Industrial Engineering: Innovative Networks

Suresh P. Sethi · Marija Bogataj
Lorenzo Ros-McDonnell
Editors

Industrial Engineering: Innovative Networks

Annals of Industrial Engineering 2011

Springer

Prof. Suresh P. Sethi
School of Management
University of Texas Dallas
PO Box 830688
75083-0688 Richardson, TX
USA

Lorenzo Ros-McDonnell
Escuela Técnica Superior de Ingeniería I
Universidad Politécnica de Cartagena
Dr. Fleming
30202 Cartagena
Spain

Marija Bogataj
Faculty of Economics
Ljubljana University
Ljubljana
Slovenia

ISBN 978-1-4471-2320-0 e-ISBN 978-1-4471-2321-7
DOI 10.1007/978-1-4471-2321-7
Springer London Dordrecht Heidelberg New York

British Library Cataloguing in Publication Data
A catalogue record for this book is available from the British Library

Library of Congress Control Number: 2011945110

Printed on acid-free paper

Springer is part of Springer Science+Business Media (www.springer.com)

Preface

We are pleased to welcome you to the CIO 2011- V International Conference on Industrial Engineering and Industrial Management and to the XV Congress of Industrial Engineering, under the motto *"Industrial Engineering: Innovative Networks"*.

CIO 2011 gives continuity to the series of annual Conferences initiated in September 1986 in La Rábida (Huelva, Spain). This conference is a very significant scientific event in Business Management and related areas. Investigators, academics, scientists and managers of diverse parts of the world have the opportunity to exchange experiences, contribute new ideas and debate topics, in those fields related with Industrial Engineering.

The V International Conference on Industrial Engineering and Industrial Management addresses the great multidisciplinary field formed by Organization Engineering, from an international point of view.

This book contains the selection of the papers that were accepted for presentation at the CIO 2011 International conference, covering consolidated and emerging topics of the conference scope.

The contributions were organized in six sections:

- Business
- Innovation & knowledge
- Management
- Manufacturing
- Logistics
- Operations Research

The papers selected for this book cover the most relevant topics in the CIO Conference scope and they may help readers gain a deeper understanding of next generation systems for Industrial engineering and innovative networks.

We will like to express our gratitude to all the authors, reviewers and the Organizing Committee for their enthusiastic work and magnificent support during CIO 2011.

Suresh P. Sethi
Marija Bogataj
Lorenzo Ros-McDonnell

Organization

Conference Chair
Lorenzo Ros-McDonnell (Spain)

Program Committee Co-chairs
Suresh P. Sethi (USA)
Marija Bogataj (Slovenia)
Maria Victoria de la Fuente (Spain)

Program Committee

Carlos Andrés	Samo Drobne	Catalina María
Mareva Alemany	Alfonso Durán	Morales Granados
Eugenia Babiloni	Sofía Estellés	Josefa Mula
Marija Bogataj	Isabel Fernández	Jesús Muñuzuri
André Boza	Enrique Flores	Carlos Ochoa
Antonio Juan Briones	Isabel García	Jordi Ojeda
José Mª Bueno	Jesús García	Luis Onieva
Francisco Campuzano	José Pedro García	Ángel Ortiz
Manuel Cardós	Nazario García	Sergio Palomero
Javier Carrasco	Susana García	Manuel Pérez
Andrés Carrion	Osvaldo Luis	Raúl Poler
Juan Gabriel Cegarra	Gonçalves	Eva Ponce
Ricardo Chalmeta	Gonzalo Grau	José Carlos Prado
Josep Coll Beltran	Gil Gutiérrez	Bernardo Prida
José Antonio	Cesáreo Hernández	Lluis Quatrecasas
Comesaña	Antonio Hidalgo	Inma Ribas
Ramón Companys	Eloy Hontoria	Diego Ros

(continued)

(continued)

Program Committee

Albert Corominas	Carmen Jaca	Rafael Ruiz
Pablo Cortes	Ana Mª Lara	Patxi Ruiz de Arbulo
Anna Mª Coves	Adolfo López	Lourdes Saiz
Francisco Cruz Lario	Antonio López	Javier Santos
Llanos Cuenca	Miguel Ángel	Mª Val Segarra
David de la Fuente	Manzanedo	Pep Simó
Mª Victoria de la	Juan A. Marín	Ana Suarez
Fuente	Carmen Martínez	Jaume Valls
Carmen de Nieves	Eva Martínez	
Ricardo del Olmo	Manel Mateo	Eduardo Vicens
Pablo Díaz de Basurto	Ana Mª Mejías	Robert Vodopivéc

Organizing Committee

Eloy Hontoria Hernández
Antonio Fernández Martínez
Maria Dolores Soler Romero
Catalina Morales Granados
Maria Candelaria Martínez Conesa

Contents

Part I
Business

Chapter 1
Smart Government's Role to Reduce Business Failure

Francisco Acosta and Antonio Juan Briones

Abstract We analyze the appropriateness of the intervention of the government outsourced through Relationship Management tutoring in order to address the new companies to reduce enterprise output rate or failure. The purpose of this research is to help explain how the government through professional associations may increase the know how of entrepreneurs or managers of new enterprises. Of the cases studied we have learned that starting a business is a difficult and bureaucratic administration that acts more of a bureaucratic intelligent tutoring that is outsourced by the AAPP but they must control the outsourced service, to be generalized the use of New Information Technologies (hereinafter, NIT) by all actors involved, both as professionals and entrepreneurs AAPP and that there is a perception that the Administration is not widely used in the NIT, that subsidies are supported but difficult to obtain and ineffective, it is proper mentoring by professionals and entrepreneurs who have experienced business failure tutoring prefer that economic subsidies, we propose a structural change in economic grants subsidies replacing no consistent economic support for the direction of new firms by professionals.

1.1 Introduction

Today's society demands constant change and modernization of the Administration, is demanding an Intelligent Management, based on new ways of organizing, thinking, innovative, flexible, adaptive, based on new models of leadership and

F. Acosta (✉) · A. J. Briones
Universidad Politécnica de Cartagena, Calle Real, 30201 Cartagena, Murcia, Spain
e-mail: Francisco.acosta@upct.es

S. P. Sethi et al. (eds.), *Industrial Engineering: Innovative Networks*,
DOI: 10.1007/978-1-4471-2321-7_1,

governance. This change requires a different conception of the organization, human capital management and knowledge processes to citizen services, technologies and systems (Martín-Castilla 2005). The introduction of ICT in public organizations has transformed and enables it to meet the speed and efficiency that our proposed model to be used by all actors involved (Salvador 2007). The information society is widespread and therefore e-government is already a reality, so now use to be successful requires the involvement of all parties, general government, professionals and entrepreneurs (Sanjuan 2010).

We propose a model that emphasizes human capital management and knowledge management, through the participation of independent professionals for tutoring to new businesses and will build this model in the use of NIT both professional direct service providers as entrepreneurs. Study as the most important change in recent years has been our general government to adopt a management model that is essentially relational focus major direct management of public services to private organizations. In it, we will hold that the PA will implement e-government but are the professional associations that offer their skills and abilities to new companies in attempting to reduce the high rate of output or business failure, knowledge transfer, establishing Cegarra-Navarro (2010).

The administration should take responsibility in the investment of public funds, combining effectiveness and efficiency with the principles of accountability, legality and transparency, using appropriate tools to manage both their human capital, and the various resources, including the administration or government knowledge (Martin-Castilla 2006). Control and quality improvement is an increasingly common goal in the AAPP, and the effectiveness of the same to manage the service in the form of aid to businesses has been demonstrated in the work may involve not only economic aid (Barbero 2009), but support in improving knowledge and policy coherence for new entrepreneurs. Therefore, efficient management in investment of public resources needed and the control thereof, it will be possible through the evaluation of the results obtained with these public resources.

1.2 Conceptual Framework, Literature Review

Intelligent Public Administration can be configured as a modern organization (Martín-Castilla 2006), which establishes quality improvement systems in the functioning of Public Administration (hereinafter PA) (Barbero 2009), must implement new management to enable continuous adaptation to the needs of citizens, in order to achieve excellence. Given the variety of contributions from different authors consulted, we consider it appropriate to propose a conceptualization of different terms that we consider critical. Intelligent Public Administration that we consider using the outsourcing organization and subject to the planning, monitoring and evaluation of outsourced public services, practice the use of standardization to exercise such control (Martín-Castilla 2006). Using it, the PAs are able to manage with a certain efficiency, achieving higher levels of quality in the provision of public services with minimal resources based on the use of NIT and the use of e-administration (Ramió 2002; Salvador 2007).

First understand outsourcing as a positive trend to decrease the work of the PA. Ramió (2000) speaks of a controlled outsourcing, where the PA will plan, monitor and evaluate private networks providing public services. However, Cosculluela (2008) is to outsource as that which is entrusted to third parties certain activities of the entity. In this regard, Ramió (2000) states that the Administration Relational arises when there is the transfer of part of the management of public services to private organizations, and Asensio (2006) defined as true when it appears Relationship Management, the unit traditional administrative reduced.

A better understanding of the managed practice of public service they are using, could improve other public services that offer general government. Sanjuán (2010) states that accelerate eGovernment in Europe is beneficial for everyone. Similar concept is the e-government, claiming that the introduction of ICT in government activities improves efficiency (Salvador 2007).

Finally, we consider the output rate is the proportion of firms that leave their activity in relation to the total number of companies that started in a given time period. We analyzed studies that show that in Spain to 80% of small and medium enterprises (hereinafter SMEs) fail before their fifth birthday. Urbano (2006a, b) proposes that because more than 50% of the companies goes out of business in the first five years of life, counseling is necessary to reduce that percentage, which suggests that PAs provide to new entrepreneurs mechanisms non-economic support.

1.3 Methodology

Casuistry conducted an investigation using scanning and matching description of the facts in a multiple case study with three blocks of businesses, based on the classification made by Montero and León (2005) with the following objectives: (1) determine whether future entrepreneurs had agreed to receive professional assistance to collect more than traditional cash subsidies, (2) ascertain the possible positioning of the newly established businesses, if they had received the tutoring or coaching; (3) understand if it was possible to avoid bankruptcy in companies that have disappeared or pending close with the assistance of a professional responsible for tutoring their economic activities.

Montero and León (2005) proposed that can be set four different levels of methodology, one of them the techniques used to collect empirical evidence, particularly well attended to seek some explanation of emergent phenomena which we compete and for the development of theories where intangibles are crucial.

We use case studies in order to have an understanding of the phenomenon studied and proposed model inquire about this contemporary phenomenon and its context in the real environment, using multiple data sources (Yin 2009). As to the scope of the research study, is limited to the Region of Murcia and interviewed and known by the investigator and others known and interviewed at the European Business and Innovation Cartagena (hereinafter CEEIC).

A version of the Pattern Matching is the examination of patterns of behavior, which is the formulation of expected behavior as a hypothesis, which will be confirmed, modified or rejected by the actual behavior, which is the tactic used. The examination of patterns of behavior resembles the presentation of results to quantitative studies (Yin 2009).

The sample for this investigation consists of three blocks:

1. Ten future entrepreneurs are Block 1 in training before starting the activity. They are going to get in two ways: first by the researcher's professional experience, which is closely related to the business world and known future entrepreneurs who are taking their first steps in the near future to develop their business ideas, second through CEEIC that prepares and provides training for them to develop their business plans.
2. Six entrepreneurs stay in the first 5 years her business. Block 2 will comprise the active entrepreneurs.
3. Five entrepreneurs who have had to close his company consequently suffer business failure. We arrived at Block 3 by knowledge of them by the researcher due to his work in consulting, as well as the research administrator and therefore be in contact with companies with serious problems of continuation of their business.

The rationale for conducting this research is the absence of a reference test case on the impact of mentoring or co-management of the new entrepreneurs with the result of lower output rate business, with which to make comparison.

The research pointed to two instruments: the three questionnaires and a matrix table to collect the information. As for the description of the questionnaires, consisting of 19 questions each, and questions we find different measurements, and these are:

- Measuring single-selection, for example the age and grade level. Are objective and unique.
- Measurement of multiple-choice questions such as "Finally, if the administration will provide professionals to assist, what would you prefer?"
- Dichotomous measure, are the majority, for example the question "Do you think the closure of his company could have been avoided if he had received support from the management as to tutoring or co-management?"
- Measuring open, for example when ordering the respondent to provide the professional who wants to assist it in the early years of the company and there is a box OTHERS.

1.4 Results

As a result common to the three blocks of cases appears to match a survey has tried to clear, fast and convenient way to respond, so that the failure of non-response has been almost nonexistent. Respondents have expressed interest in the issues, asking that they move the research findings. We have spent a block of multiple case study to future entrepreneurs, other entrepreneurs in active block and the last to companies that have closed. They all want to know first the perception of the aspects that make up the Intelligent Public Administration, through the terms that define best practices and organizational management systems towards improving service delivery by the PA. We analyze your answers:

- Efficiency: For entrepreneurs, the majority believe that the Administration does not encourage entrepreneurship and to properly manage financial resources for the promotion, therefore future entrepreneurs interviewed by the PA and energize or operate efficiently. However, for the most active entrepreneurs in the Administration does encourage entrepreneurship, considering it is energizing, but also the majority view that the Administration does not properly handle the financial resources for development, therefore there is a perception for this block of cases that the PAs do not act efficiently.
- Smart management: Most entrepreneurs believe that the future is difficult to fill and meet the permits and paperwork to start a business, and that the administration does not act intelligently to promote entrepreneurship. These majorities agree to entrepreneurs in business. As for the entrepreneurs who ceased their activity increases most saying the great difficulty of filling and know the permits and paperwork to start a business, and unanimously believe that the Administration does not act intelligently to promote entrepreneurship.
- Outsourcing: The majority of future entrepreneurs surveyed did not agree to give the promotion of entrepreneurship to private managers, and those who show their agreement, 100% said they have to control these private managers. As active entrepreneurs, 50% of the respondents agree to give up support for the creation of companies to private managers and 50% is not, so there is no majority, however there is unanimity who share control of outsourcing as 100% said they have to control these private managers.
- E-government: 90% of prospective employers thought that the general government and private operators should offer their services and supports online and you use email as a channel of communication between the parties. 100% of these cases respondents use the internet and e-mail but 40% believed that the PAs do not use and 40% say that if they use them, so no majority, but notes that there is a long way go for future entrepreneurs perceive that they are at their level of use of the NIT, they seem to hold advantages in these fields on the Administration.
- Business failure: Most entrepreneurs and future entrepreneurs active professionals believe that if help address new business, their results can be better and therefore may be less likely business failure. As for the entrepreneurs who have

experienced business failure is 100% of those who believe that if professional help address new business, their results can be better and therefore may be less likely business failure.

We find the majority of future non-university entrepreneurs without prior experience in business management. In this sense, most of the cases considered significant that the autonomous communities are the right actors to direct and plan the tutoring, so we propose it as required by professionals to bridge this gap. 100% of future entrepreneurs agree that supervises the entrepreneurs at no cost to them, however there is no majority for choosing economic subsidies or support from professionals, both get 40% back. In the case of active entrepreneurs in 100% of the respondents believe that supervises the entrepreneurs at no cost to them, but only 16.6% of the respondents preferred the grant money tutoring, even though employer only six received a monetary grant. As for the entrepreneurs who have ceased activity, 100% think that supervises the entrepreneurs at no cost to them, and most prefer the grant money tutoring, mentoring giving therefore an added value.

In the block of active entrepreneurs we find most college-educated entrepreneurs, but most also had no previous experience in business management, so that professional assistance should not be substituted for their expertise to be mainstream inexperience. As for employers who have closed their businesses, we find a high formation of the five, as we see it is not a guarantee of success, besides declaring 100% of the cases that had previously experience in business management, so business failure is not correlated in these cases the lack of business experience.

On the question of whether it supports the economic subsidies, there is near unanimity in the SI, but the question of whether they are difficult to obtain, the near unanimity of the cases say that if they are difficult to obtain, which calls into question use and effectiveness and opens the discussion of non-economic support.

The researcher the answers of the three blocks of cases we suggest a shift towards better corporate governance as a way different from directing companies through mentoring and support to management.

It is common ground in the three blocks where the view that the public should lead actor tutoring in their planning and control should be the Autonomous Community, this has transferred to it by the European Union's competence to manage the processes stimulate and improve the economic conditions of their territory and consider it suitable both survey responses and the study authors and Urbano (2006a, b).

Mentoring of entrepreneurs will be conducted through the Relationship *Management*, participating as a leading actor and professional associations to professional associations by assigning a shift system, mentoring of new businesses, as is common ground in the polls where entrepreneurs have been consulted about the appropriate professionals, and as we have studied Barbero (2009) and Martín-Castilla (2005). This public service is called "e-administration", which uses the NIT and ensure that there is a communication professional's advantageous for the entrepreneur, thus using the internet for handling OOPP (Sanjuán 2010). We recommend that there is a channel of communication between professionals OOPP and where the first will send information to the second, and another channel among professionals and entrepreneurs where the first turn information into knowledge for entrepreneurs, plus the possibility of resolve issues of good practice for the latter.

The emerging public service consisting of a tutoring business to be outsourced by the Regional Administration through the CCPP. The Regional Administration shall be reserved the powers of control, planning and evaluation, as Ramió (2000) proposed on the control of the outsourced service through professional mentoring to achieve an effective operation.

The PA act bureaucratically, efficiency suffers and instrumentation to stimulate the economy (Peña and Aranguren 2002). Through the tutoring provided by the UNJSPF companies can improve in this regard, mandated professionals in the provision of public services to reduce business failure. Criado and Ramilo (2001) state that public authorities should stimulate, coordinate, lead and build consensus between different actors in the economy. With the proposed model, we may find greater policy coherence and a likely decline in business failure, improving the economic and social planning.

We recommend a structural change in the processing of grants through a non-economic support as proposed by Urbano (2006a, b), to overcome the problems reflected by López (2001), partly replacing direct subsidies or currency, because subsidized through of the tutoring companies led by professionals for that purpose; no conditions, no prerequisites, just the fact of starting a business, and ultimately, to subsidize the employer to support management.

1.5 Conclusions and Research

Starting a business is difficult, very bureaucratic and action by government is perceived as such. However, the tutoring companies should be entrusted to professionals and the general government should control this public service to reduce business failure. In this way, should be widespread use of the NTI, both OOPP as entrepreneurs in the different modalities studied. It is unanimous opinion of respondents is appropriate tutoring by professional entrepreneurs, including the views of those who have experienced business failure are said to prefer tutoring Direct financial subsidies. Therefore, the tutoring through the implementation of professional management practices, could have prevented the closure of some companies and thus would lower the rate of enterprise output.

We believe that the proposed public service is emerging and so far has had limited research in very localized areas. We believe that this research can provide a level playing field in favor of improved public service with tremendous growth and very recent cases for the analysis of such singular, where results can be extrapolated to other areas. Find the causes recurrent failure of firms through monitoring of companies created by the Points of Advice and Processing (PAIT) to find the factors affecting the closure of businesses, and serve to define protocols that ameliorate the exit rate of firms, to improve and disseminate good management practices to outsourcing through a model of intelligent management.

References

Asensio P (2006) The book of municipal management. Ed Business and Economics, India

Barbero I (2009) Orientaciones para la implantación progresiva de sistemas de calidad en los servicios sociales locales de la CAPV. Zerbitzuan 45:39–45

Cegarra-Navarro JG (2010) GAP de Conocimiento en el sector sanitario: Aplicación práctica a UHD. Cap. VI: pp 49–59. En Apezetxea-Celaya, Martínez-Caro y Cegarra-Navarro (Coord.): Dirección estratégica del conocimiento en unidades de hospitalización a domicilio: La escalera hacia el éxito. Edita Diego Marín. Librero Editor.

Cosculluela L (2008) Manual de Derecho administrativo. Ed. Aranzadi, Pamplona

Criado JI, Ramilo MC (2001) e-administración: ¿un Reto o una Nueva Moda para las administraciones del Siglo XXI? Revista Vasca de administración pública (61):1–21

López Barja de Quiroga J (2001) Derecho penal económico. Manuales de formación continuada (14):430–495

Martín-Castilla JI (2005) La administración inteligente: un modelo de administración pública orientada al servicio del ciudadano. Revista de Auditoría Pública (36):47–60

Martín-Castilla JI (2006) Valor y valores de una administración al servicio público. Revista de Auditoría Pública (38):25–34

Montero I, León O (2005) Sistema de clasificación del método en los informes de investigación en Psicología. Int J Clinical Health Psychol 5(1):115–127

Peña-Legazkue I, Aranguren M (2002) Transferencia de conocimientos mediante acuerdos de colaboración. Economía Ind (346):67–80

Ramió C (2000) Externalización de servicios públicos y corrientes neoempresariales. Ponencia del V Congreso Internacional sobre la Reforma del Estado y Administración Pública

Ramió C (2002) e-administración y nuevos modelos de gestión pública. Ponencia del I Congreso catalán de gestión pública

Salvador M (2007) Una visión neoinstitucionalista de las iniciativas de gobierno electrónico en el ámbito local catalán. Ponencia del VII Congreso Español de Ciencia Política y de la Administración, Democracia y Buen Gobierno, pp 55–67

Sanjuán P (2010) La e-administración; ¿es común el procedimiento administrativo electrónico?, Lex Nova, la revista (59):18–21

Urbano D (2006a) La Creación de Empresas en Catalunya: Organismos de apoyo y actitudes hacia la actividad emprendedora. Ed. Addenda, Barcelona

Urbano D (2006b) Factores condicionantes de la creación de empresas en Catalunya: un enfoque institucional. Revista de Estudios de Economía Aplicada 24(2)

Yin R (2009) Discovering the future of the case study method in evaluation research. Eval Prac 15(3):283–290

Chapter 2
Outsourcing in Spain's Automotive Auxiliary Industry: Evolution, Trends, Causes, Effects and Consequences

Jose Miguel Fernández Gómez, Javier Tafur Segura and Miguel Palacios Fernández

Abstract Outsourcing is a technique implemented and consolidated through the organisation of production in the automotive sector, which consists of fragmenting the value chain into more specialised stages and deciding which activities, products and services are to be seen to within the firm and which externally. This article seeks, from among all the factors involved in the practice of outsourcing, to find indicators associated with the factors that are significant to the carrying out of outsourcing and to quantify the evolution, trends, causes, effects and consequences of its practice in the automotive auxiliary industry in Spain.

2.1 Introduction

The automotive auxiliary industry came into being from outsourcing by manufacturers (OEM). The reasons for the growing pace of outsourcing by the OEM (today over 75% of a vehicle's added value) are: specialisation, particularly as regards the technological contribution; lower costs; conversion of fixed cost into variable cost, which means running a lower risk; and improved profitability of capital employed for vehicle manufacturers (Kimura and Ando 2005). This sector covers a broad industry spectrum, ranging from steel to electronics and including chemistry, textiles, mechanics, glass and of course services (financial, insurance, etc.). Altogether it is thought that 65–75% of a vehicle's value comes from firms that manufacture automotive equipment and those that manufacture components, and hence the huge importance that suppliers have for the automotive sector.

J. M. F. Gómez (✉) · J. T. Segura · M. P. Fernández
University of Industrial Engineering, Madrid Technical University,
Calle José Gutiérrez Abascal 2, 28006 Madrid, Spain
e-mail: josemiguel.gomez@alumnos.upm.es

S. P. Sethi et al. (eds.), *Industrial Engineering: Innovative Networks*,
DOI: 10.1007/978-1-4471-2321-7_2, © Springer-Verlag London Limited 2012

Outsourcing or vertical disintegration is a production organisation technique that has been implemented and consolidated in the automotive sector. It consists of fragmenting the value chain into more specialised stages and of deciding which activities, products and finished or semi-elaborated services are to be done within the firm and which are to be outsourced, regardless of whether the outsourced part is carried out within the same country or outside it (offshoring) and who the owner is (self or third parties) (Curzon 2001). Outsourcing is carried out for several reasons: to turn part of fixed costs into variable costs, to reduce costs, to lower investments for performing activity, to boost flexibility and technological knowledge, to improve the final product/service, to enhance productivity through specialisation and so on (Hitt and Holcomb 2007; Sturgeon et al. 2009).

2.2 Description of the Model

The aim and scope of the research based on this model is the quantitative study of outsourcing in Spain's automotive auxiliary sector, analysing its evolution, trends, causes, effects and consequences. From all of the factors involved in the practice of outsourcing (costs, flexibility, productivity, competition, investment, etc.), we seek to find indicators associated with the factors described previously that are significant with the practice of outsourcing. This article breaks new ground in that it quantifies the causes, effects and consequences of outsourcing in the automotive auxiliary industry in Spain.

Two initial hypotheses have been taken into consideration in this model:

- Technological changes or technical advances have not altered the consumption of intermediate inputs per unit of final product.
- There have not been disparate alterations between the prices of intermediate inputs and prices of the final goods, a condition which has been met since the linear regression technique has been applied.

Once established the initial hypotheses, the methodology has been the following:

In the first place the values of the variables have been compiled in the Spanish Annual Industrial Company Survey (Spanish National Institute of Statistics-INE) and in Structural Business Statistics (Eurostat), according to the European classification NACE Rev 1.1 by which components manufacturers has Code 343. The series of data compiled (ranging from 1995 to 2007) has been divided by the value of the invoicing in those variables in which it has been feasible. Next we have selected the variables that are more important and with greatest importance with regard to invoicing. Following this, we have grouped these by areas: costs, productivity, purchases, added value, investment etc.

Next the outsourcing value has been calculated, taking the following definitions into account (Díaz and Gandoy 2005):

- Raw materials: Goods acquired for transformation in the production process: These are goods in which the vehicle manufacturing firm contributes the greatest added value to the final transformation, due to the fact that they have high strategic value, high technological content, final product differentiators, etc.
- Merchandise: Goods acquired to be resold without subjecting them to a transformation process. These goods are incorporated into the vehicle without any subsequent transformation process in which the vehicle manufacturing firm does not contribute further added value. They are goods in which the components industry is very specialised; they are goods that are very standardised and the final product has little differentiating quality since any vehicle manufacturer has access to them
- Jobs done by other firms (TRE): the cost forms part of the process of one's own production, and the job is commissioned to and carried out by other firms.
- External services (SE): operation costs of different types (R + D + I, repairs and maintenance, professional services, etc.).

The calculation of the value of the fragmentation value of internal production (FPI) is as follows:

$$\text{Raw materials/production value} + \text{Merchandise/production value} + \text{Jobs done by other firms/production value.} \quad (2.1)$$

The calculation of the value of outsourcing services (SE) is as follows

$$: \text{External services/Production value} \quad (2.2)$$

The calculation of the value of complete outsourcing index (OT) is the following: Fragmentation of internal production (FPI) + Outsourcing of external services (SE).

$$\text{OT} = \text{FPI} + \text{SE} \quad (2.3)$$

With the value of complete outsourcing index (OT) and the value of the variables compiled and calculated, the linear ordinary least squares regression statistical technique has been applied, and their correlation has been analysed for levels of significance of $p < 0.05$ and $p < 0.01$ and correlation coefficient r.

With the results obtained, we have analysed the effects, causes and consequences that outsourcing has in the automotive sector in Spain.

2.3 Results Obtained

The values obtained by applying the methodology presented are shown in Table 2.1.

As Table 2.1 reveals, the greatest importance of outsourcing in the auxiliary sector is shown in raw materials whose average values, variability and range have

Table 2.1 Values obtained in calculating outsourcing index

Denomination	1995	1996	1997	1998	1999	2000	2001	2002	2003	2004	2005	2006	2007
Raw materials	0.403	0.413	0.453	0.443	0.424	0.477	0.519	0.532	0.538	0.544	0.561	0.542	0.577
Merchandise	0.032	0.032	0.025	0.023	0.026	0.023	0.028	0.014	0.020	0.018	0.013	0.020	0.025
TRE	0.024	0.029	0.026	0.028	0.038	0.037	0.044	0.044	0.041	0.041	0.045	0.047	0.050
FPI [1]	0.461	0.475	0.503	0.496	0.490	0.537	0.591	0.591	0.601	0.604	0.620	0.610	0.653
SE [2]	0.108	0.100	0.101	0.105	0.106	0.105	0.111	0.113	0.114	0.112	0.111	0.115	0.117
OT [3]	0.569	0.575	0.604	0.601	0.596	0.642	0.702	0.707	0.715	0.716	0.731	0.726	0.770

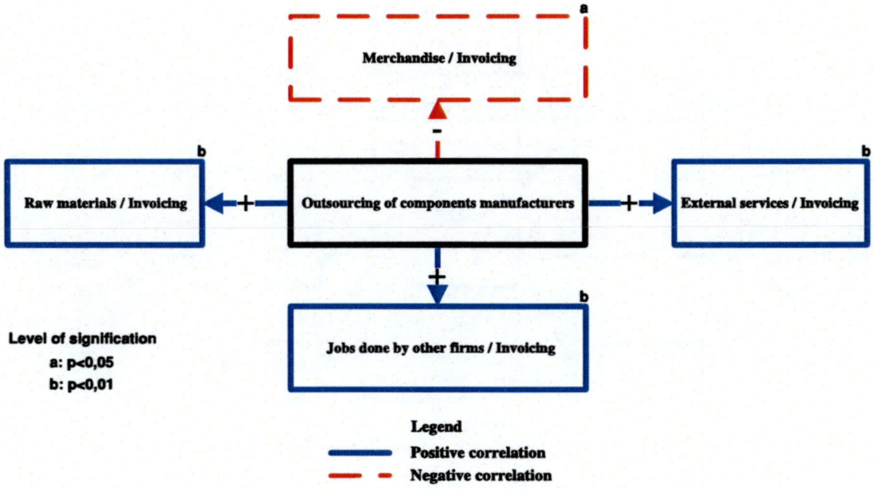

Fig. 2.1 Relation between outsourcing of components manufacturers and purchases of raw materials, merchandise, external jobs and external services

been 0.49; 0.0037 and 0.174. Below are the external services whose average values, variability and range have been 0.10; 0.00003; and 0.017 whose value over the periods of study has remained constant. The value of merchandise and of external jobs has only a very slight influence on the total value of outsourcing.

Figure 2.1 represents the results obtained from applying the linear ordinary least squares regression statistical technique (p-value and correlation coefficient r) between outsourcing index (OT) of components manufacturers and purchases of raw materials, merchandise, external jobs and external services. As can be seen in Fig. 2.1, as the value of outsourcing index grows, so do the purchases of raw materials, which are the products that greatest added value give to the product, while merchandise purchasing decreases. In turn there is growth in terms of external jobs, where the firm devotes more resources to raw materials, and external services grow meaning that the components manufacturers, due to the fact that they lack the necessary resources, acquired knowledge, people with adequate training and so on mainly stimulate cooperation with its suppliers for product/service development. The purchases of merchandise with the external jobs and external services are correlated negatively, which explains why components manufacturers invest fewer resources in cooperating with their suppliers in the development of these products. Figures 2.2, 2.3 and 2.4 represent the results obtained from applying the linear ordinary least squares regression statistical technique (p-value and correlation coefficient r) between the value of the outsourcing index (OT) and the value of the variables/invoicing, representing the significant variables and their relation with the variable of outsourcing (OT).

As can be seen in Fig. 2.2, outsourcing of components manufacturers is determined by vehicle demand. This prompts vehicle manufacturers to outsource

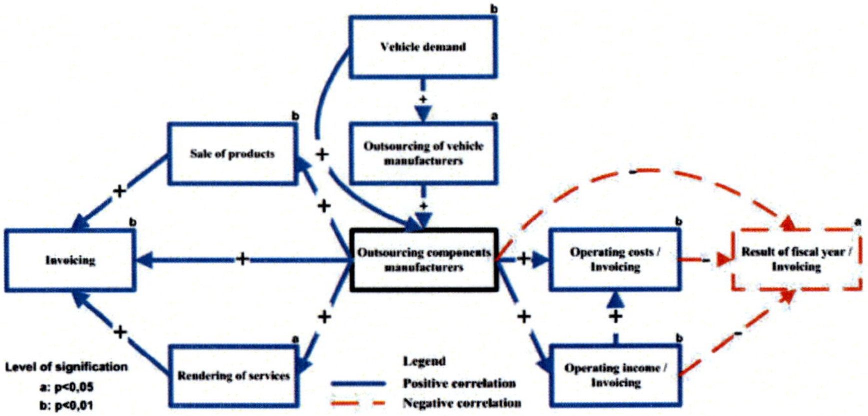

Fig. 2.2 Relation outsourcing index manufacturers-outsourcing components

in the auxiliary components industry; it causes components manufacturers to do so with their suppliers.

Because vehicle manufacturers increase their outsourcing, components manufacturers increase their invoicing owing to the increase in sales of products and rendering of services to vehicle manufacturers. By analysing Fig. 2.2 operating costs increase at a higher rate than operating income with respect to the invoicing, owing largely to the increase in the price of raw materials (more specifically to the price of steel) and of energy. Since 2003, steel prices have risen owing to the large demand for emerging countries such as China and India. This causes margins and profitability to reduce as invoicing by components manufacturers increases.

Figure 2.3 shows the relation between outsourcing index of components manufacturers (OT) and in the number of firms of the auxiliary automotive industry in Spain.

As the Fig. 2.3 shows, as the value of outsourcing index increases, the number of firms making up the sector decreases because the majority of firms making up the auxiliary sector have fewer than 20 workers. The number of firms with fewer than 20 workers decreases and the number of firms with more than 20 workers increases. This is for the most part due to three reasons:

- As has been shown, the rise in the value of outsourcing index increases invoicing yet profitability and margins are reduced, resulting in processes of purchases, acquisitions and strategic alliances in order to obtain economies of scale, reduce operating and structural costs in exchange for increasing profitability and margin and avoiding a possible temporary receivership or bankruptcy.
- The increase in the value of outsourcing index of vehicle manufacturers, raises the value of outsourcing index of the components manufacturers as represented in Graph 18 (outsourcing chain) causes cooperation among the firms to increase and firms have to have some minimum resources (know-how, staff, assets,

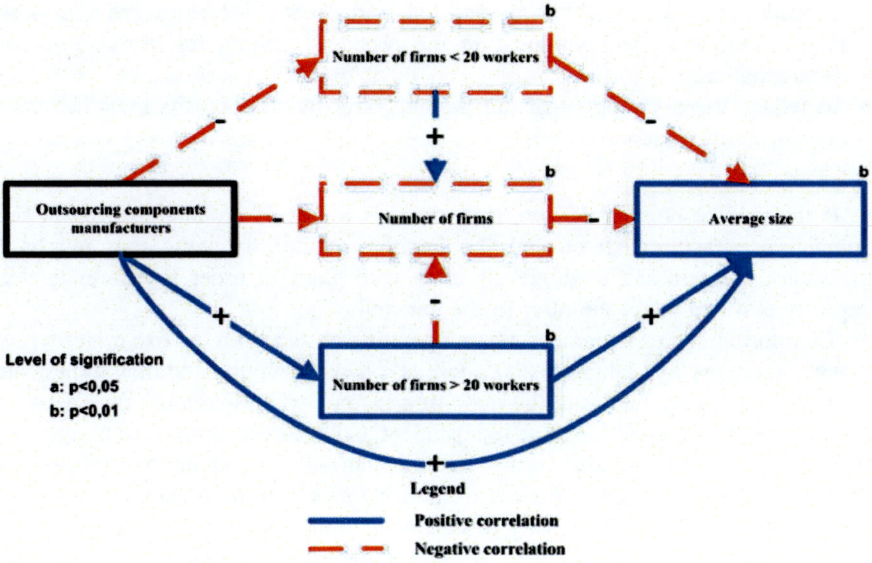

Fig. 2.3 Relation between outsourcing and number of firms

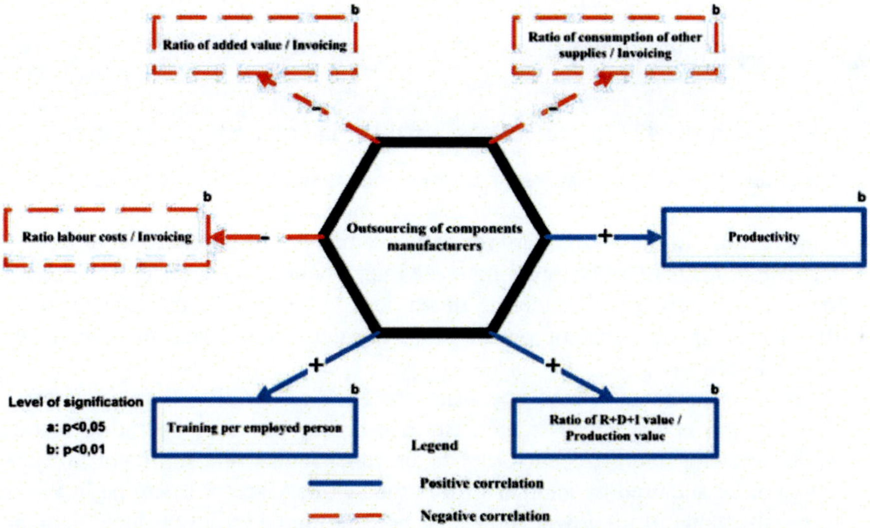

Fig. 2.4 Relation among outsourcing, cost reduction, investment and productivity

R + D + I, patents, etc.) in order to develop products/services both upstream to their buyers and downstream to their suppliers, meaning that acquisitions, purchases and strategic alliances take place in order to have these resources

available, since the level of competition in the sector increases and the firms increase the level technological of the products/services for the purpose of differentiation.
- To reduce the overlapping of the product offer by components manufacturers existing in the market.

If the vehicle demand is low, then there is a glut of production in both the vehicle manufacturer sector and the auxiliary sector, meaning that purchase processes, mergers and strategic alliances take place in order to eliminate this capacity glut and adapt the offer to the demand.

Therefore, in an environment where operating costs are on the rise (chiefly raw materials, energy and labour costs), there is a lack of supply, and operating costs increase at a higher rate than the operating income when invoicing increases, in order for an auxiliary sector to be independent and not have to resort to mergers or takeovers has to be included in a zone in which the value of the bottom limit is determined by low vehicle demand, which triggers a glut in production, and a top limit value that is determined by a high vehicle demand which causes firms to merge or be taken over in order to achieve economies of scale.

Figure 2.4 shows the relation among outsourcing index of components manufacturers, cost reduction, investment and productivity.

2.4 Conclusions

The key conclusions in the automotive auxiliary industry are the following:

- The chief cause of outsourcing of components manufacturers is vehicle demand.
- To boost the value of outsourcing, components manufacturers increase cooperation with their suppliers for the development of products/services.
- The components manufacturers increase their invoicing as a result of the boost in sales of products and rendering of services to the vehicle manufacturers.
- By increasing the value of outsourcing, the number of firms making up the sector decreases.
- In an environment in which operating costs are on the rise, there is a supply shortage and operating costs increase at a higher rate than operating income when invoicing increases. In order to be independent and not have to either merge or be acquired by another firm in the auxiliary sector has to be in a zone where the value of the lower limit is to be determined by low vehicle demand, which prompts a production overcapacity or glut, and an upper limit value that is determined by a high vehicle demand which causes firms to merge or be taken over for the consecution of economies of scale.
- Labour costs and the costs of other supplies with respect to invoicing decrease, productivity increases, staff training and the ratio value of the R + D + I/ production value increases owing to increased demand for raw materials.

References

Curzon Price V (2001) Some causes and consequences. Fragmentation: new production patterns in the world economy. Oxford University Press, Oxford

Díaz C, Gandoy R (2005) Outsourcing en la industria manufacturera Española: nuevas estrategias para el nuevo siglo. Economía Ind 358:65–77

Hitt M, Holcomb T (2007) Toward a model of strategic outsourcing. J Oper Manag 25:464–481

Kimura F, Ando M (2005) Two dimensionals fragmentation in East Asia. Intern Rev Econ Finance 13:317–348

Sturgeon T, Memedovic O, Van Biesebroeck J, Gereffi G (2009) Globalisation of the automotive industry: main features and trends. Int J Technol Lear Innov Dev 2:7–23

Chapter 3
Evaluation of the Fundamental Index's Performance in the Spanish Capital Markets From a Passive Investor's Perspective

Felipe Ruiz, Guillermo Martínez and Rafael Ruiz

Abstract A stock market index weighed by the market capitalization presents the drawback of overweighing overpriced stocks and underweighing underpriced stocks. Therefore, market capitalization-weighed indices should underperform other indices avoiding that bias, such as the "Fundamental Index" (FI). This article compares the relative performance of FI vs. IBEX-35 (main Spanish stock index), in the Spanish stock exchange, from the IBEX-35′s inception in 1992, to the end of 2010. The results of the study show a relative overperformance of the FI over the IBEX-35 of more than 1.12% compound annual rate, during the 19 years covered by the study. This excess return, in favour of the FI, reaches 5.55% compound annual rate during the 7 years, within the study period, in which the IBEX-35 showed a negative return. However, during the 12 years in which the IBEX-35 showed a positive return, the FI underperformed the IBEX-35 by 2.90% compound annual rate.

3.1 Finance Theory and Its Investment Advice

For the last 40 years, the advice from the finance theory to the investment profession has been dominated by the "Efficient Market Hypothesis" (EMH) and the "Capital Asset Pricing Model" (CAPM) (Cambell and Viceira 2002).

EMH presupposes markets in which security prices reflect information instantaneously. The strong-form version of EMH states that stock prices reflect all

F. Ruiz (✉) · G. Martínez · R. Ruiz
Escuela Técnica Superior de Ingenieros Industriales, Universidad Politécnica de Madrid, José Gutiérrez Abascal, 2, 28006 España, Madrid, Spain
e-mail: fruiz@etsii.upm.es

S. P. Sethi et al. (eds.), *Industrial Engineering: Innovative Networks*,
DOI: 10.1007/978-1-4471-2321-7_3,

information relevant to the firm, including that which can be acquired by thorough and diligent analysis of the company and the economy (Brealey et al. 2011).

In addition to those of EMH, CAPM main assumptions are, for the purpose of this paper, that all investors are price-takers and share homogeneous expectations. That is, all investors have the same beliefs about the means, variances and co-variances of the returns on all assets. If we assume that prices reflect instantaneously all relevant information and, according to CAPM, there is an equilibrium situation, then the price-formation mechanism by which information comes to be reflected upon prices must also be agreed upon by all investors, and current stock prices must be derived from unbiased estimations of their corresponding true intrinsic or fair values (Bodie et al. 2002; Lo 2007).

The main practical conclusion of CAPM is that all investors will choose their portfolio allocations based exclusively on the market portfolio and on the risk-free asset. The market portfolio holds all risky assets in the market in proportion to their capitalization weights (price per share multiplied by number of shares outstanding, divided by the total market value of all stocks). Finally, the manner in which investors should divide their money between risk-free assets and the market portfolio depends on their risk-return preferences (Cvitanic and Zapatero 2004).

Therefore, finance theory-based investment advice is for a passive investment strategy (buying a well-diversified portfolio without searching for mispriced securities to invest in). In developed markets like the USA, the historical evidence favours the strategy recommended by finance theory: (i) the returns of all funds between 1971 and 2006 were lower than the Wilshire 5,000 and the S&P 500 indexes for the that period; and (ii) out of the 138 funds that survived over the period 1971–2006, only 48 have been able to outperform the Whilshire 5,000 index, only 30 have been able to outperform it by more than 1%, only 14 by more than 2%, and only 3 by more than 4%. On the other hand, there are some exceptions in favour of an active investment strategy (trying to identify mispriced securities for investment in order to obtain superior risk-return results): a small number of investment superstars like Peter Lynch of Fidelity's Magellan Fund, Warren Buffett of Berkshire Hathaway, Francisco García Paramés of Bestinver, and a few others, have shown a consistency of superior performance that is hard to reconcile with the EMH (Siegel 2008; Swensen 2005, 2009; Graham and Zweig 1984; Goldie and Murray 2011).

3.2 Improving the Passive Investor's Performance

It follows from this that, although the finance theory-based investment advice of investing in the capitalization-weighed market portfolio is consistent with results in practice, it is based on many assumptions that do not seem very realistic. In this paper we will concentrate on the market portfolio, analyzing the implications for its optimality when market prices are not unbiased estimates of true underlying asset value (future prices are randomly distributed around current prices), but current prices are randomly distributed around true intrinsic or fair values.

Let us assume that price is the market's best guess of the intrinsic or underlying value of a company, and suppose that the errors are random and uncorrelated to the fair value of the company. In this scenario, pricing errors will be random and the chances that a company is either overvalued or undervalued will be evenly split. Therefore, when a capitalization-weighed market portfolio is constructed, more than half of it will be made up of overvalued companies and less than half will represent investment in undervalued companies. The reason is that when a security is overvalued, the capitalization-weighed index will overweigh it, and when a security is undervalued, the capitalization weighed index will underweigh it. The overweight of overvalued stocks and the underweight of undervalued stocks will cause the capitalization index to underperform.

As a consequence, it seems reasonable that an index using a set of weights other than market capitalization can eliminate the above-mentioned bias and result in improved returns for the passive investor. Although a number of different indexes that outperform those which are capitalization-weighed can be constructed (Siegel 2005; Grinold and Kahn 2000), in this paper we will follow the studies performed by Arnott et al. (2008), who have identified an index, called the Fundamental Index, which intends to measure the importance, or "footprint" of every company within the economy. This index is based on four company metrics: (i) sales; (ii) cash-flows; (iii) dividends; and (iv) book value. To obtain more stable figures, arithmetic averages covering the last 5 years are used for the first three metrics and the last year's figure is used for the fourth.

All four metrics are given the same weight, and companies are ranked by their corresponding arithmetic average. The highest-ranking companies are selected every year to become the components of the Fundamental Index.

3.3 Methodology

The methodology used to construct the Fundamental Index in the Spanish market follows six steps.

In the first step, the universe of stocks is defined as the companies included in the IBEX-35, IBEX COMPLEMENTARIO, IBEX MEDIUM CAP and IBEX SMALL CAP indices. These indices are used starting from their respective dates of inception: 1992 for IBEX-35, 1998 for IBEX COMPLEMENTARIO, and 2005 for both IBEX MEDIUM CAP and IBEX SMALL CAP. Therefore, the universe of stocks combined for this study varies accordingly: 35 stocks for the first 7 years, and an average of 75 stocks for the remaining 12 years. From that universe, 35 stocks are selected to construct the FI.

In the second step, the required data are obtained for every company in the universe of stocks: (a) accounting data from companies' annual reports: sales, cash-flows, dividends and book values; and (b) stock market data from Bolsas y Mercados Españoles, or BME (BME is a listed company that operates the stock markets in Spain): stock quotations, as well as dates and amounts of each dividend. Mergers and

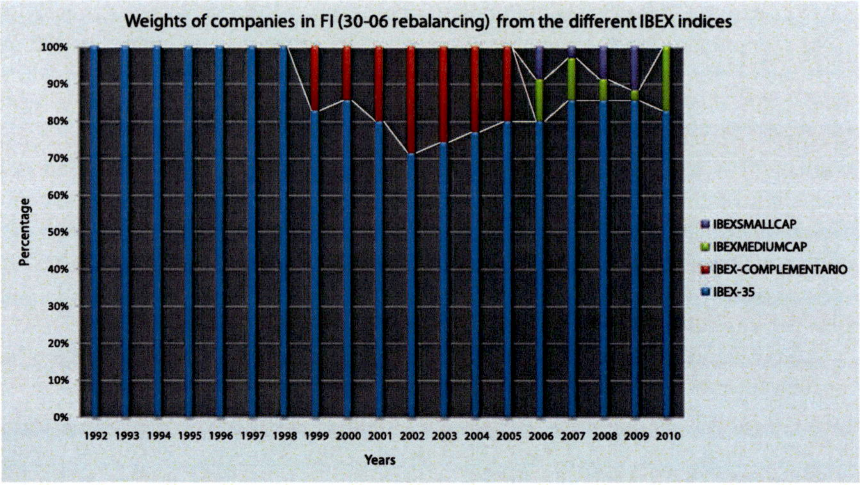

Fig. 3.1 Weights of companies in FI (30-06 rebalancing) from the different IBEX indices

acquisitions are treated in the same way as in the IBEX-35 index: depending on whether the intervening companies are part of the index portfolio or not.

In the third step, and for each year of the period being studied, the arithmetic averages of sales, cash-flows and dividends corresponding to the previous 5 years are obtained, as well as the most recent value of equity book values for each considered company in the universe of stocks.

In the fourth step, each of the concepts taken into consideration (sales, cash flows, dividends and book value) is normalized so that each concept's total adds up to 100%, and each company is assigned its percentage of contribution to each concept.

In the fifth step, all four concepts are given equal weights and averaged for each company, except in the case of lack of dividend distribution, in which the arithmetical average is obtained by using only the three remaining concepts, excluding dividends.

In the sixth step, the weights of the companies in the FI are obtained by normalizing the weights of each one of the first 35 companies, as defined in the fifth step, to add up to 100%.

The yearly re-balancing of the FI portfolio has been performed by using the data, both at 31.12 and at 30.06, for each of the 19 years of the period considered. Since most companies' annual reports are available by the end of June, the majority of results presented in the following points refer to the latter re-balancing date, thus allowing the consideration of relevant information shortly after it becomes available (Tully 2010).

In Fig. 3.1 the weights of companies in the different indices (IBEX-35, IBEX COMPLEMENTARIO, IBEX MEDIUM CAP and IBEX SMALL CAP) that are part of the construction of each year's FI, are presented.

3.4 Fundamental Index Versus IBEX-35

As seen in Figs. 3.2, 3.3, the FI overperforms the IBEX-35 index in the total period considered by 1.12% cumulative annual return; or, given a €1,000 initial portfolio, by over 21% at the end of the period.

The above-analyzed better performance of the FI versus the IBEX-35, over the 19 years of existence of the latter index, can be explained by the advantage, as indicated in Sect. 3.2, of indices based on more fundamental company data (sales, cash-flows, dividends and equity book value) over indices using market capitalization weights.

3.5 Rising and Falling Markets

In Fig. 3.4, the strong performance of the FI over the IBEX-35 can be seen during the set of 7 years of falling markets (IBEX-35 negative performance) within the period under study, amounting to a cumulative annual return of 5.55%.

However, during the twelve years of rising markets (IBEX-35 positive performance), the FI underperforms the IBEX-35 by a cumulative annual return of 2.90% (Fig. 3.5). It should be noted that the overperformance of the FI during the falling market years more than compensates for its underperformance during the rising market years, as explained in point 4 above.

The FI's relative underperformance during rising markets and its relative overperformance during falling markets can be explained by the asset pricing bubbles that tend to form during rising markets and that are usually corrected during falling markets. The capitalization-weighed indices will move more in sync with the bubble's rising prices, whereas the FI, which uses more fundamental company data, will tend to avoid them. Most explanatory theories for the developing and bursting of financial asset bubbles lie in the field of behavioural economics and behavioural finance (Pompian 2006; Bookstaber 2007; Diamond and Vartiainen 2007; Montier 2007, 2010)

3.6 Industry Weights

In Fig. 3.6, one can observe that the main differences in industry weights between the FI and the IBEX-35 are that: (a) the FI overweighs Raw Materials, Manufacturing and Construction and underweighs Telecoms and Technology, as well as Consumer Services during the majority of the study period; and (b) the FI overweighs Oil and Energy and underweighs Financial Institutions and Real Estate, as well as Consumer Goods, during the second half of the study period.

	ANNUAL RETURNS			END OF YEAR VALUE OF A €1,000 INITIAL PORTFOLIO		
	Fundamental Index with 31-12 portfolio balancing	Fundamental Index with 30-06 portfolio balancing	IBEX 35	Fundamental Index with 31-12 portfolio balancing	Fundamental Index with 30-06 portfolio balancing	IBEX 35
				1,000 €	1,000 €	1,000 €
1992	-8.04%	-5.87%	-9.94%	920 €	941 €	901 €
1993	58.08%	58.24%	61.00%	1,454 €	1,490 €	1,450 €
1994	-7.63%	-10.65%	-11.66%	1,343 €	1,331 €	1,281 €
1995	20.47%	22.01%	22.38%	1,618 €	1,624 €	1,568 €
1996	45.81%	39.60%	47.05%	2,358 €	2,267 €	2,305 €
1997	41.64%	53.26%	44.46%	3,340 €	3,474 €	3,330 €
1998	31.45%	40.01%	38.64%	4,391 €	4,864 €	4,617 €
1999	13.26%	4.80%	20.08%	4,973 €	5,097 €	5,544 €
2000	-4.45%	-6.10%	-20.54%	4,752 €	4,786 €	4,405 €
2001	0.16%	0.88%	-6.06%	4,759 €	4,829 €	4,138 €
2002	-18.70%	-13.96%	-26.50%	3,869 €	4,154 €	3,041 €
2003	32.35%	34.51%	32.19%	5,121 €	5,588 €	4,021 €
2004	20.12%	19.01%	21.09%	6,151 €	6,650 €	4,868 €
2005	24.58%	22.74%	21.98%	7,663 €	8,162 €	5,939 €
2006	36.63%	38.65%	36.04%	10,470 €	11,317 €	8,079 €
2007	8.90%	3.52%	10.71%	11,402 €	11,715 €	8,944 €
2008	-39.96%	-34.91%	-36.50%	6,845 €	7,626 €	5,679 €
2009	25.85%	19.08%	29.84%	8,615 €	9,081 €	7,374 €
2010	-12.47%	-14.25%	-12.90%	7,540 €	7,787 €	6,423 €
Geometric average / Portfolio value	11.22%	11.41%	10.28%	1,117 €	1,364 €	
Arithmetic average	14.11%	14.24%	13.76%			
FI excess geometric average	0.93%	1.12%				
FI excess arithmetic average	0.35%	0.48%				

Fig. 3.2 Annual returns of FI and IBEX-35 for the 01.01.1992–31.12.2010 period

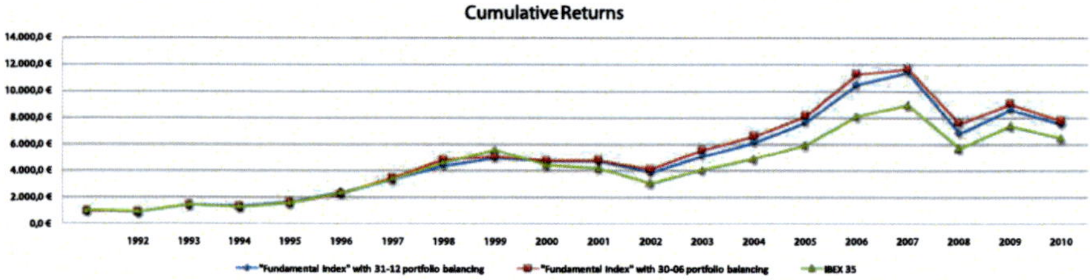

Fig. 3.3 Cumulative returns of FI and IBEX-35 for the 01.01.1992–31.12.2010 period

FALLING MARKETS	"Fundamental Index" with 31-12 portfolio balancing	"Fundamental Index" with 30-06 portfolio balancing	IBEX 35
1992	-8.04%	-5.87%	-9.94%
1994	-7.63%	-10.65%	-11.66%
2000	-4.45%	-6.10%	-20.54%
2001	0.16%	0.88%	-6.06%
2002	-18.70%	-13.96%	-26.50%
2008	-39.96%	-34.91%	-36.50%
2010	-12.47%	-14.25%	-12.90%
Geometric average	-14.02%	-12.83%	-18.37%
Arithmetic average	-13.01%	-12.12%	-17.73%
FI excess geometric average	4.35%	5.55%	
FI excess arithmetic average	4.71%	5.61%	

Fig. 3.4 Performance of the FI vs. IBEX-35 during the falling market years

RISING MARKETS	"Fundamental Index" 31-12	"Fundamental Index" 30-06	IBEX 35
1993	58.08%	58.24%	61.00%
1995	20.47%	22.01%	22.38%
1996	45.81%	39.60%	47.05%
1997	41.64%	53.26%	44.46%
1998	31.45%	40.01%	38.64%
1999	13.26%	4.80%	20.08%
2003	32.35%	34.51%	32.19%
2004	20.12%	19.01%	21.09%
2005	24.58%	22.74%	21.98%
2006	36.63%	38.65%	36.04%
2007	8.90%	3.52%	10.71%
2009	25.85%	19.08%	29.84%
Geometric average	29.24%	28.54%	31.45%
Arithmetic average	29.93%	29.62%	32.12%
FI geometric average underperformance	-2.21%	-2.90%	
FI arithmetic average underperformance	-2.20%	-2.50%	

Fig. 3.5 Performance of the FI versus IBEX-35 during the rising market years

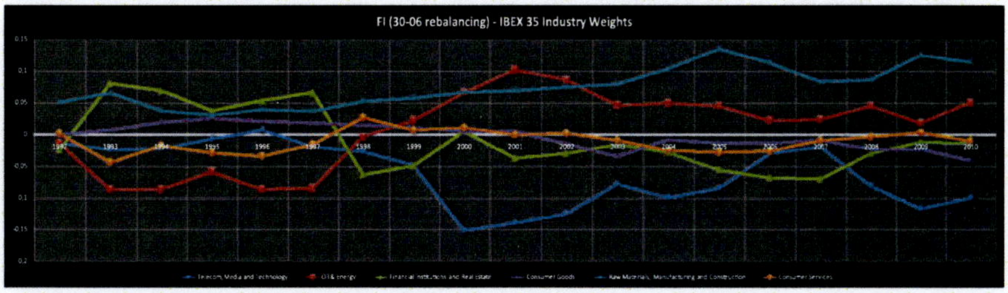

Fig. 3.6 Performance of the FI versus IBEX-35 during the rising market years

The explanation goes along the same lines as the previous point, since the FI's use of more fundamental data for obtaining its weights causes it to include a lower percentage of the relatively higher-valued companies in the Telecoms and Technology and Consumer Services industries, as well as in the Financial Institutions and Real Estate, and Consumer Goods during the second half of the study.

3.7 Conclusions

The study shows an example of overperformance of the FI, based on data reflecting the "footprint" or importance of companies within a given economy, over the capitalization-weighed IBE-35 index. The over-performance is obtained for the span of 19 years, from the inception of the IBEX-35 index to 31.12.2010.

Other significant results from this study include: (i) the overperformance of the FI during the falling market years, as well as its underperformance during the rising market years; and (ii) the lower weights within the FI portfolio of more overvalued industries such as Telecoms and Technology, and Consumer Services (including TV broadcast), as well as Financial Institutions and Real Estate (the latter from 1998 onwards, when their overvaluations reached peak values).

References

Arnott RD, Hsu JC, West JM (2008) The fundamental index: a better way to invest. Wiley, Hoboken

Bodie Z, Kane A, Marcus AJ (2002) Investments, 6th edn. McGraw Hill/Irwin, New York

Bookstaber R (2007) A demon of our own design. Markets hedge funds and the perils of financial innovation. Wiley, Hoboken

Brealey RA, Myers SC, Allen F (2011) Principles of corporate finance, 10th edn. McGraw Hill, New York

Cambell JY, Viceira LM (2002) Strategic asset allocation. Portfolio choice for long term investors. Oxford University Press, New York

Cvitanic J, Zapatero F (2004) Introduction to the economics and mathematics of financial markets. MIT Press, Cambridge

Diamond J, Vartiainen H (2007) Behavioral economics and its applications. Princeton University Press, Princeton

Goldie DC, Murray GS (2011) The investment answer. Business plus, New York

Graham B, Zweig J (1984) The intelligent investor. The definitive book on value investing. Harper Business Book, New York

Grinold RC, Kahn RD (2000) Active portfolio management. A quantitative approach for providing superior returns and controlling risk, 2nd edn. McGraw Hill, New York

Lo A (2007) The efficient market hypothesis. In: Blume L, Durlauf S (eds) The new Palgrave: a dictionary of economics, 2nd edn. Palgrave McMillan, New York

Montier J (2007) Behavioural investing: a practitioner's guide to applying behavioural finance. Wiley, New York

Montier J (2010) The little book of behavioural investing: how not to be your own worst enemy. Wiley, New York

Pompian MM (2006) Behavioural finance and wealth management: how to build optimal portfolios that account for investor biases. Wiley, New York

Siegel JJ (2005) The Future for investors: why the tried and the true triumph over the bold and the new. Crown business, USA

Siegel JJ (2008) Stocks for the long run: the definitive guide to financial market returns & long-term investment strategies, 4th edn. McGraw Hill, New York

Swensen D (2005) Unconventional success. A fundamental approach to personal investment. Free Press, New York

Swensen D (2009) Pioneering portfolio management. An unconventional approach to institutional investment. Free Press, New York

Tully S (2010) Robert Arnott's magic formula. Fortune 14 June 2010

Chapter 4
Entrepreneurial University: The Costa Rica Institute of Technology Experience

Mauricio Monge Agüero, Antonio Juan Briones Peñalver and Domingo Pérez García de Lema

Abstract This paper analyses "academic entrepreneurship" as a key factor during the creation process of academic spin offs at the Costa Rica Institute of Technology (ITCR). This initiative is the first of its kind and it includes variables such as: personality, skills, professional background and willingness to be involved in business activities. It uses the methodology of case study and key factors proposed by O'Shea R, Chugh H, Allen T (2008) Determinants and consequences of university spinoff activity: a conceptual framework. J Technol Transf 33(6):653–666.

4.1 Introduction

The process of evolution from the traditional university, focused on pedagogy and research to an entrepreneurial university committed to economic development and the needs of society (Bueno 2007), driven by internal factors, such as the need to find alternative financing sources and by external factors promoted by a Society of Knowledge and its consequences on innovative processes (Etzkowitz and Leydesdorff 2000), has its origins in the last part of the twentieth Century in American universities such as the Institute of Technology and Berkeley University. These universities are pioneers in the commercial exploitation of research results and in getting governmental support (Etzkowitz et al. 2000).

A. J. Briones Peñalver · D. Pérez García de Lema
Universidad Politécnica de Cartagena, C/. Real, n° 3, 30.201, Cartagena, Spain

M. Monge Agüero (✉)
Instituto Tecnológico de Costa Rica, Barrio Los Ángeles, Cartago, Costa Rica
e-mail: mmonge@itcr.ac.cr

S. P. Sethi et al. (eds.), *Industrial Engineering: Innovative Networks*,
DOI: 10.1007/978-1-4471-2321-7_4,

Considering the success this model has had, many countries have promoted university reforms to increase commercialization of the results of research (Wright et al. 2007) through technology transference offices, enterprise incubators, internal seed funding (Rasmussen et al. 2006) and using as a very important instrument, the creation of university enterprises (Shane 2004a, b; Wright et al. 2004).

University spin offs are companies developed by academics or doctorate students and they are the best means for transferring investigation. These companies are really innovative, greatly accepted in the marketplace and highly competitive (ANCES 2003). The increasing interest in that type of companies and the great amount of resources invested in supporting them (Lockett and Wright 2005) justify researching about how universities can facilitate their creation (Markman et al. 2008) to have a positive impact in the marketplace by introducing a new good or value added service, resulting from the advances in knowledge at the university, transferred to the marketplace for its commercialization as an innovative product or service (Laborda and Briones 2010).

This paper describes the behavior of the academic entrepreneur as a key factor in the creation of spin-offs, in this particular case study, the key factors in the process of the creation of academic spin-offs at the Costa Rica Institute of Technology (ITCR)—a public university devoted to teaching, research and extension in science and technology. This analysis will be carried out through the design and application of a questionnaire to academic entrepreneurs, public university authorities and officials from the government of Costa Rica. Its results can be compared to Ortín et al. (2007) in "La creación de Spin-off universitarios en España: Características, determinantes y resultados" and to Aceytuno and Paz (2008) in "La creación de spin-off universitarias: el caso de la Universidad de Huelva"; all of which use a methodology based on case analysis and supported on the key factors for the creation of academic spin-offs by O'Shea et al. (2007). This factors are briefly described and start with the identification of the technological opportunity, which is an indispensable requisite for the conformation of academic spin-offs (Roberts 1991), with various groups of key factors that influence the decision of creating a company, its formation process and its development. These factors are: 1. Entrepreneur's attributes and personality (Roberts 1991); 2. University organizational resources (O'Shea et al. 2005); 3. Institutional key factors (university structures and policies that promote commercialization) 4. External or environmental factors (Sorenson and Stuart 2001; Wright et al. 2004; O'Shea et al. 2007).

4.2 The Academic Entrepreneur as a Key Factor

The academic entrepreneur is a key factor because the creation of an academic spin-off is the reflection of individual actions resulting from personal characteristics such as personality, skills, professional background and willingness to engage in business activities (Aceytuno and Paz 2008).

4.2.1 The Entrepreneur

The General Monitor Entrepreneurship (GEM) Report presents the latest definition for entrepreneurship as a group of business initiatives of any type and in any sector, including self- employment, that are in the market for a period not superior to 42 months, after which the activity is considered as consolidated (Vega et al. 2010).

4.2.2 General Characteristics of an Entrepreneur

According to Pablo et al. (2004), the entrepreneur is a highly motivated person who is energetic, innovative and creative, and who has great analysis capability. An entrepreneur is a perseverant person, prone to risk and with the ability to influence others. The author groups the characteristics as follows:

1. Motivation: key factors in this aspect are: initiative, need to escape, economical benefit perception, self-confidence, resource administrator, impulse/energy and ability to get resources
2. Energy and Personal Initiative: key factors in this aspect are: hard-work, perseverance, compromise, energy, initiative, stability, self-control, New challenges
3. Analysis Capability: key factors in this aspect are: opportunity oriented, planning on deadlines, environment analysis/reflection, efficient and knowledgeable
4. Psychological Profile: key factors in this aspect are perseverance and complete commitment, leadership, tolerance to change, achievement recognition, economical benefit perception, environment analysis/reflection, responsibility.
5. Inclined towards risk: key factors in this aspect are accepts moderate risks, self- confidence, problem-solving abilities
6. Ability to relate/Influence: key factors in this aspect are networking and business knowledge

4.2.3 Types of Entrepreneurs

They can be classified by: Motivation: It includes taking advantage of an opportunity—unsatisfied need or unattended market niche-; having a good idea and working on it; enjoying what they do (Audretsch 2002). They have a need: they are unemployed or need better income (Alcaraz 2006).

According to the enterprise: A social entrepreneur seeks to produce change and benefit through his creativity, enthusiasm and non-profit work (Veciana 2005).

An entrepreneur who seeks to stand out and be a model for others. A business entrepreneur who—in a company or in his own business- can begin from scratch, buy an existing company, manufacture products, offer services or sell pre-existing products (González 2005).

4.2.4 The Academic Entrepreneur

The figure of this kind of entrepreneur has been analyzed in literature by examining aspects such as academic rank, experience, scientific productivity and social networking among others as well as the factors that influence in the probability that an entrepreneur decides to create a company (Jones 1998; Audretsch et al. 2005).

Shane (2004a, b), Franzoni and Lissoni (2006) believe that researchers create companies near the end of their careers. Levin and Stephan (1991) indicate that a researcher's publications become a requisite to have access to teniorship which provides them with the peace of mind to develop other activities and that once stability and academic recognition are gained, entrepreneurs embark in other activities such as the creation of companies. For scientists, to create a company is a means to own intellectual property and to gain access to additional means of getting funding for future investigations (Feldman et al. 2001).

Audretsch and Stephan (1999) demonstrated that academic researchers create companies at a later stage of their careers than private companies researchers. Klofsten and Jones (2000) found that most academic entrepreneurs are over 40 years old. Shane (2004a, b), Shane and Khurana (2003) state that researchers who have achieved a high academic rank have a higher probability to create a company where the scientist's position acts as an indicator of the project's quality for potential investors and tends to minimize the inherent risk when creating a company. Shane (2003) and Shane (2004a, b) point out that when at least one of the researchers involved in an enterprise holds a high status, it is easier to get funds for the project. On the contrary, Réjean et al. (2006) found that the researcher's status does not influence significantly the possibility to create a new company. Vohora et al. (2004) indicated that scientific investigation is what allows the entrepreneur to gain the necessary knowledge to be able to identify new opportunities of commercial application. Réjean et al. (2006) found that the years of researching experience influence the probability to create a spin-off. Di-Gregorio and Shane (2003) point out that older scientists show higher probability to create enterprises than younger scientists. Such is the case of "star scientists" who have an outstanding number of publications and an influential position in academic and entrepreneurial communities with experience in the handling of subvention and personnel in their labs (Franzoni and Lissoni 2006). Those scientists seek financial rentability from the accumulated knowledge. This fact has been confirmed by case studies carried out in United Kingdom by Vohora et al. (2004), who found that the founders of those companies were leading in their field of research and knowledge and that they possessed a valuable know-how and technological expertise.

Di-Gregorio and Shane (2003) found, in a study about the factors that make a university generate more enterprises than others, that there is a significant and positive relation between the university eminence and the number of enterprises created. Louis et al. (1989) found that the variables that best predicted the participation of a scientist in an enterprise were their participation in other entrepreneurial activities (consultant, financing from an industry and patents) and an environment in which entrepreneurship is the norm. Shane and Khurana (2003) found that inventors who had previously studied technology had a greater probability to create a company than the ones who had not and that experience in obtaining external financing increases the probability to create a company. Réjean et al. (2006) also state that the researchers who carried out consulting activities with private companies, government agencies or organizations, show a higher probability to create a company.

4.3 Research Methodology

The methodology used in this research is case methodology based on the study developed by O'Shea (2005), with a variable analysis as a first approach to the real study subject (Maxwell 1996) in which the answer to the question what role does the key factor academic entrepreneurship play in the creation and evolution of the academic spin offs already identified in The Costa Rica Institute of Technology is being looked for is being looked for. A bibliographic revision that included an analysis of plans, programs and research and extension policies at ITCR was carried out as well as interviews with entrepreneurs, government officials from universities and science and technology sectors. As a Pattern Matching (Yin 2009) to compare data found in ITCR, the following Studies were used: La creación de Spin-off universitarios en España: características, determinantes y resultados by Ortin et al. (2007) and La creación de spin-off universitarias: el caso de la Universidad de Huelva by Aceytuno and Paz (2008).

4.4 Discussion of Results

The motivations ITCR entrepreneurs had to engage in their enterprises: (a) 20% of them wanted to apply their knowledge and get personal fulfillment, a fact that reinforces what McClelland (1967), Roberts (1991) and Ortín et al. (2007) pointed out to be a key motivation to create a spin off: need of accomplishment; (b) 14% wanted to satisfy a market need; (c) 11% experienced working dissatisfaction, although 100% stated that they had good working conditions; therefore, this aspect is not really taken into account; (d) 9% identification of a business opportunity,

Table 4.1 Relevance of the key factors for the creation of ITCR and Spanish academic spin-offs

Item	ITCR	Spain
*To apply technical knowledge	1	2
*Business opportunity detection	2	1
*The enterprise or institute where they worked encouraged them	3	6
*Entrepreneur's prestige or status	3	4
*Self-employment advantages	3	5
*Desire to earn more than working for a salary	4	3

Source Designed from the research and from Ortín et al. (2007)

which is pointed out as a motive to create a spin-off by Saxenian (1994); (e) 5% product/service quality- which encouraged the company or institute where they worked; (f) 2% positive attitude, desire for higher income, discipline; (g) 0% better working conditions and family tradition, which contradicts what was stated by Hsu et al. (2007) about family history being a key factor in the creation of spin-offs.

Table 4.1 compares the results obtained in this research and those obtained by Ortín et al. (2007) in 70 Spanish academic spin-offs, where the options: (a) applying technical knowledge and (b) detecting a business opportunity were the two main reasons for the creation of the spin-offs. The last option among ITCR academics was (f) a desire to earn more money than working for a salary, while for the Spanish entrepreneurs it was (c) the company or institute where you worked encouraged you. Options (c), (d) and (e) obtained third place with the same value among the ITCR entrepreneurs, while options (f), (e) and (d) were third, fourth and fifth for the Spanish academics.

Besides, ITCR academic entrepreneurs show a high degree of satisfaction for deciding to begin their productive project, 90% of them mentioned feeling proud of achieving that project and 80% mention that the experience has been worthwhile in regards to personal development. Sixty percent states they are proud with the enterprise because the experience has allowed them to improve their economic situation. Spanish entrepreneurs also point out that the experience has been enriching from a personal point of view, while Ortín et al. (2007) mentions that entrepreneurs are not really satisfied in regards to economic income.

The academic entrepreneur's most important characteristics are described: educational level, gender, age, working experience as well as objectives and motivations to create the enterprise. Regarding educational level 70% of the entrepreneurs had a Masters Degree when they created the company, from those entrepreneurs, 70% had attended foreign universities. 20% of the people interview had a PhD from foreign universities and 10% had a Licenciatura Degree, evidence that is in agreement with Roberts (1991) who states that there is a positive correlation between educational level and the decision to create a spin-off. In the case of entrepreneurs with a doctorate degree, salary increase was not a factor that negatively affected the creation of the spin-off in Spain, where there are a lot of doctors participating in the creation of such companies. Regarding gender, and in agreement with Aceytuno and Paz (2008) and Murray and Graham (2007), 90% of the academic entrepreneurs are men and only 10% are women.

The age factor has been analyzed by Roberts (1991), who pointed out that the average entrepreneur is 37 years old and by Ortín et al. (2007), who indicated that entrepreneurs in Spanish universities are between 30 and 40 years old. In ITCR, the average age was 34.6. The youngest entrepreneur was 22 years old, the oldest 52 and 80% are under 40 years old. These findings are supported by Roberts (1991) and Ortín et al. (2007). Regarding working conditions or what the academic was devoted to in his university at the moment of becoming an entrepreneur, it was found that at ITCR 60% of them were devoted to teaching and 40% to both teaching and research at the same time. Sixty percent of the people interviewed stated that their working link with ITCR had a positive effect on them professionally which was also stated by Rappert et al. (1999) and Johansson et al. (2005), when they stated that universities are a source of recruitment for highly qualified professionals for companies. The companies benefit in front of their customers from the academics' image and because of their constant training due to academic demands and their teaching activities all of which enrich their entrepreneurial experience. The other 40% was not linked to ITCR anymore either because they had already retired or because they had left their practice to engage in the entrepreneurial activity.

As a conclusion, we can say that all of ITCR's entrepreneurs are restless, dynamic people who have been leaders in the institution, who have promoted enterprises oriented towards the growth and academic improvement of the university. It was not possible to link professional and institutional performance to Stuart and Ding's plan (Stuart and Ding 2006) in which they point out that the most important scientists are those who are more prone to creating spin-offs because of the facility with which they can obtain the necessary resources to begin the enterprise.

Concerning entrepreneurship commitment, 40% of the people interviewed said that they devoted more than 40 h a week to the enterprise and the other 60% from 11 to 40 h. The number of hours depended on the enterprise's size, maturity and stability, the demand being higher at the beginning and lowering in time according to what Ortín et al. (2007) stated. Family history was not a key factor among ITCR's entrepreneurs who declared to come from families with no entrepreneurship tradition.

4.5 Conclusions

The qualitative methods used in this research allowed for non-structured means to gather the information to be utilized. Such means are more flexible, of a psychological and sociological nature, within exploratory methodology and using small samples from which abundant knowledge was gotten. The limitation with these means is that the results are not quantifiable or extrapolative to the whole of the population from which the descriptions and explanations were obtained. The academic entrepreneurs from ITCR stated that felt great need of success and to apply the knowledge they had acquired as well as a high degree of satisfaction with the entrepreneurial activity developed. They also pointed out the relationship between technological innovations and market needs, which place them within international standards.

References

Aceytuno M, Paz M (2008) La creación de spin-off universitarias: el caso de la Universidad de Huelva. Economía ind,pags. (368):97–111

Alcaraz R (2006) El Emprendedor de Exito México: Mcgraw-hill

ANCES (2003) La creación de Empresas de Base Tecnológica. Una experiencia práctica proyecto para la promoción de empresas innovadoras de base tecnológica. Valencia, España: ANCES, CEIN y otros

Audretsch A, Aldridge T, Oettl A (2005) The knowledge filter and economic growth: the role of scientist entrepreneurship. Discuss Pap Entrepreneurship Growth Public Policy 1106:66

Audretsch D, Stephan P (1999) Knowledge spill-overs in biotechnology: sources and incentives. J Evol Econ 9:97–107

Audretsch DB (2002) Entrepreneurship: a survey of the literature. CEPR, Institute of development strategies, London

Bueno E (2007) La Tercera Misión de la universidad: el reto de la transferencia del conocimiento. Revista de Investigación en Gestión de la Innovación y Tecnología (41)

Di-Gregorio D, Shane S (2003) Why do some universities generate more start-ups than others? Res Policy 32:209–227

Etzkowitz H, Leydesdorff L (2000) The dynamics of innovation: from national systems and mode 2 to a triple helix of university-industry-government relations. Res Policy 29(2):109–123. doi:10.1016/S0048-7333(99)00055-4

Etzkowitz H, Webster A, Gebhardt C, Terra BRC (2000) The future of the university and the university of the future: evolution of ivory tower to entrepreneurial paradigm. Res Policy 29(2):313–330. doi:10.1016/S0048-7333(99)00069-4

Feldman M, Feller I, Bercovitz J, Burton R (2001) Understanding evolving university-industry relationships. In: Feldman MLA (ed) Innovation policy in the knowledge-based economy. Kluwer Academic, pp 171–188

Franzoni C, Lissoni F (2006) Academic entrepreneurship, patents and spin-offs: critical issues and lessons for Europe. Paper presented at the Centro di Ricerca sui Processi di Innovazione e Internazionalizzazione CESPRI, Università Commerciale "Luigi Bocconi"

González AB (2005) Definición de Triple Hélice, 2010, from http://www.triplehelice.cl/esp/qs_nombre.htm

Hsu DH, Roberts EB, Eesley CE (2007) Entrepreneurs from technology-based universities: evidence from MIT. Res Policy 36(5):768–788. doi:10.1016/j.respol.2007.03.001

Johansson M, Jacob M, Hellström T (2005) The strength of strong ties: University spn-off and the Significance of Historical Relations. J Technol Transf 30:271–286

Jones D (1998) Universities, technology transfer and spin-off activities-academia entreneurship in different European regions (Final Report). Targeted Socio-economic Research Project (1042)

Klofsten M, Jones D (2000) Comparing academic entrepreneurship in Europe. Small Bus Econ 14(4):199–309

Laborda F, Briones A (2010) Innovación y Tecnología Factores de dirección estratégica de los agronegocios en Costa Rica y la Región de Murcia. Universidad Politécnica de Cartagena, Costa Rica-España, pp 99–116

Levin S, Stephan P (1991) Research productivity over the life cycle: evidence for academic scientists. American Econ Rev 81(1):114–130

Lockett A, Wright M (2005) Resources, capabilities, risk capital and the creation of university spin-out companies. Res Policy 34(7):1043–1057

Louis K, Blumenthal D, Gluck M, Stoto M (1989) Entrepreneurs in academe: behaviours among life scientist. Adm Sci Q 34(1):110–131

Markman GD, Gianiodis PT, Phan P (2008) Full-time faculty or part-time entrepreneurs. IEE Trans Engg Manag 55(1):29–36

Maxwell J (1996) Qualitative research design. An interactive approach. California Sage Publications

McClelland D (1967) The Achieving Society. New York

Murray F, Graham L (2007) Buying science and selling science: gender differences in the market form commercial science. Ind Corp Change 16(4):657–689

O'Shea R, Allen T (2005) Determinants and consequences of university spinoff activity: a conceptual framework. Paper presented at the Babson, Kauffman Entrepreneurship Conference

O'Shea R, Allen T, Morse K (2005) Creating the entrepreneurial university: the case of MIT. Paper presented at the Academy of Management Conference, Hawaii

O'Shea R, Allen T, Morse K, O'Gorman C, Roche F (2007) Delineating the anatomy of an entrepreneurial university: the Massachusetts Institute of Technology experience. (Delineando la anatomìa de un emprendedor universitario: La experiencia del Instituto Tecnológico de Massachusetts). R&D Management 37(1):16

O'Shea R, Chugh H, Allen T (2008) Determinants and consequences of university spinoff activity: a conceptual framework. J Technol Transf 33(6):653–666. doi:10.1007/s10961-007-9060-0

Ortín AP, Salas V, Trujillo MV, Vendrell F (2007) La creación de spin-off universitarios en España: Características, determinantes y resultados, 32

Pablo ID, Santos B, Bueno Y (2004) Las dimensiones del perfil del emprendedor: contraste empírico con emprendedores de éxito. In: Piqueras UVD (Ed) El emprendedor innovador y la creación de empresas de I+D+I, Valencia, España, pp 813–830

Rappert B, Webster A, Charles D (1999) Making sense of diversity and reluctance: academic–industrial relations and intellectual property. Res Policy 28:873–890

Rasmussen E, Moen Ø, Gulbrandsen M (2006) Initiatives to promote commercialization of university knowledge. Technovation 26(4):518–533. doi:10.1016/j.technovation.2004.11.005

Réjean L, Nabil A, Imad R (2006) Why are some university researchers more likely to create spin-offs than others? Evidence from Canadian universities. Res Policy 35(10):1599–1615. doi:10.1016/j.respol.2006.09.020

Roberts EB (1991) Entrepreneurs in high technology lesson from MIT and beyond. Oxford University Press, New York

Saxenian A (1994) Regional advantage: culture and competition in silicon valley and route 128. Cambridge (Mass.) y Londres

Shane S (2003) A general theory of entrepreneurship: the individual-opportunity nexus retrieved from http://books.google.co.cr/books?hl=es&lr=&id=0FxO_Wsh30kC&oi=fnd&pg=PR9&ots=79-DtsmdJo&sig=BXbIUfHVGxsEUG4Jf6RmvYf5Pa4#v=onepage&q&f=false

Shane S (2004a) Academic entrepreneurship university spin offs and wealth creation. Edward Elgar Publishing, Londres

Shane S (2004b) Encouraging University Entrepreneurship? The Effect of the Bayh-Dole Act on University Patenting in the United States. J Bus Ventur 19:127–151

Shane S, Khurana R (2003) Bringing individuals back in: the effects of career experience on new firm founding. Ind Corp Change 12(3):519–543

Sorenson O, Stuart TE (2001) Syndication network and the spatial distribution of venture capital investments. American J Sociol 106(6):1546–1588

Stuart TE, Ding WW (2006) When do scientists become entrepreneurs? The social structural antecedents of commercial activity in the academic life sciences. American J Sociol 112(1):97–144

Veciana JM (2005) Creación de empresas: un enfoque gerencial, Vol. 33. Caja de ahorros y pensiones de Barcelona, La Caixa

Vega IDL, Cruz C, Justo R, Coduras A, González I (2010) Global entrepreneurship monitor informe ejecutivo 2009, Instituto de Empresa, España

Vohora A, Wright M, Lockett A (2004) Critical junctures in the development of university high-tech spinout companies. Res Policy 33(1):147–175. doi:10.1016/S0048-7333(03)00107-0

Wright M, Birley S, Mosey S (2004) Entrepreneurship and university technology transfer. J technol transf 29:235–246

Wright M, Clarysse B, Mustar P, Lockett A (2007) Academic entrepreneurship in Europe

Yin R (2009) Discovering the future of the case study method in evaluation research. Eval Pract 15(3):283–290. doi:10.1016/0886-1633(94)90023-x

Chapter 5
Economic Impact of Rural Tourism in Costa Rica and in the Region of Murcia, and Its Implications on Demand

Catalina María Morales, Antonio Juan Briones, Andrés Artal and Antonio García

Abstract Tourism in rural areas is a wide concept that refers to the various leisure activities that can be carried out in rural zones, which include such modalities as: rural tourism, ecological tourism, agro-tourism, adventure tourism, cultural tourism, business tourism, young person tourism, social tourism, health tourism and sport tourism. These activities, little valued in the past, have come to integrate true productive chains involving, amongst other pursuits: properties, agrarian industries, transport, restaurants and communications, all of which constitute an impeller of growth, mainly for non-agricultural activities in the rural zone.

5.1 Introduction

According to Icart and Alarcón (2006) tourism constitutes a strategic industry in the national or regional economy. It is for this reason that a competitive development of the tourist destinations is indispensable, in the face of the growth of the tourism and leisure industry at an international level.

In this sense, Verardi Fialho (2001) points out two characteristics of tourism in the rural space that complete the definition of this activity as a component of a development strategy. One of them is related to the tourist potential of rural areas, and the associated idea is that they do not need to be regions of extraordinary natural beauty, but rather to possess well-developed cultural aspects, an appreciable architecture, a characteristic gastronomy; they must be places where the

C. M. Morales (✉) · A. J. Briones · A. Artal · A. García
ETSII-Universidad Politécnica de Cartagena,
C/Dr Fleming, s/n, 30202 Cartagena, Spain
e-mail: mmgkty@hotmail.com

S. P. Sethi et al. (eds.), *Industrial Engineering: Innovative Networks*,
DOI: 10.1007/978-1-4471-2321-7_5,

Table 5.1 Topics analyzed in tourism demand

Author	Year	Topic analyzed
Cànoves, G., Herrera, L., Cuesta, L.	2005	Territorial distribution of rural tourism in Catalonia taking into account conditioning factors of both offer and demand
Lois, R., Piñeira, J., Santomil, D.	2009	Rural tourism in Galicia confronts transformations in both landscape and culture which influence the motivations of tourist demand
López, T., Millán, M., Agudo, E.	2006	Rural tourism as an instrument to achieve the development of regions where an economic imbalance exists
Lozano, A., Morales A., Navarro, M.	2006	Analysis of tourist demand in Andalucía, a substantial contribution to its regional GDP
Roig, B.	2005	A quantification of the recent evolution of the main peculiarities of Spanish demand of rural house tourism
Usach, J.	1998	Territorial analysis of internal tourist demand flows
Villalobos, D., Galdeano, E., Tolón, A.	2009	Intended demand of nature tourism by international tourists

population has preserved their habits and customs, thus making them interesting as a whole. The other feature has to do with the possibility of creating links with other activities, such as the preparation of homemade foods (breads, pastries, cheeses, cured meats, preserves, jams, sweets, honey, among others), restaurants offering typical food, arts and crafts workshops, horse-riding or cart trips or ecological trekking, among others.

One can expect that the common ground in this type of tourism be the fact that the services are rendered by the farmer's or the agricultural producer's family and that their development is concerned with the recovery and conservation of the natural, cultural and architectural heritage of the rural world.

In the same sense, Lage and Milone (2000), mentioned by Verardi Fialho (2001), point out that job creation may not be the objective of all tourist development, but it is surely one of its main results, especially if we consider its capacity to absorb and promote youth labour. Whether directly in managing and/or operating a tourist service, or derived from connected activities, employment resulting from tourist development is boosted.

Several authors have studied tourism demand in general and, more specifically, the demand for rural tourism (Table 5.1).

According to White (1999), mentioned by Icart and Alarcón (2006), rural tourism is an economic activity characterized by its development outside the urban nuclei, its fundamental orientation being to town inhabitants; by taking place in small groups, outdoors; by its wide-ranging use of both natural and cultural resources: heritage, lodgings and services, which are characteristic of the rural environment, and by contributing to local development and to tourist diversification and competitiveness.

In their examination of the economic analysis of tourism, Capó et al. (2006) clarify that two of the fields where more literature has been produced regarding the economic aspects of tourism have been the study of its demand and it's modelling.

Demand forecasting has been one of the most important areas scrutinized by investigators.

Among the reasons cited by Frechtling (1996) in order to know the determinants of the demand is the fact that the tourist product cannot be stored, so an unused offer cannot be kept for future periods of higher demand. Nor is it possible to separate the production process from the consumption.

This interaction between consumers and producers forces to offer the goods and services at the exact time when the demand takes place. It must also be considered that tourist satisfaction depends to a large extent on the complementary services. Another factor is that tourist demand is extremely sensitive to natural disasters and to political-social problems such as wars, terrorist attacks and natural disasters, amongst others. Finally, there is the added necessity to undertake long-term investments in equipment and infrastructures that are suitable for the expectations of a future demand.

Therefore, the market is the point of encounter between offer and demand, and it is controlled by the various elements and/or components of these two concepts. Each one of them has their own distinctive variables. This being the case, next we will define the dimensions of the demand that are analyzed in this article through the case study of the economics of tourism in Costa Rich and in the Region of Murcia.

Tourism activity has a strong repercussion on the quantitative economic variables (private income and employment, among others) and qualitative economic variables (standard of living and welfare, among others) of the regions and countries where it occurs; therefore, it is important to assess its positive contribution to the progress of countries and tourist destinations (Sancho 2001). Firstly, let us consider the injection of revenue for the economy of a region. At a national level, this income will initially improve the balance of payments and will later stimulate both domestic and foreign investment. In this context, it is acknowledged that the percentage of tourist expense that will remain in the local economy depends on the proportion of tourism companies, suppliers, local and foreign workers to whom each round of expenditure is destined.

Secondly, the new tourist attitude could represent an initial supplement to agriculture and/or industry, which permits a diversification of traditional sources of income. Seasonality in tourism and agricultural demands contributes to maintaining a constant source of income. A third economic effect of tourism is that it usually contributes to the creation of employment, both in the tourist sector and in non-tourist activities. Recipient regions have traditionally witnessed strong levels of employment growth, due to the basic fact that tourism services are labour-intensive.

A fourth positive effect of tourism is the creation and/or improvement of infrastructures, which is a requirement for its development. This will benefit the visitors as much as the rest of the economic sectors and the local residents. Finally, another effect of tourism is its capacity to improve the distribution

5.2 Demand Indicators

The alluded problem of finding the exact definition of tourist product and its quantitative determination leads to consider indicators that help us to approach the definition of rural tourism demand. Particularly, in this essay several indicators of demand are considered: the foreign currency derived from foreign tourist spending, the arrival of tourists, the number of overnight stays and the number of travellers.

A traveller is defined as a person visiting any kind of enterprise classed as a rural tourism establishment. Overnight stay is defined as every night that a traveller is put up at a rural establishment.

5.3 Case Studies

5.3.1 The Economics of Tourism in Costa Rica

Tourism is importantly related to the external macroeconomic situation, so it is fair to expect that in years of world expansion, tourism grows. In the same way, a global income reduction will affect the sector and the benefits derived from it.

In addition to this, the entrance of foreign currency associated with tourism totalled 5.38% of the final consumption figure (both Costa Rican households and the general government) in 1991. Later, this percentage increased to the region of 9%, where it remained steadily between 2004 and 2007. The number of visitors to the country has risen by approximately 10.33% annually between 1986 and 2007. In connection with the flow of such visits, Costa Rica has been relatively successful in comparison with the rest of Central American countries. During the years 2005–2007, according to data released by the World Tourism Organization, the number of guests arriving in Costa Rica has been the largest in all of Central America, ranking at 10th place for the whole of the continent.

The entrance of foreign currency due to tourism has grown in importance when compared with other sources of foreign exchange. For instance, the entrance of foreign currency due to tourism represented only 82% of the entrances related to coffee and bananas in 1998. This rate has increased in such a way that by 2007 tourism accounted for twice as much as the combined national revenue of both coffee and bananas. Similarly, since 2000 tourism has supplied as much foreign currency as exports by manufacturers and corresponds to approximately 40% of all free trade zone exports.

In 2007, 48.18% of all visitors to Central America came from North America. Costa Rica, the first tourist destination in Central America, is also largely dependent on North American tourism, which totalled 40% in 2007. After the USA, the countries from which more visitors are received are Nicaragua and Canada.

This dynamic has made tourism one of the most important sectors in terms of generation of external revenue.

5.3.2 *The Economics of Tourism in the Region of Murcia*

According to a report by the Spanish National Institute of Statistics, in 2007 Spain was the second country in the world in terms of revenue obtained from international tourism, with France ranking first, and second only to the United States in the number of international tourist arrivals. In 2008 Spain kept its second place in terms of revenue obtained from international tourism, again surpassed by France, but it was third in the number of international tourist arrivals, after France and the United States.

349.4 million overnight stays took place in Spain during 2009, with hotels claiming 72.10% of the total. If we look at 2008, the number of overnight stays was higher: 375.7 million. Again, hotels account for most of them (71.40%). This percentage made 2009 slightly better. The evolution of these figures in the last 10 years has been a positive one, although in the new recessive situation, a decline can be observed since 2008. In February of 2011 a total of 2.8 million international tourists visited Spain, which means a growth of 4.30% as compared with the same month of the previous year. This positive evolution is parallel to that of the first month of 2011.

The two main issuing markets, the United Kingdom and Germany, have reduced their tourist figures, as opposed to the growth registered by other tourist-issuing markets, especially the Nordic countries. If we analyze this on a regional basis, without question it is in the Canary Islands where the best evolution can be seen. So far this year 5.5 million international tourists have been received in Spain, 235,000 more than in the same period of 2010, which is a 4.5% increase.

5.4 Discussion of Results

In Costa Rica as described previously, the arrival of tourists in the country has grown by 10.33% annually from 1986 to 2007. In connection with such a flow of visitors, Costa Rica has been relatively successful in comparison with other Central American countries. For the period 2005–2007, according to the World Tourism Organization, the number of tourist arrivals in Costa Rica has been the largest in all Central America. The country ranks in 10th position within the American continent. The same applies to the entrance of foreign currency in dollars.

Despite being the leading tourist destination in Central America, Costa Rica largely relies on visitors from the United States: up to 48.18% of all visits to the country in 2007. This percentage translates in nearly one million visitors (953,812) from the USA, followed by Central America and Europe with 592,840 and 271,631 visits respectively.

Within North America, the United States generate the largest amount of visitors, with more than 790,000 visits since 2007, which represents nearly 40% of the

Table 5.2 Comparative analysis of the variables studied

Comparative analysis			
Variable studied	Costa Rica	Region of Murcia	Relevant data
Age	Age of tourists in Costa Rica ranges from 30 to 45	The 26–35 bracket is where most tourists are included	Tourists visiting both countries are mostly young adults
Stay at destination	Average stay is 10 nights	Tourists stay, on average, 3–4 days	The farther the destination, the longer the stay
Reason for trip	Main reason is holidays	Tourists enjoy natural surroundings, sports and excursions	Relaxation and holidays are the main reasons in both destinations
Average expenditure	Average expenditure is approximately 70 Euros per day	Rural tourists spend approximately 50 Euros per day	Average daily expenditure at the destinations studied is 60 Euros per day
Holiday periods	High season is January–March, July and December	Most tourist arrivals occur in July, August and September	Dry season or summer are the months with the most visitors

total number of visits to Costa Rica, followed by Nicaragua and Canada, with 379,222 and 102,061 visitors respectively.

This dynamic has turned tourism into one of the most important sectors in the generation of external revenue. For example, focusing in the main products exported by each country, it can be observed that the tourism industry has got to "equate or surpass exports of oil, manufactured food or automobiles". In this sense, tourism in Costa Rica has come to be twice the value of the combined exports of coffee and bananas, and tends to be on a par with the exports of the manufacturing industry. It has not managed to gain ground on free trade zone exports in recent years, but still it amounts to 40% of the total value of exports.

It must be noted that these revenues have grown not only in what concerns exports of certain goods or sectors, but also in connection with other related macroeconomic issues, such as Gross Domestic Product, national income and consumption. In this way, the tourist sector has experienced a significant growth in the last years and has become an important variable in the Costa Rican economy, in so much as it finances a considerable part of the running account deficit and therefore of the consumption. Moreover, it has allowed a reduction in the level of dependence of traditional exports (coffee and bananas) and its exports have come to contribute in similar terms to those of the manufacturing industry.

A survey carried out in by the ICT 2007 among foreign visitors arriving at the Juan Santamaría Airport revealed that 68% of the interviewees entered the country for the purpose of vacation, leisure, recess and pleasure. However, 52% of Central

American visitors stated business and professional reasons as the reason for their visit to Costa Rica (Table 5.2).

According to the same survey, "non-resident visitors stay in Costa Rica for an average of 12.4 nights, whereas visitors residing in Central America are those with the smallest amount of overnight stays: 8.2 nights. On the contrary, European visitors stay the longest, with an average of 16.9 nights, followed by residents in Canada with 16 nights, the Caribbean with 14.3 and the Americans with 11.4 nights."

It is also evidenced that "the most visited area is the Central Valley, mentioned by 80% of the interviewees, followed by the Mid-Pacific area and the North Zone, with 30 and 29% respectively. Nevertheless, the place where most tourists spend the night in Costa Rica is the Central Valley, which averages 7 nights, followed by South Guanacaste with 6.4, Puntarenas with 6 nights, North Guanacaste with 5.7 and the Osa Peninsula with 5.6; the areas of Mid-Pacific and South Caribbean also feature as the places where tourists stay practically as long, with approximately 5 nights. On the contrary, the North Area and North Caribbean, obtain, respectively, 3 and 2.5 nights on average.

The main difference between rural tourism and rural community tourism is that the latter is an economic activity that has been planned by the communal organization and it is the residents of the communities who participate in a direct way in the administration of the enterprises and in their profits.

In the case of the Region of Murcia, the tourist activity is, "without question, one of those which have experienced a greater dynamism in the last decades", as a study carried out by the Economic and Social Council (CES) of the Region of Murcia affirmed in 2010.

During 2009 349.4 million overnight stays occurred in Spain, 72.1% of which were in hotels, while the same data for 2008 show a larger amount, with overnight stays reaching 375.7 million and again hotels claiming most of them (71,4%). These data point to a small improvement for 2009. The numbers of rural tourism lodgings have seen a light ascent, since they pass from 2.1% of the total tourist lodgings in 2008 to 2.3% in 2009. In terms of overnight stays, the increase is from 7.8 to 7.9 million.

The evolution of the number of travellers and their overnight stays in the last 10 years has been a positive one, although given the current financially recessive circumstances, a decrease can be perceived, starting from 2008.

This idea is confirmed in the Region of Murcia by reports elaborated at the Regional Council of Culture and Tourism in 2008 and 2009: tourist affluence to rural lodgings diminished, as did the number of travellers staying at camping sites. In 2008, 132,187 visitors made se of camping facilities, but this figure was reduced by 7,253 in 2009, so the total number of campers was 124,934. Occupation decreased in rural lodgings by over 10,000 visitors in 2008 and 2009.

In 2009 there were fewer overnight stays by non-residents than in the previous year in all kinds of tourist lodgings, in comparison with camping sites and rural tourism lodgings, where the number of residents increased (1.2 and 2.2% respectively) (Table 5.2).

The reduction in the number of overnight stays in rural lodgings has been considerable, though not in the amount of incoming travellers. This means that tourists have continued to visit the Region of Murcia but have not included in their plans the option of staying at this type of lodging.

Camping sites on their own show little variation in their total numbers of guests, but similarly to rural lodgings, have seen a decrease in the total number of overnight stays.

It is important to realize that the accommodation offer did not undergo major changes in 2009 in comparison with the previous year. The most significant variations were in country house rentals, with 2008 registering 486 lodgings and 3,000 users, while in 2009 those figures increased to 500 and 3,092 respectively.

The average length of stay of rural lodgings visitors is approximately 3 days in Spain, while in the Region of Murcia this increases significantly in 2008, reaching its peak: three and a half days approximately. In 2007 and 2009 it was approximately 3.25 days. The daily expenditure per person in this type of lodging, at least in 2008, came to slightly over 70€.

5.5 Conclusions

Performance in the tourism sector has declined, possibly due to three basic factors: the crisis, which has led us to be more careful with our money and to make sacrifices; secondly, reservations are practically inexistent, but price reductions will only worsen the situation, since rural tourism also needs to be competitive; finally, the world situation holds back the entrance of foreigners, which could re-activate this sector. Both Costa Rica and the Region of Murcia should keep in mind that this new rural tourism product is complementary of other more traditional, consolidated type of tourism products.

References

Blanco M (2006) "Potential analysis of rural tourism in the cantons of Upala, Los Chiles, Guatuso y La Cruz."

Capó J, Riera A, Roselló J (2006) "A revision of the economic analysis of tourism." *Principios* magazine, no 5

Icart B, Alarcón A (2006) "Quality and authenticity of rural tourism." *Estudios Turísticos* magazine (168)

Frechtling DC (1996) Practical tourism forecasting. Butterworth Heinemann, Oxford

National University of Litoral. II World Forum on agrotourism and rural tourism, Santa Fé, 7 to 10 November 2002 (online). Available at: www.foroagroturismo.unl.edu.ar

Riveros H, Blanco M (2003) "El agroturismo, una alternativa para revalorizar la agroindustria rural como mecanismo de desarrollo local". Documento Técnico. IICA-Prodar, Lima (Perú)

Rodríguez A, Saborío M (eds) (2009) Lo rural es diverso, evidencia para el caso de Costa Rica. Instituto Interamericano de Cooperación para la Agricultura, San José (Costa Rica)

Sancho A (dir) (2001) Apuntes de metodología de la investigación en turismo, Organización Mundial del Turismo. Madrid, España

Spanish National Statistics Institute. "Spain in figures 2009" (2009) Madrid, Spain http://www.ine.es

Verardi Fialho M (2001) "Rural tourism and non-agrarian rural employment. The case of the colonial route Dos Irmaos, Rio Grande do Sul." (online). Available at http://www rimisp.cl/dfid/colonial-es.pdf

Chapter 6
Does the Accessibility of CE Regions Influence Investment Attractiveness?

Marija Bogataj, Milan Domborsky, Vinko Vidučić and Robert Vodopivec

Abstract This paper presents some factors influencing investment decision of investors in smaller industrial site location, where the accessibility of the site where industrial activities will take place is of the main concern. There was a need for empirical analysis to identify factors which influence location of investments in Central European Countries and to evaluate how the accessibility to a central place and to the other activities influences investment attractiveness. Empirical study of factors has been completed in Czech Republic, Slovenia and Croatia and the hypotheses about similarities and differences among regions have been tested. The paper is the result of complementary study to the core of ATTREG ESPON project, which strive to achieve a better understanding of the contribution of European regions' and cities' attractiveness to economic performance. Here we have to differentiate two groups of factors influencing migrations: (a) internal factors which influence the quality of places and (b) those which influence accessibility to region.

M. Bogataj (✉)
University of Ljubljana, and MEDIFAS, Mednarodni prehod 6, Šempeter pri Gorici, Slovenia
e-mail: marija.bogataj@guest.arnes.si

M. Domborsky
University of Economics, W. Churcill Sq. 4, Prague, Czech Republic

V. Vidučić
POMORSKI FAKULTET U SPLITU, Zrinsko-Frankopanska 38, 21000 Split, Croatia

R. Vodopivec
MEDIFAS, Mednarodni prehod 6, Šempeter pri Gorici, Slovenia

S. P. Sethi et al. (eds.), *Industrial Engineering: Innovative Networks*,
DOI: 10.1007/978-1-4471-2321-7_6, © Springer-Verlag London Limited 2012

6.1 Introduction

Extensive spending has taken place under ERDF, Cohesion Fund and ISPA to reduce disparities among regions. One of the prominent initiatives in the European Union in this respect is the development of the Trans-European transport networks (TEN-T), where also investments in Central European (CE) networks took place among priority list, which is based on the accessibility index value. This paper presents some factors influencing investment decision of investors in smaller industrial site location, where the accessibility of the site where industrial activities will take place is of the main concern. There was a need for empirical analysis to identify factors which influence location of investments in Central European Countries and to evaluate how the accessibility to a central place and to the other activities influences investment attractiveness. Empirical study of factors has been completed in Czech Republic, Slovenia and Croatia and the hypotheses about similarities and differences among regions have been tested. The paper is the result of complementary study to the core of ATTREG ESPON project, which strive to achieve a better understanding of the contribution of European regions' and cities' attractiveness to economic performance. Here we have to differentiate two groups of factors influencing migrations: (a) internal factors which influence the quality of places and (b) those which influence accessibility to region.

In 2010 new ESPON project ATTREG was starting, where we study the flows between regions dependent of endowment factors of regions which are (Russo et al. 2010, Drobne and Bogataj 2011): Environmental Capital, Anthropic Economic, Social & Cultural Capital and Human Capital but also ***Institutional Capital*** variables measured also by accessibility potential of regions. These variables are directly related to the actions of public bodies at European, national or local/regional level. Each source of territorial capital consists of a number of indicators covering the whole territory of analysis (endowment factors) which differ in places and in time. Those people who are more skilled and educated, whose affluence permits more travel, and who are familiar with different parts of the country, tend to migrate more If we divide the flow of human resources to two groups: (a) top-workers and creative class and (b) opportunity-driven immigrants, we can assume that different endowment factors from the groups of capital listed above are differently important for this two classes. We also find that an accessibility to central places influence strongly in-migrations. Here we wish to investigate, how accessibility influence investment in regions.

6.2 Accessibility Index

Accessibility plays a significant role in European policy related to the development of territorial units. In several European policy documents as it is European Commission Green Paper on Territorial Cohesion or Territorial Agenda of the EU involving all EU Member States, accessibility is seen as key factor in improving attractiveness of Members States, their regions and cities. Accessibility and mobility are prerequisites for territorial economic development. Regions having a high accessibility to raw materials, suppliers and markets are in general economically more successful regions enjoying a more competitive position in the global market. If so, transport infrastructure improvement might be an important policy instrument to promote regional economic development (ESPON 2009). Accessibility is a measure of ease of access and in all of our previous research works it was assumed to be symmetrical. Converting distance to time spending or some other actual costs of access, we get a measure that may provide a changed ranking of localities by centrality. By accessibility in general we want to describe an integral view of localization qualities that result from nonlocal influences.

The interest of industrial real estate owners is to attract activity cells in recovered supply chains to participate in the future growth. Regional economic integrations in EU had provided an important stimulus to trade and to investments before economic crisis has appeared, and these integrations could positively influence further development of economic activities of EU accessed regions. Accessibility to central places of EU states or states in accession processes has been improved by investments in EU corridors and by removing the barriers between states.

Existing corporate and industrial locations had therefore changed localization characteristics. New location decision process can appear to be an important challenge for analysts and decision makers. It could be efficient if a logical and systematic procedure is followed. When such an approach is adopted, the chance of finding a location that maximises the firm's ability to realise critical objectives is increasing.

Theories of Smart Growth (Turner 2007) usually point at accessibility as the most significant factor. In most cases higher accessibility is influencing mobility as a measure of the vehicle-miles or person-miles involved in travel. Finding efficient road configurations for sustainable developments in towns and regions is basically an accessibility problem. If road configurations and modes are made in such a way that they provide effective intermodal access to majority of the human resources and activities, it will make developments more sustainable in the future. The goal of transportation accessibility is therefore to find the most efficient way to connect origins and destinations, provide road designs that promote different modes of transportation, reduce congestion and create livable, developed regions with sustainable transportation.

While in its narrow sense localization is modifying or adapting a production to fit the requirements of a particular locale, and while transportation accessibility participate to design and develop a production to function in connection of multiple locales (internationalization), in transportation and physical planning we understand localization broadly as a process of making decisions in space and to adopt it. Each place offers certain resources and each economic activity is characterised by certain needs. The best location of a socio-economic activity is where optimum resources exist for it. Resources and needs vary over time and, as a result, the optimal localization of socio-economic activities changes on time horizon.

Localization is among the most important issues studied in the research areas focused on the utilisation of space (socio-economic geography, spatial economics etc.). This is one of the first topics of regional development. People have always been interested in where is the best to locate their activities. Von Thünen (1726, 2009) was the first theorist who addressed the localization issue in details and many modern theories based on this approaches (Fujita 2010).

6.3 Accessibility in the Modern Monocentric and Polycentric City Models

How important accessibility is also in modern urban society, was shown first in Alonso's (1964) book Location and land use where von Thünen's approach, replacing agricultural products with urban activities in the model of localization, was used. Here direct distance and the existence of one centre are the only criteria of accessibility and transport costs. Like in von Thünen model the costs of transport are supposed to grow proportionately to distance from the centre while rentals are declining proportionately to this distance. The distance is assumed to be Euclidian, but in modern theory the accessibility is often measured by time spending or cost distance.

According to Alonso people and enterprises on the demand site can be divided into groups based on preference to smaller rent and those who prefer smaller transport costs. Though there exist different new theories about localization (Fujita 2010), all these contributions have not included study of accessibility to different activities as an important factor influencing significantly the investment attractiveness of region regarding global supply chains matrices, having activities in different regions. Though the value-added approach in cluster and supply chain management studies (SCM) is promising and seems to be good direction to study connection between capacity of activities in supply chain and regional environment (attractiveness of region), only first steps towards embedding accessibility in SCM problems connected with regional environment have been done [see Bogataj (1996), Bogataj and Usenik (2005, 2009), Bogataj et al. (2011a, b), Bogataj and Grubbström (2011), Bogataj and Bogataj (2011)]. In all this studies it was able to see how the location of individual activity cells in a global supply chain could significantly influence net present value of activities in the supply chain and especially in reverse activities.

6.4 Empirical Results

The key issue of all location theories and research of localization is the determination of factors influencing location decision. Above mentioned theories relatively exactly determine main localization factors but the importance of these

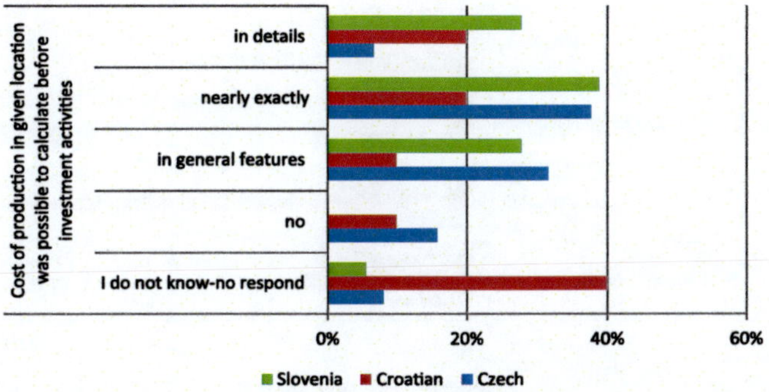

Fig. 6.1 How have been able to calculate costs of productions before investment activities

factors differ from theory to theory. The question appears which of these factors more significantly influence investment attractiveness of regions in the CE Regions (see: Damborský and Jetmar 2008). Therefore comparative study among the Czech Republic, Slovenia and Croatia was prepared.

297 firms have participated in empirical research in the Czech Republic. 19.8% of firms are in the micro-firm category (1–9 employees), 57% of firms are in the small firm category (10–49 employees), 22.4% of firms in the medium firm category (50–249 employees) and 0.8% of firms in the large firm category (250 and more employees). 3.7% of firms are agricultural, 56% are industrial and 40.3% are services. In Slovenia we took small sample of 18 units, 50% of them are working in logistics and the others mostly in production. We have a sample of 20 units from Croatia.

Location decision of decision makers in Czech companies has been making more intuitively in more than 50% of cases and before capital investment (choosing one location among available) nearly one half of Czech companies would be able to "exactly specify"or "nearly exactly specify" differences in costs of production as a result of differences in location accessibility before capital investment (Fig. 6.1). This percentage does not seem to be significantly different in Croatia (P-value of association test for Croatia is around 0.2) but for Slovenia percentage of firms which have been able to make more detailed calculation of differences in localization costs (in details or nearly exactly, see Fig. 6.2) is higher (with probability higher than $P = 0.01$). Decision makers from Croatia and Slovenia were trying to choose location less intuitively than those from the Czech Republic ($P < 0.01$).

We wanted to get answers on the following questions:

1. What were the main costs influencing location decision;
2. How had accessibility (transportation costs) to consumers influenced location decision;
3. How was geographical distance of competitive companies and companies of the same branches important for location decision and how the decision makers are inclined to clustering;

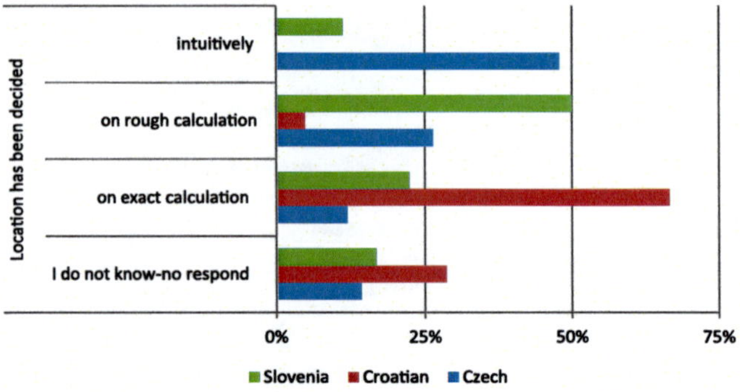

Fig. 6.2 How intuitively location has been chosen

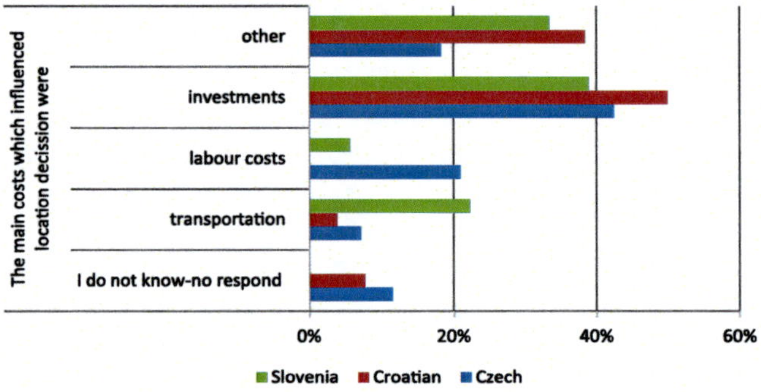

Fig. 6.3 The main costs which influenced location decision

4. How had distance from an important city, which is able to provide enough services (financial, education, cultural, business institutions) influenced final location decision;
5. Did the decision makers prefer regions with higher economic growth;
6. How did accessibility to suppliers influence decision of location;
7. How did accessibility to existing managers of enterprises or accessibility to owner's home influence location decision;

Using Pearson's Chi-square test of association we have found out the following:

1. Transportation costs have been the main cost for location decision of investment only in around 42% of Czech companies and the percentage of Slovenian and Croatian firms has been significantly lower as presented at Fig. 6.3 (P-value is less than 0.01) though in Slovenia this percentage seems to be a little higher than in Croatia (P-value = 0.15). Also our previous study on the influence of distance to the CBD and the land price confirm these results (Bogataj et al. 2011a, b).

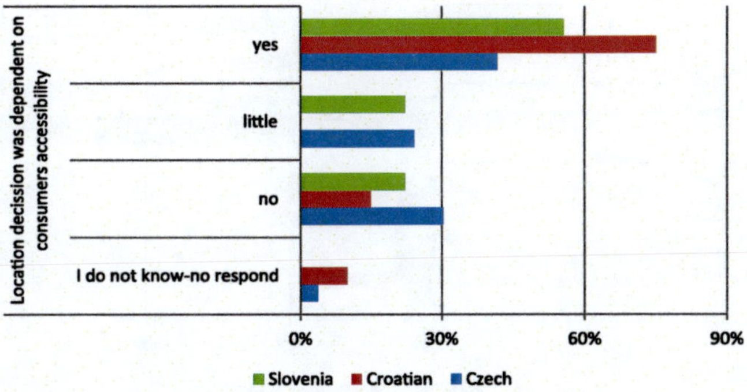

Fig. 6.4 The influence of customer accessibility

2. For more than 42% of Czech companies the accessibility to consumers was important for location decision. The percentage of positive answers was significantly higher in Slovenia and in Croatia (see Fig. 6.4). Here, the difference between Slovenian and Croatian decision makers was not significant.
3. Geographical distance of competitive companies was important for location decision only in less than 20% of Czech enterprises. The percentage of such enterprises in Croatia and in Slovenia is not significantly different. Only 12% of enterprises were trying to locate close to the area where other firms of the same branches are located. Also they did not try to avoid the region where competitors are located in majority of Czech, Croatian or Slovenian firms.
   In all three countries the firms are not trying to locate close to the firms of the same branches, it means that they are not inclined to clustering in this sense. While Slovenian and Czech firms prefer to locate in regions which are attractive also for other enterprises, this is not the case in Croatia.
4. Following von Thünen theory, Alonso developed the hypotheses that the distance from an important city (central business district, CBD), which is able to provide enough services (financial education, cultural, business institutions), influences the most the final location decision. Marginal accessibility value for different activities is different and is not constant (see: Bogataj et al. 2011a, b). The sample of Czech firms does not confirm this theory for high part of enterprises. Only 41.4% of decision makers have followed this rules of importance to be close to central business district. The percentage of Croatian decision makers with similar evaluation of accessibility impact was not significantly different, but for Slovenian decision makers we can say that higher percentage of them are following Alonso's, von Thünen's or modern Krugman's theory (P-value <0.05) and agree the distance to CBD is the most important factor for localization (Fujita et al. 1999).
5. Regional growth influenced localization in 36.4% Czech firms and there is not significantly different percentage of Slovenian firms influenced by regional growth. This percentage seems to be higher in Croatia. That regional growth influenced location decision in Croatia often than in Czech Republic (significance level $P < 0.005$).

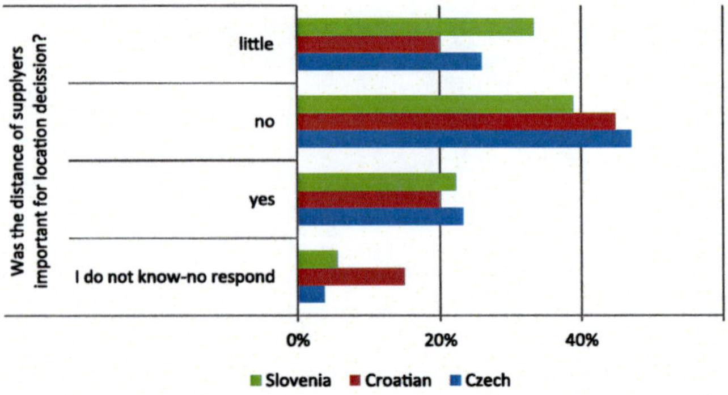

Fig. 6.5 The distance to supplier seems to be of little or no importance to decision makers

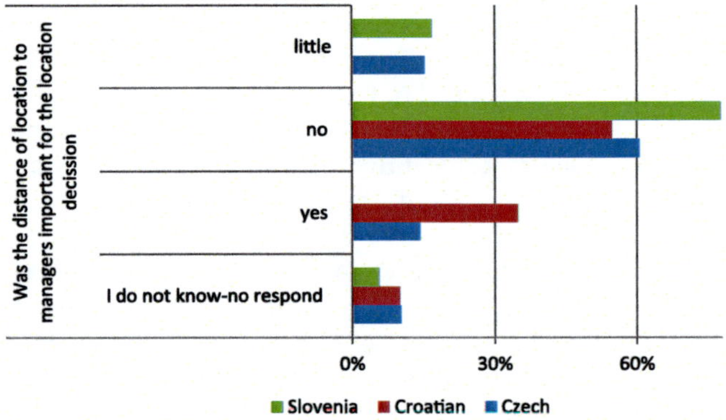

Fig. 6.6 The influence of accessibility to managers' home

6. Accessibility to suppliers (Fig. 6.5) influenced decision where to locate in nearly 23% of Czech enterprises. The percentage of such enterprises in Slovenia and in Croatia was nearly the same as in the Czech Republic. The Chi-Square value of association tests was 0.1 or lower in both cases or *P*-values have been close to 1 in both cases.
7. An accessibility to home of existing managers of enterprises (Fig. 6.6) or the accessibility to owner's home did not influence localization significantly in any of studied countries.

6.5 Conclusion

The cohesion policy of the European Commission influences development of new transport infrastructure and it is important support to regional economic development. Extensive spending has taken place under ERDF, Cohesion Fund and ISPA to reduce disparities among regions by investments in the countries with law accessibility index. One of the prominent initiatives in the European Union in this respect is the development of the Trans-European transport networks (TEN-T), where also investments in CE networks took place among priority list, which is based on the accessibility index evaluation. We can see that the accessibility to the costumers and to the business districts influence decision in many but not in all cases, but the main costs which influence location decision are costs of investment in the industrial site and not net present value of the future transportation costs. We know that the costs of transport which are the results of accessibility influence net present value in any supply chain, but the maximum net present value of activities connected in a supply chain does not seem to be the main goal of new investors in Central European Countries like Czech, Slovenia or Croatia.

Acknowledgements The paper and it's dissemination was partly financed by ATTREG—The attractiveness of European regions and cities for residents and visitors. ESPON Applied Research 2013/1/7 and ARRS: J5-4279-2734.

References

Alonso WA (1964) Location and land use: towards a general theory of land rent. Harvard University Press, Cambridge

Bogataj M (1996) Inventories in spatial models. Int J Prod Econ 45(1):337–342

Bogataj D, Bogataj M (2011) The role of free economic zones in global supply chains—a case of reverse logistics. Int J Prod Econ 131(1):365–371

Bogataj M, Grubbström RW (2011) On the representation of timing for different structures within MRP theory. Int J Prod Econ. doi: 10.1016/j.ijpe.2011.04.016

Bogataj M, Usenik J (2005) Fuzzy approach to the spatial games in the total market area. Int J Prod Econ 93:493–503

Bogataj M, Grubbström RW, Bogataj L (2011a) Efficient location of industrial activity cells in a global supply chain. Int J Prod Econ 133(1):243

Bogataj M, Tuljak Suban D, Drobne S (2011) Regression-fuzzy approach to land valuation. Cent Eur J Oper Res. doi:10.1007/s10100-010-0188-x

Damborský M, Jetmar M (2008) Localization decision—making of the firms as dynamic process. Regional studies association annual conference 2008

Drobne S, Bogataj M (2011) The accessibility and the flow of human resources between Slovenian regions, (MEORL, serial no. 11). Ljubljana: Faculty of Civil and Geodetic Engineering, Šempeter pri Gorici, Medifas, Slovenia

ESPON (2009) Territorial dynamics in Europe, TRENDS in Accessibility, TO2, Luxembourg. http://www.espon.eu/main/Menu_Publications/Menu_TerritorialObservations/trendsinaccessibility.html. Accessed 1 July 2010

Fujita M, Krugman P, Venables AJ (1999) The spatial economy: cities, regions and international trade. The MIT Press, Cambridge

Russo AP, Smith I, Atkinson R, Servillo L, Madsen B, van der Borg J et al (2010) ATTREG—The attractiveness of European regions and cities for residents and visitors. Applied Research 2013/1/7, Interim Report

Turner MA (2007) A simple theory of smart growth and sprawl. J Urban Econ 61(1):21–44

Von Thünen J H (1726) Der isolierte Staat in Beziehung auf Landwirtschaft und Nationaloekonomie. Fischer G (1910), Jena

Von Thünen J (2009) The isolated state in relation to agriculture and political economy (trans: Tribe K). Tünengesellschaft, Tellow

Chapter 7
Private Sector Initiative for the Development of Industrial Land. La Palma Industrial Park

Diego Ros-McDonnell

Abstract Modern Industrial urban development requires that location and land use are adapted to the functions and needs of the productive sector. The Spanish legislation on urban development has established that the way municipal land is used is via the General Plans of Urbanization. The General Plan of Urbanization of the municipality of Cartagena, passed in 1987, classified some of the areas for industrial use. Afterwards the initial forecasts were proved to be insufficient by the development of economic activities in the district of Cartagena, so successive modifications of the Plan of Cartagena were necessary in order to enlarge the area destined to be used by industrial units. This article describes the plans of the private sector initiative in the reversal of the shortcomings of urban development plans; in particular, the activity of fruit and vegetable industries in order to create an appropriate area for their production facilities in the Cartagena region.

7.1 Background

Cartagena is located near the Mediterranean Sea, in the southeast part of the Iberian Peninsula. The importance and relevance of Cartagena has long been a result of the many uses of the area, as a military base, as a strategic stronghold dating back to Roman times; as a port where a variety of goods have passed through on their way to Europe and further afield; as a mining and metallurgic town with auxiliary industries; and as a regional seat of Government and Lawmaking.

D. Ros-McDonnell (✉)
Universidad Politécnica de Cartagena, Paseo Alfonso XIII, 30205 Cartagena, Spain
e-mail: diego.ros@upct.es

S. P. Sethi et al. (eds.), *Industrial Engineering: Innovative Networks*,
DOI: 10.1007/978-1-4471-2321-7_7, © Springer-Verlag London Limited 2012

By the middle of the twentieth century the district of Cartagena had benefited from a noticeable increase in the network of well established companies (Gallardo 1958). After the remarkable industrial expansion of the 1960s the manufacturing base of Cartagena had two clearly defined strands. On the one hand, the indigenous manufacturing industry with its roots in arts and crafts from previous centuries and, working alongside it, international businesses with investment from foreign capital, based in large industrial parks. Since then the planning and promotion of industry, and consequently the pattern of increased industrial development, have followed two different but not contrary paths: an industry based on the strategic position of the port, run by large firms with an ambitious outlook, alongside traditional industry in clear disadvantage with it (Andres Sarasa 1982). The former can be seen in the mining activity and mineral transformation works, the shipyards, the arsenal, the chemical industry, the oil refinery and the thermal power station. The latter include all kinds of privately owned and run industries characteristic of the fruit and vegetable trade of Cartagena.

Generally speaking, the large productive establishments settled on a few specific sites and tend to be concentrated in the Escombreras valley and in the areas surrounding the Cartagena port. In fact, most of the industrial units of the region were located in these places. It was also logical, in those years, to forecast that the highest demand for industrial space would take place in Escombreras, the Cartagena port and the routes connecting both places. The rest of the industries, whether light or medium, did not have their own industrial parks or were located in specific areas. Productive activities coexisted with residential use of the municipal land.

7.2 The General Plan of Urban Ordination of Cartagena

General Plans of Urban Ordination are the instruments used for the integral design of the territory and may be inclusive of one or several complete municipal areas. Their classification makes different types of land subject to various applicable legislation. They define the fundamental elements of the general structure to be adopted by the urban planning of the territory, establish the programme for the development and completion of works, as well as their minimum term of validity (TRLS 1976). These plans must detail development in two phases of 4 years, coordinate activities and investment, both public and private, and the programmes of the different Ministerial Departments.

The General Plan of Urban Ordination of Cartagena, from now on known as the GPUO of Cartagena, established seven areas to be developed in its first four years of application, 1987–1991. Five were to be residential areas; two of them were classed as industrial areas.

The GPUO of Cartagena established another seven areas to be developed during the second four years, 1991–1995: six residential and one industrial.

Figures 7.1 and 7.2 verify the similar initial importance, in terms of ground allocated by the GPUO of Cartagena, for areas destined to become residential and

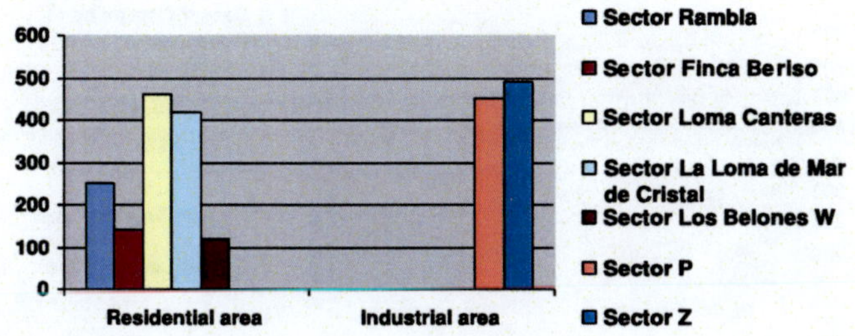

Fig. 7.1 Soil forecasts for the first four years in hectares

industrial. However, the territorial distribution of these two kinds of urban developments is very different. The residential area is spread over eleven different zones, with their total ground span ranging from 68 to 460 hectares. Industrial land is concentrated in two places, sector P and sector Z.

In addition to this, the General Plans of ordination must anticipate new developments after the first two 4-year periods. These would take place in unused urban land, called Non-Programmed Urban Soil, from now on known as NPUS. The GPUO of Cartagena established a bigger number of sectors destined to be industrial areas, and made a distinction between two categories: land destined to become light and medium industry and land destined to become heavy industry. It becomes apparent that this classification intentionally or unintentionally responds to the two types of existing industries in Cartagena, as shown in the first point above. As for the sectors of NPUS:

- Light and medium industry:
  - Cartagena East, CE-1, 44.00 ha.
  - Cartagena East, CE-2, 54.66 ha.
  - Los Dolores, LD-3, 7.54 ha.
  - La Aljorra, AJ-1, 100.00 ha.
  - Vista Alegre, IU-1, 39.02 ha.
  - Santa Lucía, SL-1, 10.80 ha.
  - La Palma, LP-1, 32.27 ha.
- As for heavy industry, it only forecasts one area: Escombreras, EC-1, 854 ha.

By using two sets of regulations, Vi and Ai, for the building of warehouses or industrial facilities, the urban development ordinances of the GPUO of Cartagena also differentiate two categories of industry, the object of each of them being (PGOUC 1987):

- Vi, Industrial street: These norms apply to those areas of urban land in villages and boroughs of the Municipal Area which may accommodate workshops and low-disturbance industries, as well as warehouses adjacent to housing in these areas. They were basically destined to industries compatible with homes.

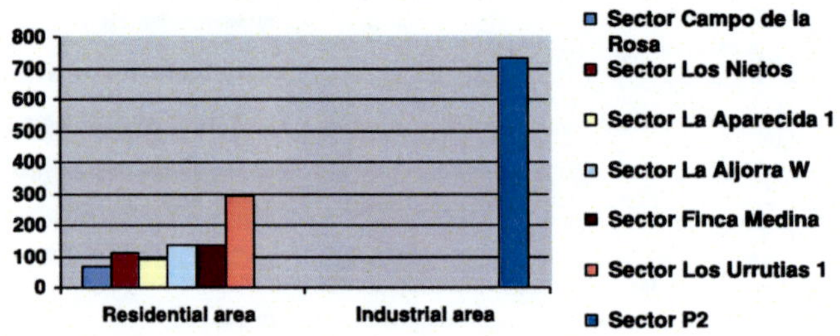

Fig. 7.2 Soil forecasts for the second four years in hectares

- Ai, Isolated industrial area: These norms apply to those areas of urban land destined to the implementation of the "industrial activities that must be distant from residential areas."

7.3 Amendments to the General Plan of Cartagena

The initial allocations of land to be given to industry of the GPUO of Cartagena were either proved insufficient in the face of new industrial needs or because of poor planning, or else could not be carried out due to urban management issues. In order to resolve the problems, successive modifications of the plan were made to give an appropriate location to the productive sector. Among the most relevant corrective actions were:

- Amendment N. 15. Casa Grande. La Aljorra.
- Amendment N. 20. La Palma. Industrial Park.
- Amendment N. 50. El Albujón. Industrial Park.
- Amendment N. 68. La Aljorra North, area surrounding the GEP® site.
- Amendment N. 99. La Concepción borough, lot in road to La Algameca changed from industrial to residential use.
- Amendment N. 109. Cabezo Beaza North.
- Amendment N. 115. Escombreras basin, General Port System.
- Amendment N. 124. Escombreras basin, urban development norms.

These modifications were caused by:

- The Establishment of the General Electric® industrial site in La Aljorra.
- Demands of the fruit and vegetable sector in the areas of the Cartagena region. La Palma and El Albujón.
- Development of service sector in the Cabezo Beaza Industrial Park.
- Enlargement of the Cartagena port.

Fig. 7.3 GPOU of Cartagena. Sector La Palma, LP-1. (GMUC 2011)

Amendment N. 20, dealing with the demarcation and building of the La Palma Industrial Park, is the most important modification made by the private initiative of the traditional fruit and vegetable sectors in the municipality of Cartagena during the period of validity of the GPUO of Cartagena.

7.4 The Industrial Park in La Palma

The "Campo de Cartagena" is an area covering the municipal terms of Cartagena, La Unión, Los Alcázares, Fuente Álamo, San Javier, San Pedro del Pinatar and Torre Pacheco. The village of La Palma is located in the middle of this "Campo de Cartagena".

As explained in the previous point, among the forecasts of the GPUO of Cartagena was the definition of a NPUS area in the surrounding area of La Palma destined to become light- and medium-scale industry. As shown in Fig. 7.3, its location between the villages of La Palma and Pozo Estrecho, relatively far away from both, its design around the existing railroad, which would split the industrial park in two halves, the layout of nearby property and the existence of facilities and buildings in the demarcation entailed excessive difficulties for the management and completion of the sector.

Fig. 7.4 Aerial photograph of La Palma. 1981. (CARTOMUR 2011)

At the same time, the obvious shortage of industrial areas for livestock and agricultural activities in the Campo de Cartagena and the delicate condition of the existing companies, with houses transformed into workshops and warehouses, including fertilizer and chemicals stores amid the residential areas, demanded an allocation of land for productive activities.

In the late 1980s an increase in the vitality of the private sector in Cartagena was experienced, fundamentally as a consequence of the increase in agricultural production due to the spread of the irrigation system facilitated by the availability of water from the Canal connecting the rivers Tajo and Segura, and the settling of multinational industries.

This entailed a combined meeting of all the production and labour elements of the population, initiated by the management of the "Association of Neighbours of La Palma", which was the origin of the non-profitable organization Community of Proprietors of the Industrial Park of La Palma, created with the sole purpose of promoting and developing an industrial area in which companies would have suitable businesses premises, while freeing the residential areas of the characteristic inconvenience of those industries already active in the village (Moral Solana et al. 1989).

Consequently, most of the industrial firms, manufacturers and producers from the different sectors of the population of La Palma, supported by the Association of Neighbours of La Palma, formed the "Community of Proprietors of the Industrial Park of La Palma", a corporation under the legal form of community property. This Community was to become the catalyst of a whole series of urban development projects; the advocating of the drawing of documents and planning instruments such as the project to amend the General Plan, the Programme for Urban Activities, the Partial Plan of Ordination, the project of Compensation, the

Fig. 7.5 Aerial photograph of La Palma. 2011. (CARTOMUR 2011)

project of Allotment and projects of Urbanization; they undertook the tasks of managing and monitoring legal procedures with the various public administration organizations: The City Council of Cartagena, the Regional Government (*Comunidad Autónoma*) of Murcia, the National General Directorate for Roads, the Irrigators' Association (*Mancomunidad de Regantes*), assigned and financed the works. Once the building process had been finalized, the resulting lots were assigned to the different community members on the grounds of the rights held by each of them, and the members, independently, built their own premises as required by the needs of their industrial sectors.

The area marked off had an extension of 32.5 ha and was situated east and south of the village, stretching over one kilometre along the regional road N-311, with a variable width of 280–430 m, most of its area comprised between the said regional road and the irrigators' service lanes S-XV.t-5-4 and S-XV.t-3-4-2 of the YRYDA. The area licensed for industrial building came to 162,500 m^2 and the total lot surface amounted to approximately 179,000 m^2 (Fig. 7.4).

The period of time used in the construction of the industrial park, including the legal procedures ordained to the issuing of administrative licenses for the urban development documents, and the completion of the urbanization works was 7 years, from 1989 to 1995. The completion of the Industrial Park of La Palma has greatly enhanced the social, productive and economic growth of the zone. Since the urbanization was finished 15 years ago, most of the ordinance area has been built upon, as shown in Fig. 7.5.

The timely construction of all the lots licensed for industrial buildings in the Industrial Park of La Palma has confirmed that it was necessary to have industrial areas in the zone and, what is more, shows the need to enlarge its area. The new General Municipal Plan of Ordination of Cartagena (Seguí 2007), the current name of the planning outlining the urban growth of a municipality, initially approved in

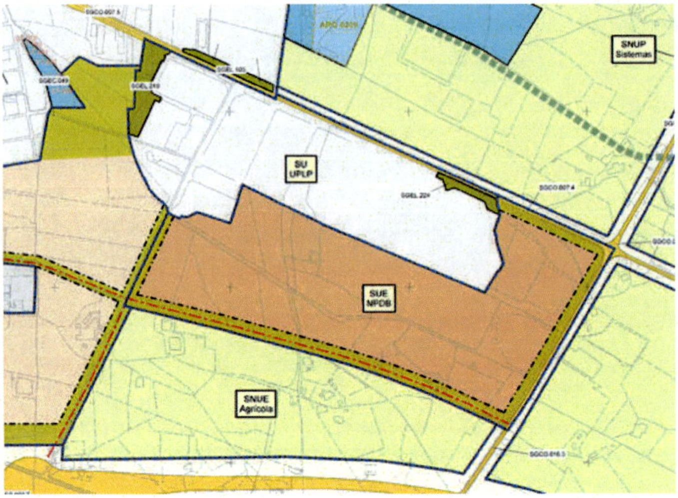

Fig. 7.6 General municipal plan of ordination of Cartagena. 2009. (GMUC 2011)

2009, has taken note of these important facts and has defined a wide extension of the current Industrial Park (Fig. 7.6).

7.5 Conclusions

The necessity of making successive timely modifications of the General Plan of Urban Ordination of Cartagena passed in 1987, some of them initiated shortly after its approval, in order to adapt the content of the plan to the demand of land shows that the forecasts were insufficient or inadequate for the development of the existing industrial networks in the zone.

The initial concentration of industrial land in two large areas, the port of Cartagena and Escombreras, and the diffusion and integration of smaller-scale productive facilities, small and medium industry, into the residential area stopped or hindered the development of the traditional industry.

The construction of industrial estates has permitted the building of new facilities and the quick growth of productive activities, and has helped to increase the local economy.

Private initiative promoted the drawing up of documents and planning instruments, and acted as agent, administrator and financer of the works. This was decisive in propelling the urban development process in order to make the necessary changes to compensate for the lack of industrial land in the Cartagena area.

The current Town Hall administration shares the view of industrial areas located around the town of peripheral population as the site for development of traditional industry. The new Plan of Ordination expands the industrial parks existing in small villages in the Cartagena area, which have used up all the available pockets of space, and there is an increasing demand for more land to build on.

References

Andres Sarasa JL (1982) Demographic growth and industrial development. Regional Provincial Council of Murcia, Murcia, p 213

CARTOMUR (2011) Infrastructures of referential land data of the Region of Murcia. IDERM. http://www.cartomur.com/visorcartoteca/. Accessed 2 May 2011

Gallardo F (1958) Industrial Cartagena. Publicaciones españolas, Madrid

GMUC (2011) Gerencia Municipal de Urbanismo de Cartagena (Cartagena Municipal Urban Planning Management) http://www.gemuc.es/urbanismo/. Accessed 2 May 2011

Moral Solana G, Moreno Alfaro J, Martínez Fortun G, Sánchez Vidal M, y García Inglés M (1989) Report study of La Palma Industrial Park. Cartagena

PGOUC (1987) Most honourable city council of Cartagena. General plan of urban ordination of Cartagena. Urban Development Norms

Seguí J (2007) General municipal plan of urban ordination of Cartagena. Cartagena

TRLS (1976) Merged Text of Law on applicable Legislation to Soil and Urban Ordinance passed by Royal Decree 1346/1976, of 9 April

Chapter 8
Statistical Data Mining and Artificial Neural Networks: A Case of Study in Financial Modeling

Pedro J. García-Laencina, M. Ángeles Varela-Jul, José L. Roca-González, Carmen de Nieves-Nieto and Joaquín Roca-Dorda

Abstract Nowadays, an organization or institution works with a huge amount of information about itself and its environment. This data has the potential to predict the evolution of interesting variables or trends in the outside environment. Data mining is the process that uses a variety of data analysis tools to discover meaningful patterns, trends and relationships in data that may be used to make valid predictions. In the last decades, artificial neural network-based technology stands out as one of the most suitable approaches. The goals of this work are to give a comprehensive analysis of the data mining process, to present the last advances on neural networks and its application for modeling financial data. In particular, an efficient neural network model is constructed for modeling the return on assets from other financial variables.

8.1 Introduction

By the most reliable estimate, 12 exabytes-plus of data representing the history of the world was generated by the year 2000. An exabyte is a unit of computer storage meaning one billion billion bytes. Within these masses of data lies hidden information of strategic importance, although raw data are rarely of direct benefit. For example, in the business field, innovative organizations use the most relevant information from their databases to locate to higher-value customers, to change their product offerings to increase sales and to reduce losses due to error or fraud.

P. J. García-Laencina (✉) · M. Ángeles Varela-Jul · J. L. Roca-González ·
C. de Nieves-Nieto · J. Roca-Dorda
Centro Universitario de la Defensa de San Javier
(University Centre for Defence at the Spanish Airofrce Academy),
MDE-UPCT, Murcia, Spain
e-mail: pedroj.garcia@cud.upct.es

S. P. Sethi et al. (eds.), *Industrial Engineering: Innovative Networks*,
DOI: 10.1007/978-1-4471-2321-7_8,

In recent years, computing and statistics researches have recently led to the development of flexible and scalable procedures that can be used to analyse large databases. This emergent research area is known as "Data Mining" (DM) (Giudici 2003). Simply stated, DM refers to extracting or "mining" useful knowledge from large amount of data by means of intelligent data analysis methodologies (Larose 2004). Whereas, the term KDD refers to the overall process of "Knowledge Discovery in Databases". While some people treat DM as a synonym for KDD, some others view it as a particular step in this process involving the application of specific algorithms for extracting patterns from data (Giudici 2003; Larose 2004).

In general, a dataset **X** can be defined as $N \times n$ matrix which is composed of N records or input samples and each record is multi-dimensional vector of n variables or input features. For example, consider a database of N industrial companies which contains different financial indicators over the last years, such as total debt, number of employees, amount of dividends,... In this case, each record of the dataset gives the financial information of a company for a specific year.

$$\mathbf{X} = \{\mathbf{x}_k\}_{k=1}^{N} = \begin{bmatrix} \mathbf{x}_1 \\ \vdots \\ \mathbf{x}_N \end{bmatrix} = \begin{bmatrix} x_{11} & x_{12} & \cdots & x_{1n} \\ x_{21} & x_{22} & \cdots & x_{2n} \\ \cdots & \cdots & \cdots & \cdots \\ x_{N1} & x_{N2} & \cdots & x_{Nn} \end{bmatrix} \quad (8.1)$$

The main part of DM is concerned with the data analysis and the use of intelligent software methods for finding patterns and regularities in datasets. This analysis starts with a reference dataset, known as *training set*. An intelligent data processing method extracts knowledge from the training dataset. Once knowledge has been acquired, this can be extended to larger sets of data working on the assumption that the larger dataset has similarities to the training data. This process is widely known as "*learning from data*" and two different approaches can be considered (Giudici 2003; Bishop 2006): *supervised* and *unsupervised* learning from data.

- *Supervised learning*. The goal is to use the available training data for building a predictive model that describes one particular target variable, t, in terms of the d input variables. For example, in the previous financial dataset, consider that there is additional information about the net profit of each company. In this case, the target task may be to construct a predictive model for the net profit. Then, the model can be used to estimate the impact on the net profit for different new corporate policies. This work is focused on supervised learning approaches.
- *Unsupervised learning*. This approach attempts to find similarities among group of records without the use of a particular target variable. In the previous example, different clusters/subgroups of industrial companies may be defined in order to compare their financial results in an accurate way.

Many of the methodologies used in data mining come from two branches of research, one developed in the machine learning community (Bishop 2006) and the other developed in the statistical community. In the first branch, artificial neural

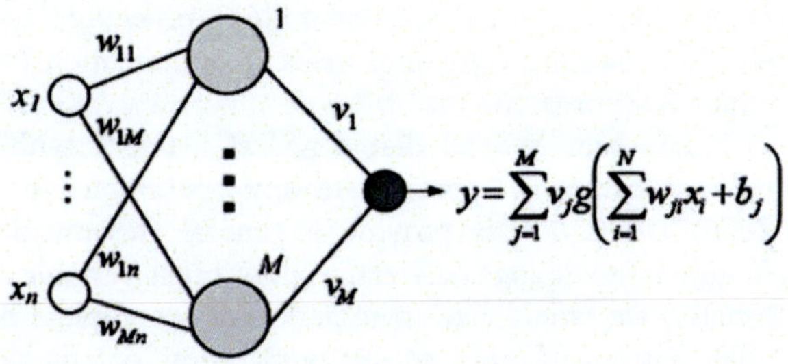

Fig. 8.1 Standard neural architecture with single layer of *M* neurons

network-based technology stands out as one of the most suitable approaches (Haykin 1998) and it has been successfully used in many different applications, such as fraud detection, demand forecasting, targeted marketing, credit scoring, cryptography, target classification, biometric recognition, biomedical engineering and control systems.

The aim of this work is to introduce the last recent advances on artificial neural networks and to shown how these methods can be applied for modeling a real financial dataset. Section 8.2 introduces the basic notions on artificial neural networks and presents its most emergent techniques. In Sect. 8.3, we first analyze a real financial dataset, related to two hundred companies during a period of five years, using statistical procedures in order to measure the relevance of 29 financial indicators with respect to the Return On Assets (ROA), which is the target variable. Then, we implement a predictive model using the neural network methods in this financial problem. Finally, Sect. 8.4 gives the main conclusions.

8.2 Artificial Neural Networks

An artificial neural network (ANN) is a computational model that is inspired by the structure and/or functional aspects of biological neural networks. This model is an adaptive system which consists of an interconnected group of artificial neurons by using *synaptic weights*. Each artificial neuron receives n inputs and computes a single output from n inputs by forming a linear combination according to its *input weights* (w_{ji}). The weighted sum, plus an additional bias term (b_j), is passed through a non-linear function $g(\cdot)$, which is known as *activation function*, in order to produce the *neuron output*. An appropriate activation function is the sigmoid. After that, these neuron outputs are linear combined by output weights (v_j) for computing the *network output* (y). Figure 8.1 shows a scheme of feed-forward architecture with single layer of M neurons.

The output network is a function of the values of synaptic weights that have to be adjusted in order to minimize an error between the output network and the desired outputs. This error is computed using the training dataset. The way these parameters are found is known *learning* algorithm (Haykin 1998; Bishop 2006). Then, knowledge is acquired by the network through a learning process (i.e., training error minimization) and the synaptic weights are used to store this knowledge. The most popular training procedure is Back-Propagation algorithm, which is based on gradient descent optimization (such as the Scaled-Conjugate Gradient approach).

It has been proved that a neural network with enough number of sigmoid neurons can well approximate any continuous function (Haykin 1998; Bishop 2006). Then, ANNs provide a general framework for representing non-linear functional mappings between a set of input variables and a set of output variables. Finally, we would like to remark that a complete description of the ANN technology is out of scope of this work. More details can be found in (Haykin 1998; Bishop 2006), which are very good reference books.

8.2.1 Emergent Neural Network Methodology

For past decades, gradient descent-based methods have mainly been traditionally used in various learning algorithms of ANNs. However, it is well known that gradient descent based learning methods are generally very slow due to improper learning steps or may easily converge to local minima. Moreover, many training repetitions may be required by such learning algorithms in order to obtain an accurate predictive model.

Lately, a new learning algorithm called *Extreme Learning Machine* (ELM) has been introduced to solve the speed limitation in the learning process (Huang et al. 2006; Garcia-Laencina et al. 2011). The ELM algorithm is based on the concept that if the input weights of ANNs are randomly assigned then an ANN can be considered as a linear system and the output weights can be analytically determined through simple generalized inverse operation of the hidden layer output matrices. This random assignment can be done if the activation functions in the artificial neurons are infinitely differentiable. In general, the neural network learning has to find appropriate weight parameter values in order to provide accurate estimates for the target variable, i.e.,

$$y_k = \sum_{j=1}^{M} v_j g\left(\mathbf{w}_j^T \mathbf{x}_k + b_j\right) \approx t_k \quad k = 1, \ldots, N; \tag{8.2}$$

where v_j and $\mathbf{w}_j$ are, respectively, the j-th output weight and the jth input weight vectors. This equation can written compactly: $\mathbf{Hv} = \mathbf{t}$, where $\mathbf{H}$ is the hidden output matrix:

$$\mathbf{H}_{N\times M} = \begin{bmatrix} g\left(\mathbf{w}_1^T \mathbf{x}_1 + b_1\right) & \ldots & g\left(\mathbf{w}_M^T \mathbf{x}_1 + b_M\right) \\ \vdots & \vdots & \vdots \\ g\left(\mathbf{w}_1^T \mathbf{x}_N + b_1\right) & \ldots & g\left(\mathbf{w}_M^T \mathbf{x}_N + b_M\right) \end{bmatrix} \tag{8.3}$$

It has been proven that a fast and accurate neural network model can be obtained by (8.1) the random assignation of input weights w_{ji} and biases b_j and, after that (2) computing the output weights according to:

$$\mathbf{v} = \mathbf{H}^{\dagger}\mathbf{t} \tag{8.4}$$

where $\mathbf{H}^{\dagger}$ denotes the Moore–Penrose generalized inverse of $\mathbf{H}$. The ELM algorithm tends to provide good generalization performance at extremely fast learning speed. Besides, the ELM learning algorithm looks much simpler than most learning algorithms for ANNs.

8.3 A Case of Study in Financial Modeling

Data mining and neural networks have attracted many attentions and they have been successfully applied in many financial and business areas. This work applies last advances on neural networks for modeling a dataset from financial field. The dataset is related to 200 industrial firms listed on the Paris Bourse (currently it is known as Euronext) during 1991–1995. The following table shows the real meaning in financial field about all the variables Table 8.1.

In particular, this dataset is composed of 650 samples and 29 variables are used, which are financial indicators that are measured every year. The target variable is ROA, which is defined as the ratio between the net income and the total assets. Note that this financial dataset has been previously defined in (Yu et al. 2007). Before modeling the target data, the dataset is randomly divided into training and test subsets: 2/3 samples for training and the remaining ones for testing. All simulations have been carried out in MATLAB 7.11 environment running in the same machine with 4 GB of memory and 2.67 GHz processor.

8.3.1 Statistical Analysis

Before building the predictive neural network model, a previous statistical analysis is done. First, a linear regression model has been implemented using the training set and, then, its prediction accuracy is measured in terms of the Root Mean Square Error (RMSE) over the test set: **1.004**. After that, we analyze the relationship between input variables and target data. The following table shows the correlation coefficient (C_i) of each input variable (x_i) and the target variable (t). Besides that, those C_i with a p-value lower than 5%, which means that they are statistically significant, are shown in bold face Table 8.2.

Whereas, it is important not to confuse correlation with usefulness. In general, correlation expresses the degree that, on an average, two variables change correspondingly. Nevertheless, a useful feature helps to build good predictors for the target data. In general, a variable is useful if, when added to a subset of other variables, it results in a prediction performance improvement, or if when removed

Table 8.1 List of financial indicators (input variables) of the dataset

Index	Variable
x_1	Industry sector
x_2	Number of shares exchanges during the year
x_3	Number of employees
x_4	Security turnover rate
x_5	Amount of dividend for one share during the year
x_6	Sales
x_7	Other assets
x_8	Dotations on other assets
x_9	Property, plant and equipment
x_{10}	Dotations on property, plant and equipment
x_{11}	Fixed assets
x_{12}	Stocks or inventories
x_{13}	Accounts receivables
x_{14}	Cash in hands and at banks
x_{15}	Total of current assets
x_{16}	Total of capital group
x_{17}	Account payables
x_{18}	Financial debt
x_{19}	Debt whose maturity is inferior to 1 year
x_{20}	Debt whose maturity is superior to 1 year
x_{21}	Total Debt
x_{22}	Cost of workers
x_{23}	Dotations on amortizations
x_{24}	Operation income before tax
x_{25}	Interests taxes
x_{26}	Financial income
x_{27}	Operating income before tax + Financial income
x_{28}	Extraordinary item
x_{29}	Taxes from state

it results in a performance loss. Indeed, the reduced linear model, which has been built by using only the most correlated variables, gets worse prediction results (**1.025**). As it is shown in the next section, the subset of selected useful features for the ANN is not the same than the most correlated input variables with the ROA. It is worthy to note that the usefulness depends on the predictive model, i.e., a variable can be useful to learn the target data with a model and the same variable can be useless using another different predictor.

8.3.2 Predictive Model Using Artificial Neural Networks

This section shows obtained results of ANN models for modeling financial data. Firstly, we have evaluated two neural network design procedures: (I) the standard ANN training based on BP algorithm; (II) the emergent ANN training on the ELM algorithm. A comparative evaluation between both procedures is performed in

Table 8.2 Correlation coefficients between each input variable and ROA

	x_1	x_2	x_3	x_4	x_5	x_6	x_7	x_8	x_9	x_{10}	x_{11}	x_{12}	x_{13}	x_{14}	x_{15}
C_i	**+0.124**	−0.007	**+0.121**	**+0.125**	−0.036	−0.003	+0.034	−0.001	−0.017	−0.020	−0.013	−0.050	−0.049	−0.022	−0.045

	x_{16}	x_{17}	x_{18}	x_{19}	x_{20}	x_{21}	x_{22}	x_{23}	x_{24}	x_{25}	x_{26}	x_{27}	x_{28}	x_{29}
C_i	+0.034	−0.021	−0.049	−0.049	−0.047	−0.053	−0.041	−0.002	**+0.093**	−0.038	−0.041	**−0.120**	**−0.100**	**+0.074**

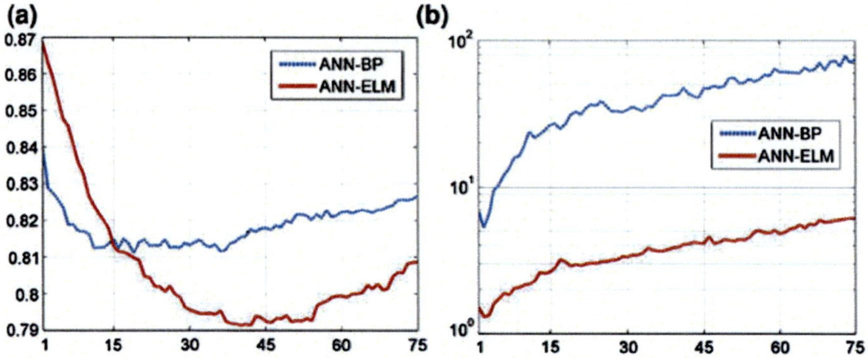

Fig. 8.2 In **a** it is shown the evolution of the Root Mean Square Error (RMSE) in the test set with respect to the size of the ANN which is designed using the BP training approach (gradient-based optimization) and the ELM algorithm. These results are averages of 30 trials. Meanwhile, in **b**, it is shown the total training time (in seconds) vs. number of neurons

order to show the clear advantages of the ELM approach. Figure 1 shows the evolution of the RMSE (in test set) and the total training time (in seconds) with respect of the number of neurons for both procedures. This experiment considers all input variables. The network size is from 1 to 75 and the number of training epochs is 500. Note that 30 different trials (different weight initializations) are done.

From Fig. 8.2a, we can observe that both ANN procedures outperform the obtained results of the standard linear regression model (around a 20% of enhancement in the prediction accuracy). Besides that, we can observer that the standard BP training performs better when the network size is small. The performance obtained by ELM is better to BP when the number of neurons is higher than 14 and of course with less training time. The best prediction (**0.791** $\pm$ **0.021**) is achieved using an ELM network with 41 neurons. As it is depicted in Fig 8.2b, the ELM algorithm runs around 10 times faster than the BP training approach. This enhancement may have a great positive impact (in terms of computational efforts) when larger datasets are analyzed. In general, it is straightforward to see that the training time increases as the network size is larger.

Once the ROA has been modeled using all input variables, we perform an incremental forward feature selection procedure (Garcia-Laencina et al. 2011), in which features are sequentially added to an empty candidate set until the addition of further features does not decrease the cross-validation (leave-one out approach) error of the ELM network. For each iteration in this exhaustive search of 'best/useful' features, a feature is added to the candidate subset if it helps to obtain a better prediction of ROA (i.e. it helps to obtain a better ELM model). The selected features are the following: x_{27}, x_{29}, x_{21}, x_{13} and x_{15}. Note that these five variables are ordered increasingly, and then, the most relevant feature results x_{27}, which is "Operating income before tax + Financial income". The remaining features are discarded in terms of usefulness for implementing the ELM network. The obtained model provides a predictive accuracy of $\mathbf{0.765 \pm 0.016}$ (mean and standard deviation of 30 trials) and its averaged size—number of neurons—is equal to 23, which are less than the selected size for the ELM network with all input variables. Therefore, we have enhanced the prediction of ROA using a simpler neural network model, which has been implemented with an appropriate set of relevant financial indicators.

8.4 Conclusions

Data mining (DM) provides useful and reliable tools, such as artificial neural networks, for business and financial applications. This work has introduced the most important aspects of statistical data mining and artificial neural networks (ANNs), doing a special effort on the last emergent procedure for implementing ANNs: the Extreme Learning Machine (ELM) methodology. Statistical approaches and neural network-based procedures are applied in a financial context: the prediction of ROA (return on assets) from 29 financial indicators of 200 industrial companies. Obtained results, in terms of prediction accuracy, show the usefulness of neural network-based technology and the importance of performing an accurate selection of the relevant information in databases.

References

Bishop C (2006) Pattern recognition and machine learning. Springer, New York

Garcia-Laencina PJ, Bueno-Crespo A, Sancho-Gómez J-L (2011) Design and training of neural architectures using extreme learning machines. Neurocomputing: learning, architectures and modeling. Nova Science Publishers

Giudici P (2003) Applied data mining: Statistical methods for business and industry. Wiley, West Sussex

Haykin S (1998) Neural networks: a comprehensive foundation. Prentice Hall, New York

Huang G-B, Zhu Q-Y, Siew C-K (2006) Extreme learning machine: Theory and applications. Neurocomputing 70(1–3):489–501

Larose DT (2004) Discovering knowledge in data: An introduction to data mining. Wiley, New Jersey

Yu Q, Séverin E, Lendasse A (2007) Variable selection for financial modeling. In: 13th International conference on computing in economics and finance, Montréal, Canada

Chapter 9
An Empirical Analysis of the Most Influential Components of an Income Statement

Roberto Alcalde Delgado, Lourdes Saiz Barcena and Miguel Ángel Manzanedo

Abstract This paper presents an empirical analysis of income statements in 1,000 Spanish companies. Firstly, the relation between the components of an income statement is studied, focusing on the operative profit/loss. Then, the most influential components in the operative profit/loss are identified and their effects measured. Finally, we conclude that they provide business groups with higher predictable profits.

9.1 Introduction

Earnings Before Interest and Tax (EBIT) is an appropriate indicator for measuring the profit of operational activities (Bohuslava et al. 2011). It is calculated as revenue minus expenses, without taking into account tax and interest. Hence, it is related to sales, the sum of raw material costs, other operating costs (human resources, depreciation, etc.). EBIT is also referred to as "operating earnings", "operating profit" or "operating income".

R. Alcalde Delgado (✉) · L. Saiz Barcena (✉) · M. Á. Manzanedo (✉)
Area of Organization. Higher Polytechnic School, University of Burgos,
C/Francisco de Vitoria, s/n, 09006 Burgos, Spain
e-mail: roberto.alcalde.delgado@gmail.com

L. Saiz Barcena
e-mail: lsaiz@ubu.es

M. Á. Manzanedo
e-mail: mmanz@ubu.es

S. P. Sethi et al. (eds.), *Industrial Engineering: Innovative Networks*,
DOI: 10.1007/978-1-4471-2321-7_9, © Springer-Verlag London Limited 2012

Table 9.1 Variables use in the model

Variable	Ratio	Indicator of percentage of
$EBIT_S$	EBIT/S	Profitability
CM_S	CM/S	Cost of material
PE_S	PE/S	Personnel expenses
DA_S	DA/S	Depreciation and amortization of fixed assets
PL_S	PL/S	Changes in provisions for losses and non-payments

Equation 9.1 defines EBIT by using the following variables: sales (S), cost of materials (CM), personnel expenses (PE), depreciation and amortization of fixed assets (DA), and changes in provisions for losses and non-payments (PL).

$$\mathrm{EBIT} = \mathrm{S} - \mathrm{CM} - \mathrm{PE} - \mathrm{DA} - \mathrm{PL} \tag{9.1}$$

The result of dividing Eq. 9.1 by sales is Eq. 9.2. It is written with new variables contained in Table 9.1. Because these variables are related to sales, it facilitates the comparison between companies.

$$\mathrm{EBIT_S} = 1 - \mathrm{CM_S} - \mathrm{PE_S} - \mathrm{DA_S} - \mathrm{PL_S} \tag{9.2}$$

9.2 Sample

The sample for this study is composed of 1,060 Spanish companies. The data corresponds to the annual financial statements system between 2001 and 2005. Companies with unconsolidated financial statements and those business areas (with the same four-digit CNAE[1] code) integrated by no more than two companies are excluded.

The dataset of the sample contains 5 years of the following indicators: operating profit, sales, cost of materials, personnel expenses, depreciation and amortization of fixed assets and changes in provisions for losses. After that, we work out the average indicators of the five-year period for each company, which are then exported to an SPSS® statistical tool.

9.2.1 Descriptive Analysis of the Sample

Table 9.2 shows a descriptive analysis of the indicators obtained. According to it, cost of materials has the highest impact on profit (56%), followed by the amount

[1] CNAE stands for National Code for Business Activities, the Spanish system for classifying business entities according to their productive activity.

Table 9.2 Descriptive statistics of sample

Variable		CM_S	PE_S	DA_S	PL_S	$EBIT_S$
	N	1060	1060	1060	1060	1060
	Range	0.99	0.89	0.75	0.89	1.71
	Min	0.00	0.00	0.00	0.01	−0.98
	Max	0.99	0.89	0.75	0.90	0.74
	Mean	0.5259	0.1783	0.0432	0.1833	0.0709
	Std. dev.	0.23446	0.13755	0.05917	0.13109	0.09904
	Variance	0.055	0.019	0.004	0.017	0.010
	1st Qu.	0.3853	0.0835	0.0100	0.0983	0.0252
	Median	0.5600	0.1460	0.0280	0.1529	0.0541
	3st Qu.	0.6995	0.2344	0.0531	0.2244	0.1022
	Asymmetric	−0.485	1.871	4.771	1.981	−0.053
	Kurtosis	−0.433	5.030	34.337	5.410	20.656
Standard error	Asymmetric	0.075	0.075	0.075	0.075	0.075
	Kurtosis	0.150	0.150	0.150	0.150	0.150

allocated to provisions for losses and non-payment (15.3%), the cost of expenses on personnel (14.6%) and depreciation (2.8%). EBIT medians are approximately 5.4% of sales.

9.2.2 Correlation Between Independent Variables

There is a high inverse correlation between personnel expenses and cost of materials (Table 9.3). Therefore, labour-intensive companies (for example, consulting firms) have lower raw materials costs.

Additionally, companies with higher cost of materials have lower provisions for losses and non-payment. This can mean that a higher cost of raw material implies a better risk management. In other words, one of the most pressing tasks for these companies is to preserve their most valuable asset: their raw materials.

9.2.3 Regression Analysis of the Sample

The use of statistical regression is a standard feature in data analysis. We use statistical linear regression for Eq. 9.3. As it has to be equal to Eq. 9.2, carrying out regression tests has never been easier.

$$\mathrm{EBIT_S} = \mathrm{a} + \mathrm{b\,CM_S} + \mathrm{c\,PE_S} + \mathrm{d\,DA_S} + \mathrm{e\,PL_S} \tag{9.3}$$

The significance of Analysis of Variance between groups (ANOVA) data test is 0.000. Hence, the medians of independent variables are different.

Table 9.3 Correlation between independent variables

Variable		CM_S	PE_S	DA_S	PL_S	$EBIT_S$
CM_S	Pearson correlation	1	−0.669(**)	−0.433(**)	−0.700(**)	−0.252(**)
	Sig. (bilateral)		0.000	0.000	0.000	0.000
	Covariance	0.055	−0.022	−0.006	−0.022	−0.006
PE_S	Pearson correlation	−0.669(**)	1	0.162(**)	0.193(**)	−0.159(**)
	Sig. (bilateral)	0.000		0.000	0.000	0.000
	Covariance	−0.022	0.019	0.001	0.003	−0.002
DA_S	Pearson correlation	−0.433(**)	0.162(**)	1	0.267(**)	−0.151(**)
	Sig. (bilateral)	0.000	0.000		0.000	0.000
	Covariance	−0.006	0.001	0.004	0.002	−0.001
PL_S	Pearson correlation	−0.700(**)	0.193(**)	0.267(**)	1	−0.065(*)
	Sig. (bilateral)	0.000	0.000	0.000		0.034
	Covariance	−0.022	0.003	0.002	0.017	−0.001
$EBIT_S$	Pearson correlation	−0.252(**)	−0.159(**)	−0.151(**)	−0.065(*)	1
	Sig. (bilateral)	0.000	0.000	0.000	0.034	
	Covariance	−0.006	−0.002	−0.001	−0.001	0.010

(*) Significant at 0.05 (bilateral)
(**) Significant at 0.01 (bilateral)

Table 9.4 Regression analysis of sample

		Coefficients		Sign	Confidence interval at 95%	
Model		Coefficient	Stand. error		Lower bound	Upper bound
1	Constant	0.962	0.006	0.000	0.951	0.974
	CM_S	−0.960	0.007	0.000	−0.973	−0.947
	PE_S	−0.968	0.008	0.000	−0.983	−0.953
	DA_S	−0.982	0.013	0.000	−1.006	−0.957
	PL_S	−0.937	0.008	0.000	−0.953	−0.921

$EBIT_S$ Dependent variables. R Square is 0.955

The result of the statistical regression is very revealing because of higher R-squares (Table 9.4) and the absolute values of coefficients are close to one (Eq. 9.4). Moreover, it is predictable because the regression correspond to Eq. 9.2.

$$EBIT_S = 0.962 - 0.960\,CM_S - 0.968\,PE_S - 0.982\,DA_S - 0.937\,PL_S \qquad (9.4)$$

9.3 Model

The financial components are encoded as categorical variables by the median value. These variables can be zero or one as can be seen in Table 9.5.

In line with this, four categorical variables are introduced in the regression model (Eq. 1.5).

Table 9.5 Categorical variables used in the model

Variable	Value = 0	Value = 1
$CM_{S\ DUMMY}$	$CM_S <= 0.56$	$0.56 < CM_S$
$PE_{S\ DUMMY}$	$PE_S <= 0.146$	$0.146 < PE_S$
$DA_{S\ DUMMY}$	$DA_S <= 0.028$	$0.028 < DA_S$
$PL_{S\ DUMMY}$	$PL_S <= 0.152$	$0.152 < PL_S$

Table 9.6 Regression with independent and dummy variables

Model		Coefficients		Sign	Confidence interval at 95%	
		Coefficient	Stand. error		Lower bound	Upper bound
1	Constant	0.961	0.006	0.000	0.949	0.974
	CM_S	−0.953	0.007	0.000	−0.967	−0.938
	PE_S	−0.967	0.009	0.000	−0.984	−0.949
	DA_S	−0.979	0.014	0.000	−1.006	−0.952
	PL_S	−0.937	0.009	0.000	−0.955	−0.920
	$CM_{S\ DUMMY}$	−0.005	0.002	0.036	−0.009	0.000
	$PE_{S\ DUMMY}$	−0.001	0.002	0.699	−0.005	0.003
	$DA_{S\ DUMMY}$	−0.001	0.002	0.521	−0.004	0.002
	$PL_{S\ DUMMY}$	0.000	0.002	0.859	−0.003	0.004
2	Constant	0.962	0.006	0.000	0.950	0.974
	CM_S	−0.953	0.007	0.000	−0.967	−0.938
	PE_S	−0.967	0.009	0.000	−0.984	−0.949
	DA_S	−0.980	0.014	0.000	−1.007	−0.953
	PL_S	−0.937	0.008	0.000	−0.953	−0.920
	$CM_{S\ DUMMY}$	−0.005	0.002	0.031	−0.009	0.000
	$PE_{S\ DUMMY}$	−0.001	0.002	0.708	−0.005	0.003
	$DA_{S\ DUMMY}$	−0.001	0.002	0.534	−0.004	0.002
3	Constant	0.962	0.006	0.000	0.950	0.974
	CM_S	−0.953	0.007	0.000	−0.968	−0.939
	PE_S	−0.968	0.008	0.000	−0.983	−0.953
	DA_S	−0.980	0.014	0.000	−1.007	−0.953
	PL_S	−0.937	0.008	0.000	−0.953	−0.921
	$CM_{S\ DUMMY}$	−0.005	0.002	0.033	−0.009	0.000
	$PE_{S\ DUMMY}$	−0.001	0.002	0.465	−0.004	0.002
4	Constant	0.961	0.006	0.000	0.949	0.973
	CM_S	−0.953	0.007	0.000	−0.968	−0.939
	PE_S	−0.969	0.008	0.000	−0.984	−0.954
	DA_S	−0.984	0.013	0.000	−1.008	−0.959
	PL_S	−0.937	0.008	0.000	−0.953	−0.921
	$CM_{S\ DUMMY}$	−0.004	0.002	0.042	−0.009	0.000

$EBIT_S$ Dependent variables. Square R is 0.955

$$EBIT_S = a + b\,CM_S + c\,PE_S + d\,DA_S + e\,PL_S + f\,CM_{S\,DUMMY} + g\,PE_{S\,DUMMY} + h\,DA_{S\,DUMMY} + i\,PL_{S\,DUMMY} \tag{9.5}$$

9.3.1 Regression Analysis of the Model

As a result of the regression analysis, all categorical variables except $CM_{S\ DUMMY}$ are excluded.

$$\begin{aligned} EBIT_S = {} & 0.961 - 0.953\,CM_S - 0.969\,PE_S - 0.984\,DA_S - 0.937\,PL_S \\ & - 0.004\,CM_{S\,DUMMY} \end{aligned} \tag{9.6}$$

According to Eq. 9.6, companies where the cost of materials was higher than 56% of sales experienced a reduction in operative profit quantified by 0.4% of sales.

With reference to Table 1.3, a higher ratio of material costs to sales is related to a lower ratio of personnel expenses to sales and vice versa. Despite this, the categorical variable of $PE_{S\ DUMMY}$ is not significant in the regression analysis, but it is the last variable to leave Table 9.6.

9.4 Conclusions

The aim of this study is to offer some insight into the position and future of the operative profit and loss, and the competitive advantage of an enterprise, based only on information on the main components of an income statement.

The main conclusion of this study is that those companies with a material cost higher than 56% of sales experience a reduction in operative profit quantified by 0.4% of sales.

Finally, a higher ratio of material cost to sales means trends of better risk management of raw material and lower competitive advantages compared to wide value chain companies.

Reference

Bohuslava K, Tomas K et al (2011) EBIT criterion: financial analysis' issues. Int J Math Model Methods Appl Sci 5(3):499

Part II
Innovation and Knowledge Management

Chapter 10
Five Information-Based Key Factors for Innovative Companies

Lourdes Canós-Darós, Cristina Santandreu-Mascarell, Julio J. Garcia-Sabater and Juan A. Marin-Garcia

Abstract In this paper we propose five common points for companies to innovate and adapt to constant change. For this, we study the interdependence between the organizational structure and information flows in generating ideas. It has been observed that to practice innovation, a company must develop an efficient and effective information management, considered as a principal resource or value for making decisions. This research is based in grounded theory which allows, through the collection and analysis of qualitative data, to build new theory.

10.1 Introduction

Currently, we speak about the information society. In history, the change of industrial societies to post-industrial and knowledge ones is recognized, where progress is essential knowledge. This new society, that includes organizations based on learning, is based on an unprecedented technological development. Then, it is the point at which large companies plan their products based on human resources management and the feasibility for obtaining them not only in a good quantity but in a good quality.

An organization is considered a system comprised of people, material resources and information. Information determines order and chaos between individuals, resources and both of them. For this reason, organizations should be seen as information systems. Information can be used to modify behaviours in the organization.

L. Canós-Darós (✉) · C. Santandreu-Mascarell ·
J. J. Garcia-Sabater · J. A. Marin-Garcia
ROGLE Group, Dpto. de Organización de Empresas, Universidad Politécnica de Valencia, Valencia, Spain
e-mail: loucada@omp.upv.es

S. P. Sethi et al. (eds.), *Industrial Engineering: Innovative Networks*,
DOI: 10.1007/978-1-4471-2321-7_10,

A good information management is a fundamental tool for decision making, staff training, product evaluation, identification of errors and control processes. In consequence, information is a vital resource for organizational development. Nowadays, the intangible nature of information makes difficult its analysis.

The impact of economic, political, cultural and technological changes has led to a revolution in information management in organizations. Rules, concepts, procedures, behaviours, and products and services were transformed. It is clear that the new management models are based on information management, because it deals with the final results in the company.

It is clear that an interdependence of an organization with its main elements (as information system) exists. For instance, an Integrated Model of Organizational Structure and Management of Ideas (IMOSMI) that shows the relationship between the structure of the company and the generation of ideas is presented in Canós and Santandreu (2010). Obviously, information is the basis to generate ideas in order to get and maintain competitive advantages. In order to handle the flow of information through information systems we must also define the relationship between organization and systems. Organizations are social systems composed of different variables with very complex relationships. Any way, managers have to construct models that represent reality designed to facilitate the understanding and treatment of complex systems with the purpose of facilitate the complexity of relationships (Loewe and Dominiquini 2006).

The new paradigm is characterized by autonomous factors, people, information and highly visible economic pressure. The need to review the current understanding of business and change organizational structures is introduced as a strategic imperative. Through incorporating the management of strategic assets, the goal to achieve is to increase added value offered to customers in new products or services. In this context, the company is an organization that specializes its functions, that are, among others:

- Ability to identify, trace and capture relevant information (creativity, ideas) for purposes.
- Assessment capacity and analysis of the information for the production of new knowledge.
- Diffusion capacity, cooperation and action to share appropriately information within and outside the organization.
- Distributed decision making processes (efficient and effective decisions).
- Purchase or develop advanced communication technologies.
- Have modern and configurable distributed information systems.

In this paper we link all the previous concepts and analyse their relationships. Then, we determine common points based in information for innovative companies and we check them in some Spanish companies by using grounded theory, a qualitative technique to obtain results from social agents.

10.2 Information Management in Innovative Companies

Ideas, knowledge and information are linked; they depend on people. The availability of information is the first step to create an information-based company.

Companies have to adapt its structure to the appropriate characteristics for the development of an information management system. Once the equipment has been purchased, the staff is hired, and policies and procedures have been established, the organization becomes more resilient to change (Miller et al. 1988; Robbins 1990). The success of an organization does not depend on the mere existence of a structure, but the adjustment between the structure and business strategy (Oke et al. 2009), including information.

This evolution has been driven by new information technologies that have allowed even to the specialists of the economy to see how theoretical assumptions are closed to reality by increasing the available information. It confirms how companies need to adapt their organizational structures from traditional hierarchical models oriented to vertical control (appropriates for industrial structures) to more horizontal chain of command for its efficiency in managing information, being a fundamental tool the technology and the necessary input information, specific to the knowledge society (Drucker 1988). When we refer to this evolution, we have to think about changes. That is, where the focus must be set to identify and clearly specify what aspects of the organization how and when are going to be affected; for instance, changes in administrative structures, control and new ways of managing information (receive, distribute, store, process and transmit).

Some reasons to justify the change in a structure are, for instance, the emergency creativity of people, the fact that the responsibility for decision making and management can not be based in a single person, the increase of the amount of processed information, the ever-increasing competition or the new markets opened as a consequence of globalization. In this context, the use of information and communication technologies (ICT) is essential for any company and this leads to a change in the structures, processes and organizational culture, which means new models of management and leadership framed in the contemporary knowledge society.

On the other hand, there is not a homogeneous set of characteristics that differentiate innovative organizations (Tornatzky et al. 1983; Wolfe 1994). At the empirical level, how innovation affects various aspects of organizational structure and the existence of work teams have been compared (Chiesa et al. 1996; Rothwell 1992; Souitaris 2002). Also, other concepts have been compared in the same way, as communication channels (Rogers and Shoemaker 1971) or the age of the company (Huergo and Jaumandreu 2004) assuming that this is a factor that influences the development of routines and reflects the knowledge gained through experience.

As a consequence of having free access to all corporate information, rotation among different technology areas and across departments can be improved, and this is positive to generate ideas. Rotation allows employees to know the company from multiple perspectives and not to develop just routines but creative tasks; duplication of knowledge is generated (Ortt and Smits 2006). In this context, the value of an employee's contribution is not determined by their position in the organizational structure, but the importance of the information provided to the entire knowledge-creating system. This does not mean that there is no differentiation in the roles and responsibilities; the creation of new knowledge is the product of a dynamic interaction between functions.

In most companies, the ultimate test to determine the value of new economic knowledge is increase efficiency, reduce costs and get higher return on sales. But in the knowledge-creating company other qualitative factors are considered equally important. Managers can ask: Does the idea incorporate the vision of the company? Is it an expression of strategic objectives and aspirations of senior management? Do you have ability to widen the range of organizational knowledge of the company? Qualitative evaluation criteria are essential to give a sense of direction to the activities of the company that creates value. Even more, it is important to note that the vision of the company must also be open to different interpretations and even contradictory ones. At first glance, this can seem a contradiction: Does the company's vision must be clear, consistent and unambiguous? The answer is yes, but if the vision is not absolutely ambiguous, there will be an instruction or an order not to encourage this high level of personal commitment that lies in the effective knowledge creation.

All these ideas implied new ways to work, by considering works teams/workgroups, the age of the company, communication channels, strategy of rotation, free access to information and the vision of the company.

10.3 Methodology and Description of the Research

The methodology used to develop our research is the grounded theory, a qualitative technique. Grounded theory is a general methodology for developing theory from data that are systematically captured and analyzed. It implies that from some observations of the real phenomenon, a general concept is reached. In qualitative research we deal with understanding in depth more than with exactitude. Qualitative methods not only provide us with the means to explore complex and chaotic situations in real life, but provide many methodological options on how to approach a field of study in accordance with the problem and the long-term objectives of research. Any qualitative method starts from a real event from which a person want to make one concept [an applied example can be seen in Marin-Garcia et al. (2009)]. The event to be studied is the starting point of the research, while the determination of their properties is the goal (Eneroth 1984).

Grounded theory was originally developed by Glaser and Strauss (1967) and was expanded in different directions by Glaser (1978, 2000, 2002) and Strauss (1987). Although there are many points of affinity in which grounded theory is identified with other approaches of qualitative research, the main differentiation is its emphasis on theory building.

By applying a conceptual-inductive model based on grounded theory we have to define the phenomenon to study. Then, we ask some general questions: What this phenomenon is? What are its main characteristics? What qualities make it different from other ones? In our case the event can be described as the circuit that follows the information in the company according to its organizational structure and management of ideas in order to identify what features or forms of work can be observed in innovative companies.

The second step is to contact the cases to study in order to observe the basis of the concept (Hobday 2005). Then, we contacted 15 companies, which are the entities where the phenomenon occurs. These companies were chosen because of their innovation capacity. In concrete, we focus in a region called La Safor, Gandia, in the Valencia Community (Spain). It is the commercial head of which orbit the populations of the region, close to two hundred thousand inhabitants. The main base of local economies is trade and services small companies.

To develop our research, we create a semi-structured interview that implies a flexible and dynamic style of questioning, with the aim of understanding the meanings of human experiences from the interviewee perspective (Pace 2004). During these interviews, researchers get information without condition or restrict the answer of the interviewed people. Due to the depth of the interview and the average length (3 h), the interviews have been recorded in order to make later a correct transcription taking into account not only the information supplied by the manager but also the impressions and experiences he or she has conveyed to the researcher, as this is an important part of the grounded theory.

Once the data collection and the transcription is finished, qualitative data analysis is done by using Atlas.ti. The main advantage of this software is that it establishes relationships between concepts, categories and families, and indicates the moment of saturation in which data always lead to the same conclusions.

10.4 Discussion and Result

As we can see in the previous section, by using grounded theory the phenomenon under study has to be defined first. In our case, the event or phenomenon we want to analyse can be described as the information-based key factors related with organizational structure and management of ideas in order to identify which of them can be observed in innovative companies. To this effect, general questions such: How is the company's organizational structure? How does information circulate through it? Who has access and how to that information? What is innovative? What kinds of innovations have been carried out? Who is creative? Who has ideas for innovation?

The five items identified from the previous described interviews are: shared vision, generation of flexible project teams, team-building with freedom and

Table 10.1 Information-based key factors from a theoretical and an empirical point of view

Theory	Company
Works teams/Workgroups	Project teams/Team-building
Age of the company	
Communication channels	Open access to all information
Strategy of rotation	Rotation
Free access information	Open access to all information
Vision of the company	Shared vision

independence, rotation, and open access to all information of the company. All these concepts have been related with information flows, systems, management, etc., and some authors demonstrated its importance in innovation within the company, for instance, work teams (Chiesa et al. 1996; Rothwell 1992; Souitaris 2002) or duplication from rotation (Nonaka et al. 2000).

In Table 10.1 we can see that key factors described in literature are reflected in real business world through ideas provided by managers in the developed interviews. Concretely, we can observe the close relation between 'communication channels' and 'free access information', reflected both in business world through the factor 'open access to all information', according to interviewees' opinion. In this way, we can deduct that a company generates innovation and is continuously adapting to the environment when its information flow. Notice that 'age of the company' is not relevant for interviewed managers. Then, the five information-based key factors identified for innovative companies are: work teams/workgroups, communication channels, strategy of rotation, free access information and the vision of the company.

10.5 Conclusions

In this paper we first present some definitions related with information in companies. In order to survive in the environment, managers have to make decisions and to achieve and maintain competitive advantages. People in the company are able to generate, transform, share and use information. Then, they can be considered as a source of competitive advantage. With this aim, we propose some ideas for innovative companies based in these concepts.

Then, we contrast the model in some Spanish companies, in the region of Valencia Community, by using grounded theory, a qualitative technique to obtain results from stakeholders, in our case, human resources (managers). A research can not contain all the real factors, but some important ones that can explain phenomena in order to extrapolate the obtained results and generate findings useful for academic and professional fields. The results show that we can identify five information-based key factors common in innovative companies. We think these ideas can be useful for human resources management and leads to the improvement of social and economical results in companies.

Acknowledgments This work has been partially supported by the projects TIN2008-06872-C04-02/TIN and CORSARI MAGIC DPI2010-18243.

References

Canós L, Santandreu C (2010) An integrated model of organizational structure and ideas. INBAM Conference, Valencia

Chiesa V, Coughlan P, Voss CA (1996) Development of a technical innovation audit. J Prod Innov Manage 13:105–135

Drucker P (1988) The coming of the new organization. Harvard Business Review, 1–19

Eneroth B (1984) Hur mäter man vackert?. Natur and Kultur, Göteborg

Glaser B, Strauss A (1967) The discover of grounded: strategies for qualitative research. Aldine, Aldine

Glaser B (1978) Theoretical sensitivity: advances in the methodology of grounded theory. Sociology Press, Mill Valley

Glaser B (2000) The discovery of the grounded theory. Sociology Press, Mill Valley

Glaser B (2002) Conceptualization: on theory and theorizing using grounded theory. International Journal of Qualitative Methods, 1 (2). Article 3. http://www.ualberta.ca/~ijqm/

Hobday M (2005) Firm-level innovation models: perspectives on research in developed and developing countries. Technol Anal Strateg Manag 17(2):121–146

Huergo E, Jaumandreu J (2004) How does probability of innovation change with firm age? Small Bus Econ 22:193–207

Loewe P, Dominiquini D (2006) Overcoming barriers to effective innovation. Strategy Leadersh 34(1):24–31

Marin-Garcia JA, García-Sabater JP, Perello-Marin MR, Canós-Darós L (2009) Propuesta de competencias para el Ingeniero de Organización en el contexto de los nuevos planes de estudio. Intangible Cap 5(4):387–406

Miller D, Dröge C, Toulouse JM (1988) Strategic process and content as mediators between organizational context and structure. Acad Manag J 31(3):544–569

Nonaka I, Toyama R, Nagata A (2000) A firm as a knowledge-creating entity: a new perspective on the theory of the firm. Ind Corp Chang 9(1):1–20

Oke A, Munshi N, Walumbwa F (2009) The influence of leadership on innovation processes and activities. Organ Dyn 38(1):64–72

Ortt JR, Smits R (2006) Innovation management: different approaches to cope with the same trends. Int J Technol Manag 34(3/4):296–318

Pace S (2004) A grounded theory of the flow experiences of web users. Int J Hum Comput Stud 60(3):327–363

Robbins SP (1990) Organization theory: structure, design and applications, 3rd edn. Prentice Hall, Englewood Cliffs

Rogers EM, Shoemaker FF (1971) Communication of innovations: a cross-cultural approach. Free Press, New York

Rothwell R (1992) Successful industrial innovation: critical factors for the 1990s. R&D Manag 22:221–239

Souitaris V (2002) Technological trajectories as moderators of firm-level determinants of innovation. Res Policy 3(6):877–898

Strauss A (1987) Qualitative analysis for social scientists. Cambridge University Press, New York

Tornatzky L, Eveland J, Boylan M (1983) The processes of innovation: analyzing the literature. National Science Foundation, Washington
Wolfe RA (1994) Organizational innovation: review, critique and suggested research directions. J Manag Stud 31:405–431

Chapter 11
e-Loyalty Towards ICT-Based Healthcare Services: A Patients' Perspective

Eva Martínez-Caro, Juan Gabriel Cegarra-Navarro and Marcelina Solano-Lorente

Abstract Public health institutions are making a great effort to develop patient-targeted ICT-based services in an attempt to enhance their effectiveness and reduce expenses. However, if patients do not use those services regularly, public health institutions will have wasted their limited resources. Hence, patients' e-loyalty is essential for the success of ICT-based healthcare services. In this research, an extended Technology Acceptance Model (TAM) is developed to test e-loyalty towards ICT-based healthcare services from a sample of 256 users. The results obtained suggest that the core constructs of TAM (perceived usefulness, ease of use and attitude) significantly affect users' behavioural intentions (i.e. e-loyalty). This study also reveals a general support for patient satisfaction as a determinant of e-loyalty in ICT-based healthcare services. Finally, the implications of the findings are discussed and useful insights are provided on what policy to follow to establish the appropriate conditions to build patients' e-loyalty.

11.1 Introduction

In most countries, healthcare is a very expensive service. For countries with a finite budget, the development of the healthcare sector should call for a search for new, more efficient ways of providing healthcare services (Lanseng and Andreassen 2007). Information and Communication Technologies (ICT) can play a significant role in the solution of that problem.

E. Martínez-Caro (✉) · J. G. Cegarra-Navarro · M. Solano-Lorente
Business Management Department, Universidad Politécnica de Cartagena,
C/Doctor Fleming s/n, 30202 Cartagena, Spain
e-mail: eva.martinez@upct.es

S. P. Sethi et al. (eds.), *Industrial Engineering: Innovative Networks*,
DOI: 10.1007/978-1-4471-2321-7_11,

The applications of ICT in the healthcare sector are numerous. This research refers specifically to online health services offered by public health institutions and therefore targeted to all citizens, such as making a doctor's appointment, checking waiting lists, asking physicians, making a complaint, suggestion or query, sharing information with others with the same health problems, and reading on-line information about public health centres resources (e.g. medical staff, medical and care services, medical specialities, medical facilities), organizational issues (e.g. timetables, location, instructions for admissions, rules), or health information (e.g. health news, events, health topics).

Public health institutions are making a great effort to develop such services in an attempt to enhance their effectiveness. These services are more convenient to the patients because they save the patients' time, require less effort from the patient and are more accessible. In addition, they offer faster methods of communication between healthcare professionals and patients. In this way, healthcare personnel can save patient support hours and can also be employed in performing other tasks. In short, the use of ICT-based healthcare services can lead to cost savings and patient satisfaction increments.

However, those improvements are possible only if patients use ICT-based healthcare services regularly, by integrating them into their daily lives. If their use of the services is discontinued, public health institutions would be wasting their limited resources. Hence, patients' e-loyalty is the key for ICT-based healthcare services success. According to the marketing literature, loyalty is present when favourable attitudes for a brand are manifested in repeat buying behaviour (Keller 1993). Adapting that definition to the healthcare sector, e-loyalty can be defined as the favourable attitude of the patient towards an ICT-based healthcare service that results in repeat use behaviour.

There is a considerable volume of work related to technology acceptance. What we employ in this paper is an application of the Technology Acceptance Model (TAM) (Davis 1989). Despite the amount of academic research dedicated to examining the determinants of information technology acceptance, and to TAM in particular, very little research has been conducted on health-related research areas (Aggelidis and Chatzoglou 2009). In addition, while e-loyalty has been widely studied in the information technology area, the few studies focused on the healthcare sector are limited to the analysis of health information websites (e.g. Kim and Chang 2007).

Hence, the primary aim of this research is to use the core concepts of TAM to test the e-loyalty of patients towards ICT-based healthcare services offered by public health institutions. To achieve this goal, a modified TAM is developed and tested by using the Structural Equation Modelling (SEM) approach.

11.2 Background

The Technology Acceptance Model (TAM), first introduced by Davis (1989), is one of the most frequently employed models for research into new information technology acceptance. In TAM, user acceptance is evaluated by assessing the user's beliefs, attitudes and intentions. Attitude toward using a technology (A) was defined by Davis (1989) as "the degree of evaluative affect that an individual associates with using a system in his or her job". Attitude is determined by a function of two beliefs: Perceived Usefulness (PU) and Perceived Ease Of Use (PEOU). PU was defined as "the degree to which a person believes that using a particular system would enhance his or her job performance". PEOU is "the degree to which a person believes that using a particular system would be free from effort" (Davis 1989). PU and PEOU create belief among potential users and subsequently form their attitude. The attitude, in turn, leads to user Behavioural Intention (BI) to use an information technology. Finally, actual system use is determined by BI.

The research model built was based on TAM, while introducing several modifications which were not in TAM. Firstly, satisfaction construct was introduced in the model. Satisfaction construct represents the degree to which a user's perceived personal needs and the need to perform specific tasks satisfactorily are met by a system (Goodhue and Straub 1991). Therefore, user satisfaction is conceptualized as affective reactions of individuals and it can be defined as an attitudinal construct. According to Guimaraes and Igbaria (1997), users with a more positive attitude towards this type of technology are likely to be more satisfied with it. Hence, attitude was considered here as an antecedent of satisfaction.

Secondly, BI and actual system use was replaced by e-loyalty. BI was excluded because it refers to the probability of using a technology in the future, but this conceptualization does not include repeat use behaviour. On the other hand, initial system usage and intentions of repeat use can differ (Karahanna et al. 1999). This study suggests that e-loyalty, that is, patients' intention of repeat use, should be a closer measure of ICT-based healthcare services success than patients' initial or probable usage of them. The relationship between satisfaction and loyalty seems almost intuitive, and several researchers have found that satisfaction leads to loyalty in their studies (Bhattacherjee 2001; Wang 2007). Therefore, in this research, satisfaction was considered as an antecedent of e-loyalty.

According to these arguments, two groups of hypotheses arise: (1) Hypotheses for the relations between the TAM variables as in the original TAM framework: PEOU will positively affect PU (H1). Furthermore, enhanced PU and PEOU will increase Attitude (H2 and H3). (2) Hypothesis between patient satisfaction and e-loyalty: Attitude will positively affect Satisfaction (H4). In turn, enhanced Satisfaction will positively affect E-Loyalty (H5).

Figure 11.1 represents the research model proposed.

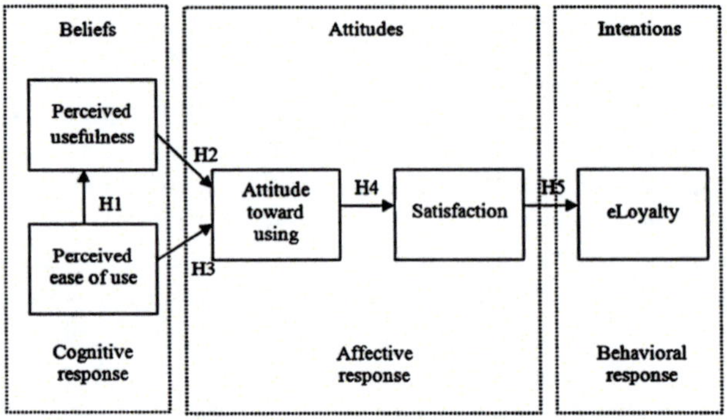

Fig. 11.1 Research model

11.3 Methods

A questionnaire was developed to be the instrument for data collection. All items were measured using a seven-point Likert-type scale with anchors from "Strongly disagree" to "Strongly agree". Question items used were mainly adapted from prior relevant studies. In this paper, the items used to measure perceived usefulness, perceived ease of use and attitude were adapted from prior work by Davis (1989). Scale of satisfaction was measured by items derived from Wang (2007), Oliver (1980) and Spreng et al. (1996). Items measuring E-Loyalty were adapted from previous efforts of Roca et al. (2006) and Wang (2007).

Data used to test the research model were gathered from a sample of users of ICT-based healthcare services. The participants were picked randomly from different parts of a region located in the South–East of Spain. All in all, the sample was diverse and the demographic differences were well covered to obtain a realistic picture of the people living there. Respondents were first asked whether they had ever used the above mentioned services; if they replied in the affirmative, they were asked to participate in the survey. Of the 277 questionnaires handed out, 256 were completed and usable questionnaires. The subjects' average age was 33.57 years, ranging from 18 to 81 years old. Approximately 52% were male.

11.4 Results

11.4.1 Assessment of the Measures

In order to get a more robust evaluation of the quality of the measurement model, a confirmatory analysis was completed by using the covariance matrix as input via LISREL 8.50 (Jöreskog and Sörbom 2001) robust maximum likelihood method. With regards to the measurement model, we began by assessing the individual

Table 11.1 Construct summary: confirmatory factor analysis and scale reliability

Constructs	Items	Standardized loading	T-value	Reliability (SCR[a], CA[b])
USE (perceived usefulness)	USE1	0.79	16.66	SCR = 0.73
	USE2	0.86	19.90	CA = 0.85
	USE3	0.83	22.67	
ATT (attitude)	ATT1	0.85	25.03	SCR = 0.73
	ATT2	0.82	22.95	CA = 0.84
	ATT3	0.84	21.52	
SAT (satisfaction)	SAT1	0.74	17.12	SCR = 0.77
	SAT2	0.80	19.53	CA = 0.81
	SAT3	0.83	21.09	
ELO (e-loyalty)	ELO1	0.85	22.12	SCR = 0.76
	ELO2	0.81	14.09	CA = 0.80
	ELO3	0.74	16.11	
EOU (perceived ease of use)	EOU1	0.75	14.70	SCR = 0.75
	EOU2	0.88	22.83	CA = 0.82
	EOU3	0.79	17.10	

The fit statistics for the measurement model were: Satorra-Bentler $\chi^2_{(80)}$=126.20; CFI = 0.94; IFI = 0.94; RMSEA = 0.049

[a] SCR = Scale Composite Reliability (SCR) of $p_c = (\Sigma\lambda_i)^2 var(\xi) / \left[(\Sigma\lambda_i)^2 var(\xi) + \Sigma\theta_{ii}\right]$ (Bagozzi and Yi 1998)

[b] CA = Cronbach's Alpha

The asymptotic covariance matrices were generated to obtain the scaled Chi-square (Satorra and Bentler 1988) and robust estimation of standard errors

item reliability (Table 11.1). The indicators exceed the accepted threshold of 0.70 for each factor loading. From an examination of the results shown in Table 11.1, we can state that all of the constructs are reliable as they present greater values for both Cronbach's alpha coefficient and composite reliability than the value of 0.7, as required in the early stages of research.

The fit statistics for the resulting 15 items, which are summarized in Table 11.1, indicate a reasonable data fit, with $\chi^2_{(80)} = 126.20$; comparative fit index [CFI] = 0.94; incremental-fit Index [IFI] = 0.94; root mean square error of approximation [RMSEA] = 0.049. The fit index of RMSEA is below 0.08, and indices of CFI and IFI are above the common standard of 0.9 (Hair et al. 1998). Although a significant Chi-square value indicates that the model is an inadequate fit, the sensitivity to sample size of this test confounds this finding, and makes rejection of the model inappropriate on the basis of evidence alone (Bagozzi 1980). However, a ratio of less than three ($\chi 2/\text{df} < 3$) indicates a good fit for the hypothesized model (Carmines and McIver 1981). In the interest of thoroughly discriminant validity, an additional test was examined, which supported this assumption, since the confidence interval (±2 standard errors) around the correlation estimated between any two latent indicators never includes 1.0 (Anderson and Gerbing 1988). This condition was satisfied for all pairwise correlations in the measurement model.

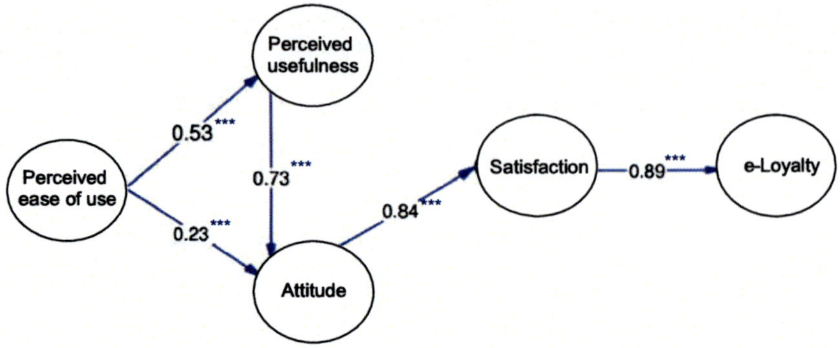

Fig. 11.2 Structural model (***$p < 0.01$ Satorra-Bentler $\chi 2/\text{df} = 1.79$; CFI = 0.92; IFI = 0.92; RMSEA = 0.054)

11.4.2 Structural Model Results

The fit of the structural model is satisfactory (CFI = 0.92; IFI = 0.92; RMSEA = 0.054). It is also interesting to note that the Parsimonious Normed Fit Index (PNFI) is above the common standard of 0.6 (James et al. 1982), thereby suggesting that the nomological network of relationships fits our data—another indicator of support for the validity of these scales (Churchill 1979). The results of the hypothesis tests are also shown in Fig. 11.2, referred to in the text as H1, H2, H3, H4 and H5. By testing our hypotheses, Fig. 11.2 shows that in all cases the relationships are highly significant and the structural model explains 79% of the variance in e-loyalty (ELO). Therefore, this analysis provides full support for H1, H2, H3, H4 and H5. The managerial implications of the relationships observed between the factors that constitute the hypothesised relationships are discussed in more detail in below.

11.5 Discussion

The goal of this study was to empirically extend research in the area of technology acceptance to understand why patients continue to use ICT-based health services. With this aim, TAM was modified by introducing e-loyalty as a postcedent variable. In consequence, the first contribution of this research is to analyze patients' acceptance of ICT-based health services by using TAM. The core constructs of TAM (perceived usefulness, ease of use and attitude) significantly affect user behavioural intention in our study (i.e. E-Loyalty), which implies that the main aspects of TAM apply to this context as well. In addition, perceived usefulness was found to be the most significant effect on attitude, which suggests that a patient's belief in usefulness is a decisive antecedent of affective variables (i.e. attitude and satisfaction), and consequently, of behavioural variables (i.e. e-loyalty).

The second contribution is to extend the basic TAM towards the postcedents by adding the variables satisfaction and e-loyalty. Modelling ICT-based healthcare services acceptance is very useful to public health institutions but understanding why patients build e-loyalty towards them is crucial. The research model tested provides deeper insights into the process of patients' loyalty build-up. Furthermore, satisfaction was included in the model acting as a link between TAM variables and e-loyalty. Consistent with prior studies, there was a significant positive impact of satisfaction on e-loyalty.

The third contribution of this research is to test the TAM in a patient context. Previous research on the TAM has mainly been conducted in workplace settings and, in particular, with hospital personnel. In this type of environment, people's attitudes, intentions, and behaviours, as well as their interrelationships are likely to be shaped by formal authority and directives (Lanseng and Andreassen 2007). This research has empirically supported the core concepts of TAM in a patient context, where respondents are free to form their own beliefs, attitudes, and intentions, as the theoretical foundation of the TAM assumes. Thus, the results contribute to the general validity of the model.

From a managerial point of view, these findings suggest that policy makers should focus on the following issues: (1) Since perceived usefulness is the most important antecedent of affective variables, managers should increase patients' E-Loyalty by improving their beliefs of how ICT-based healthcare services can offer advantages to their experience. Informative actions (guides, leaflets, posters, promotional tasks, etc.) must focus on the usefulness of using ICT-based services. (2) While usefulness can be communicated, ease of use must be designed into the services. Policy-makers must instruct ICT-based services developers to focus on ease of use. Since population have different levels of technology readiness, reliable, user-friendly services, with good user interface consistency, should be designed. (3) It is essential to get a large part of the population to use the ICT-based services and to ensure e-loyalty of use in the majority of citizens. Policy makers should develop promotional campaigns to publicize the ICT-based healthcare services between the users with potential interest in using e-services. These people could be the medium harnessed to increase the use of ICT-based healthcare services among the general public.

This research has several limitations. First, individual variables (such as age, gender or level of education) may affect the results. In futures studies, TAM could be extended towards the antecedents by incorporating these variables into the research model. In addition, other external variables (such as technological or organisational factors) that influence on perceived easiness of use and usefulness could be analyzed. In this way, the model's predictive power could be improved. Secondly, the research model was tested in one geographical area. Caution is necessary when generalizing the findings in relation to other population areas.

References

Aggelidis VP, Chatzoglou PD (2009) Using a modified technology acceptance model in hospitals. Int J Med Inform 78:115–126

Anderson JC, Gerbing DW (1988) An updated paradigm for scale development incorporating unidimensionality and its assessment. J Mark Res 25(2):186–192

Bagozzi RP (1980) Causal models in marketing. Wiley, New York

Bagozzi RP, Yi Y (1988) On the evaluation of structural equation models. J Acad Mark Sci 16(1):74–94

Bhattacherjee A (2001) An empirical analysis of the antecedents of electronic commerce service continuance. Decis Support Syst 32(2):201–214

Carmines EG, McIver SP (1981) Analysing models with unobserved variables: analysis of covariance structures. In: Bohrnstedt GW, Borgatta EF (eds) Social measurement; current issues. Sage, Beverly Hill, pp 65–115

Churchill GA (1979) A paradigm for developing better measures for marketing constructs. J Mark Res 16(1):64–73

Davis FD (1989) Perceived usefulness, perceived ease of use, and user acceptance of information technology. MIS Q 13(3):319–340

Goodhue D, Straub D (1991) Security concerns of system users. Inf Manag 20(1):13–27

Guimaraes T, Igbaria M (1997) Client/server system success: exploring the human side. Decis Sci 28(4):851–876

Hair JFJ, Anderson RE, Tatham RL, Black WC (1998) Multivariate data analysis, 5th edn. Prentice-Hall, Englewood Cliffs

James LR, Muliak SA, Brett JM (1982) Causal analysis assumptions, models and data. Sage, Beverly Hills

Jöreskog K, Sörbom D (2001) LISREL 8.50. Scientific Software, Chicago

Karahanna E, Straub DW, Chervany NL (1999) Information technology adoption across time: a cross-sectional comparison of pre-adoption and post-adoption beliefs. MIS Q 23:183–213

Keller KL (1993) Conceptualizing, measuring and managing customer-based equity. J Mark 57: 1–2

Kim D, Chang H (2007) Key functional characteristics in designing and operating health information websites for user satisfaction: an application of the extended technology acceptance model. Int J Med Inform 76:790–800

Lanseng EJ, Andreassen TW (2007) Electronic healthcare: a study of people's readiness and attitude toward performing self-diagnosis. Int J Serv Ind Manag 18(4):394–417

Oliver RL (1980) A cognitive model for the antecedents and consequences of satisfaction. J Mark Res 17:460–469

Roca JC, Chiu C-M, Martínez FJ (2006) Understanding e-learning continuance intention: an extension of the technology acceptance model. Int J Hum Comput Stud 64:683–696

Satorra A, Bentler PM (1988) Scaling corrections for Chi-square in covariance structure analysis. In: Proceedings of the business and economic statistics section of the American statistical association, pp. 308–313

Spreng RA, MacKenzie SB, Olshavsky RW (1996) A reexamination of the determinants of customer satisfaction. J Mark 60:15–32

Wang Y-S (2007) Assessing e-commerce systems success: a respecification and validation of the DeLone and McLean model of IS success. Inf Syst J 18:529–557

Chapter 12
Visualizing the Scientific Landscape Using Maps of Science

Gaizka Garechana, Rosa Rio, Ernesto Cilleruelo and Javier Gavilanes

Abstract The use of tech-mining tools in Science and Technology databases can provide vital information to understand and foresee innovation processes. Some of that information can be visualized by using a wide variety of maps, which gives fast, intuitive comprehension of the social and cognitive structure of the analyzed topic. Among these, the maps of science are used to reflect graphically the structure, evolution and main actors of a given scientific field. Institutions can use maps of science with various purposes, ranging from benchmarking to the assessment of strategic choices about their position and direction in R&D efforts. The aim of this paper is to give a brief state-of-the-art review of science maps as visualization tools.

12.1 Introduction

A large portion of the scientific production worldwide gets in bibliographic databases, whether it be articles, conference proceedings or patents. When talking about reputable databases, stored information must fulfil some requirements concerning its quality and structure, making these databases a reliable and easily-searchable source of worldwide knowledge. Several authors claim that a proper

G. Garechana (✉)
University of the Basque Country, calle Elcano 21, 48030 Bilbao, Spain
e-mail: gaizka.garechana@ehu.es

R. Rio · J. Gavilanes
University of the Basque Country, calle Nieves Cano 12, 01006 Vitoria, Spain

E. Cilleruelo
University of the Basque Country, Alameda Urquijo s/n, 48030 Bilbao, Spain

S. P. Sethi et al. (eds.), *Industrial Engineering: Innovative Networks*,
DOI: 10.1007/978-1-4471-2321-7_12,

analysis of the information contained in science and technology databases can contribute to R&D processes with valuable knowledge, the main problem being the vast amount of textual data to be processed in the analysis. Text-mining tools come to solve this problem, automatically "reading" thousands of fields and allowing the analyst to extract not only trends of bibliometric elements such as authors, articles or keywords, but also relationships between elements like terms or authors, creating what we call the "second generation" indicators.

The phenomena of co-citation and co-occurrence are the very pillars of the second generation indicators (Porter and Cunningham 2005).

- Co-citation takes place when two authors are cited together in the same document. Co-citation can also be found when talking about articles. For example, two articles being cited together by a third article. It is assumed that co-citation implies a cognitive relationship between the items being co-cited, a relationship derived from the opinion of the researchers working on that particular topic (Moya-Anegón et al. 2004).
- Co-occurrence takes place when two or more terms appear together in bibliometric fields such as title, abstract, classification codes, or even in the free text of the article. The assumption here is that terms that frequently appear together must be cognitively related. Depending on the text-mining tool utilized, the analysis can be fine-grained looking for the co-occurrence of terms within the same two phrases, three phrases, the same paragraph, and so on (Porter and Cunningham 2005).

Bibliometric maps are the easiest and most intuitive way to display the relationships derived from co-citation or co-occurrence. They are two- or three-dimensional representations of bibliographic items, arranged in such a way that the ones which are cognitively related to each other are positioned in each other's vicinity, whilst the ones that are not or hardly related are distant from each other. The degree of relationship is given by the number of times that the items are co-cited (or the times they co-occur) together. The transformation of this co-citation or co-occurrence frequency into physical distances is made by statistical techniques such as Multidimensional Scaling (MDS). The map of science is a particular type of bibliometric map (Noyons 2005).

12.2 The Map of Science

The map of science is used to reflect the cognitive structure of a scientific field, to monitor its evolution and the main actors within. This section explains some of the main approaches available to construct a science map, whether it be a local science map (focused on a particular scientific area) or a global science map.

12.2.1 Precedents

The first attempts started by seeking to define research fronts (Price 1965) with the use of document citation data, or manually generate a historical-topological science map on the basis of citation of DNA research (Garfield et al. 1964), but the big step was taken by Henry Small using document co-citation as a variable of study. Small believed that, in highly cited documents, citations were more potent symbols conceptually, than words to represent the structure of a scientific field, and he applied this principle to the visual representation of the cognitive structure of Particle Physics (Small 1973). The following year he succeeded in creating a global map of natural sciences, starting from a collection of highly cited documents and clustering them by the single-linkage method (Small and Griffith 1974).

Another major effort to represent the global structure of science was the elaboration of the Atlas of Science, a project that took 6 years to materialize and started with Garfield in 1981, mapping the fields of Biochemistry and Molecular Biology. Document co-citation was the basis, but a new spatial positioning system was used by applying Multidimensional Scaling techniques. An example is given (see Fig. 12.1), where highly cited scientific documents were clustered by co-citation analysis criteria, and their co-citation links represented by lines (Small 1999).

In the 1980s and 1990s the progress of computer graphics technologies made it possible to use the maps as an interface to explore the results and underlying data of the analysis. This is the aim of Small in the picture below, right (Small 1999), where 36,720 highly cited documents were grouped in 35 clusters, allowing the detailed exploration of the internal structure of each cluster by zooming into the object. Labels are based on a frequency analysis of article titles and journal category names. Clusters have been spatially positioned by MDS.

Subsequent efforts have been directed to allow the location of research bodies of prominent authors or institutions in the science maps, as well as to facilitate the construction of maps for different time intervals, permitting the analysts to track temporal changes in the scientific field subject to study. The enhanced availability of data and the development of visualization tools have been widely exploited in this sense, greatly contributing to the readability and usefulness of science maps.

12.2.2 Main Types

12.2.2.1 Document Co-Citation Maps

The number of times that a document has been co-cited with another (see Sect. 12.1) is the starting datum for the building of these maps. The objective is to define the groups of documents that form the core literature of a particular scientific field and to identify the researching fronts composed by the citing

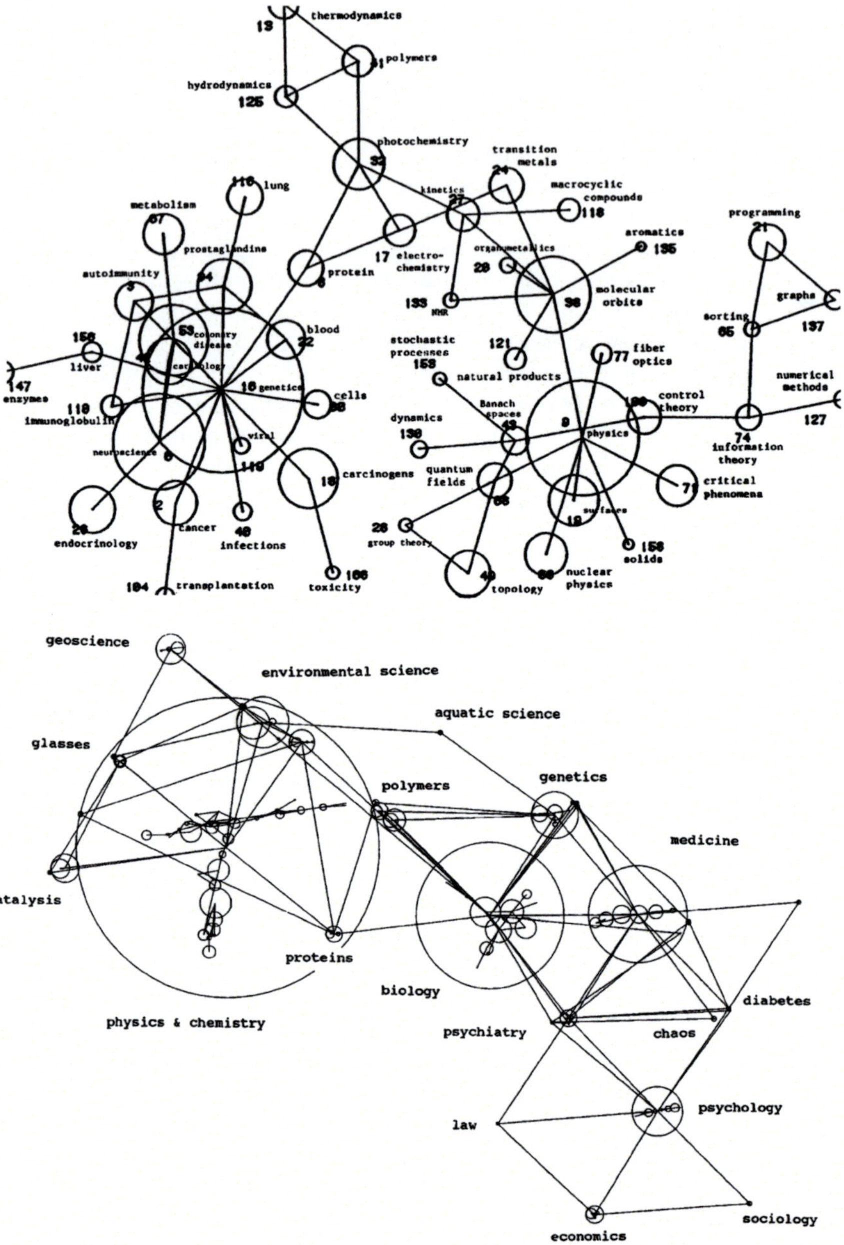

Fig. 12.1 *Top*: An early co-citation map of science showing the disciplines Biology, Chemistry and Physics. *Bottom*: map of science, lines represent co-citation links between clusters (Small 1999)

documents around them. The core documents usually correspond to intellectually specialized individuals while the citing documents can be associated to multiple core groups, giving valuable information about the degree of interrelation between different scientific fields. The Atlas of Science developed at ISI, with Henry Small as the main researcher involved, stands as a landmark in the field of document co-citation maps (Small 2003). Two visualizations of this type are given in Fig. 12.1.

12.2.2.2 Author Co-Citation Maps

Data for the construction of these maps are supplied by the counting of the number of times that two authors are cited together (see Sect. 12.1). The resulting map shows identifiable author groups, describing at the same time the social and intellectual structure of a scientific field (Rostaing 1996). The Fig. 12.2 show two author co-citation science maps. The first corresponds to the Information Science field (White and Griffith 1981) and has been built from a collection of the most representative authors according to the researchers' criteria. The second takes co-citation data from the articles resulting from a query oriented to Domain Visualization Science and presents the data in a 3D visualization. The height of the bars gives additional information about the number of publications corresponding to the author. Both pictures are given together for the reader to perceive the improvement made in readability and interpretation richness thanks to the development of visualization techniques.

12.2.2.3 Co-Classification Maps

Some databases assign a content-based classification code to documents. These codes are built by de-composing science in separate sectors of activity and assigning a univocal code to each sector. This indexation labour sometimes takes place in the form of keywords assigned to the document, but an alphanumerical code is often preferred because of the absence of synonyms and polysemy issues, and its resulting accuracy (Rostaing 1996). The data to build a co-classification map are given by the co-occurrence of these classification codes for each represented item. The represented items may be prominent authors, institutions or even single documents.

An example is the work of Spasser (1997) representing the structure of the scientific field "Pharmacy". The classification code utilized is the "Section Heading" code given by the IPA[1] database (there are 25 different codes altogether), and the input data to build the map is the co-occurrence of two "Section Heading" codes in the fields "Primary Heading" and "Related Heading". When a large number of documents share two classification codes, a cognitive relationship between these subsectors can be inferred. Clusters are made to show the main

[1] International Pharmaceutical Abstracts, provided by Thomson ISI.

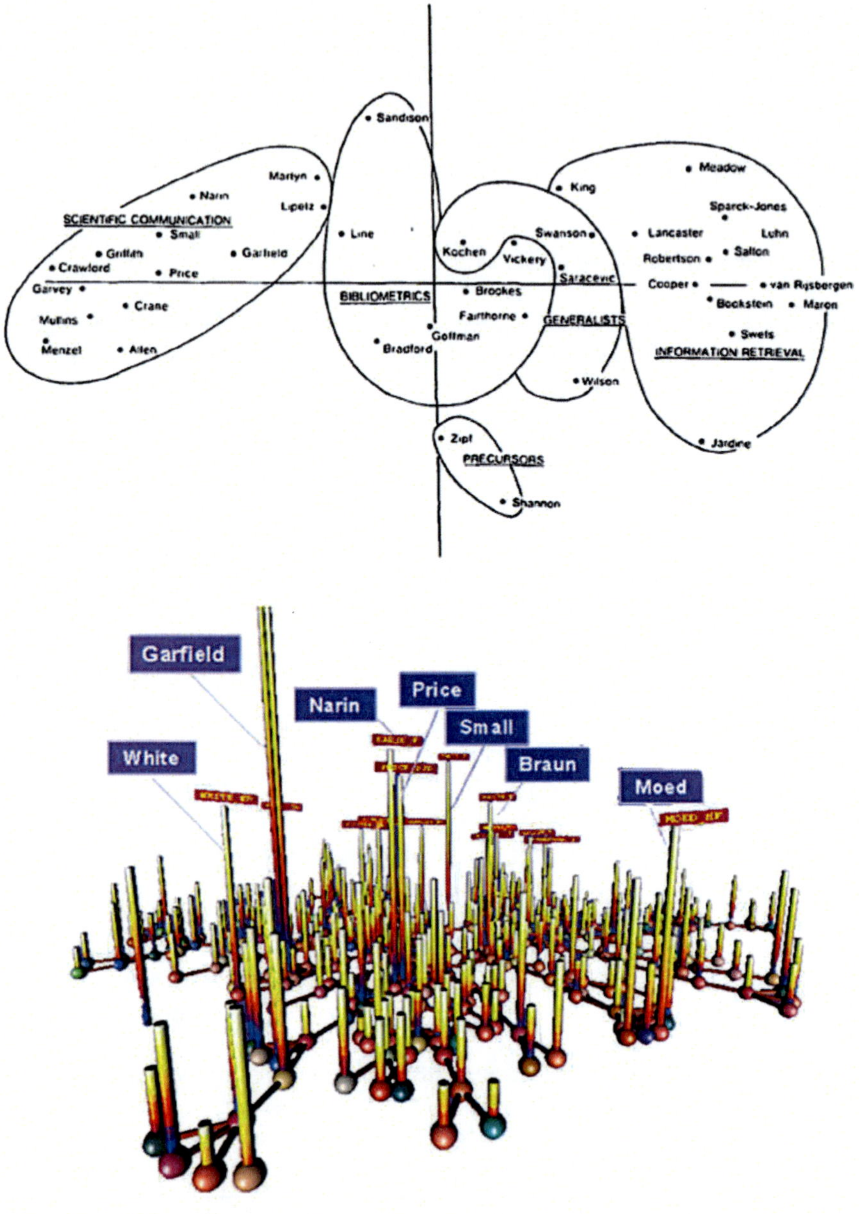

Fig. 12.2 Two author co-citation maps representing Information Science (*top*) (White and Griffith 1981) and Domain Visualization Science (*bottom*) (Börner et al. 2003)

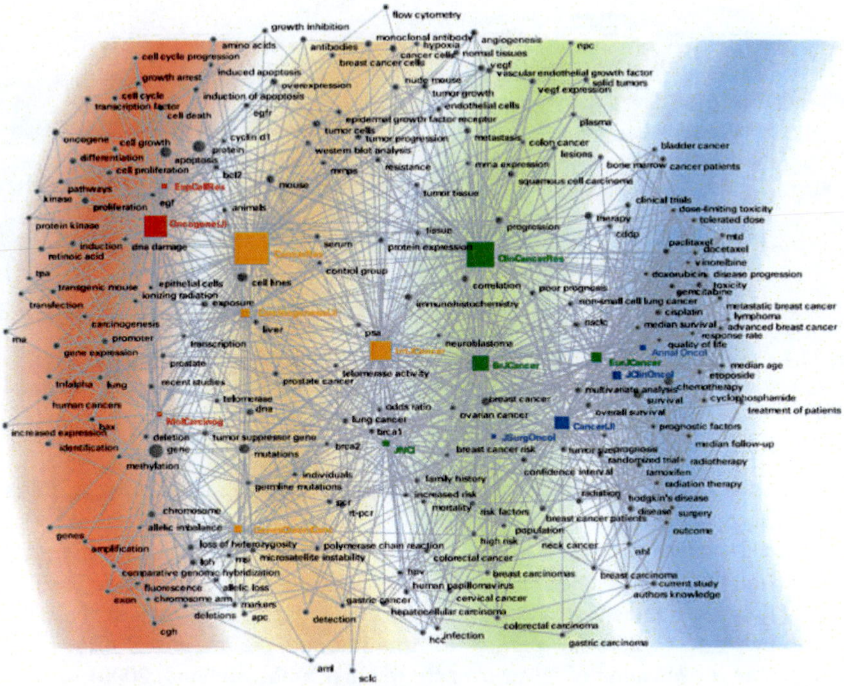

Fig. 12.3 Co-word science map reflecting the cancer research field (Cambrosio et al. 2006)

cognitive subsectors of the studied field. The advantages of this kind of maps depend on the accuracy and availability of the classification codes, which specifically define the cognitive framework of the article. The disadvantage is generated by the reliance on the discernment of indexers.

12.2.2.4 Co-Word Maps

Co-word mapping is based on measuring the co-occurrence of words in such fields as title, keywords, abstract or even in the free text of the document. Text-mining tools can detect the co-occurrence of any given words inside the same phrase or a given number of phrases. Figure 12.3 reflects the cognitive field of cancer research. Input data were extracted from articles related to cancer research, retrieved from the SCI[2] by exploiting a filter using title keywords. Titles and abstracts were text-mined and the top 200 semantic concepts extracted. Links reflect the relationship between represented concepts and between these and the main journals. Main journals were

---

[2] Science Citation Index, provided by Thomson ISI.

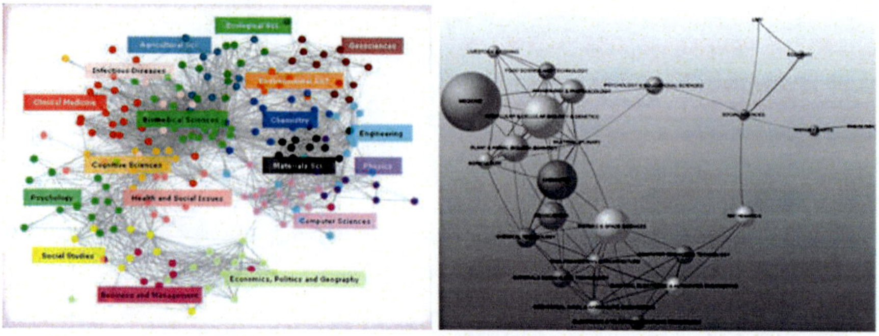

Fig. 12.4 *Right*: Spanish science map based on ANEP classification (Moya-Anegón et al. 2004)] *Left*: Global map of science based on ISI's Subject Categories (Leydesdorff and Rafols, 2009)

superposed within the resulting overall semantic network. Colours reflect different research level categories.

Co-word analyses are also known as semantic maps, and their objective is to facilitate the understanding of the cognitive structure of a scientific field. The existence of polysemy and synonyms calls for the development of a complex thesaurus, and the mapping of different periods of time presents the additional difficulty of instability in the meaning of terms over the years (Cambrosio et al. 2006).

12.2.2.5 Journal Co-Citation Map

Input data for the building of these maps are the number of times that articles pertaining to two journals are cited together in the articles contained in a third journal. The work of McCain (1998) representing the journal co-citation map of the neural network research field is given as an example. The maps obtained show cognitive subsectors defined by statistical clustering methods.

12.2.2.6 Citation and Co-Citation of Classes and Categories

Principles of citation and co-citation can be applied to any classification available in the database. For some analyses, a broader unit of visualization can be desirable, in order to obtain a better image and a simpler map. The accuracy of the criteria used in the classification made by the indexers must be checked before building the map. Figure 12.4 (right) gives an example of a science map reflecting the scientific production in Spain, taking the co-citation between scientific articles with Spanish addresses as input data, and the ANEP[3] classification of science as a basis to build the visualization (Moya-Anegón et al. 2004).

Citation between categories can be the data upon which to build a proper science map. Citation between ISI's Subject Category classification has been used as a robust classification of science upon which to construct a global map of science (Leydesdorff and Rafols 2009). Figure 12.4 (left) reflects the global structure of science as contained in JCR database, and items represented are ISI's Subject Categories. The labels in the figure are the names of ISI's 18 macro-disciplines that group Subject Categories at a wider level (items of the same colour represent subject categories that pertain to the same macro-discipline).

12.3 Conclusions

An effective mapping of scientific fields, whether it be at a global level or at a specific research level, allows researchers or interested institutions to identify the emerging research fronts, revealing the interests of researchers now at work and permitting to associate authors' names to each front (Leydesdorff and Rafols 2009). Some authors emphasize the utility of science maps to track the temporal change of scientific fields, to enhance the benchmarking collaborative activities and to explore the cross-disciplinary relations in emergent technologies, since it is a well-known fact that the majority of present-day scientific activities cannot be easily included into a single discipline (Rafols et al. 2010). Others look for applications in the field of information retrieval and classification (Börner et al. 2003). The progressive advances being made in the availability and quality of information, as well as in the tools for text-mining and visualization, will undoubtedly make maps of science a concept closer to what Derek Price defined as "War maps of science" (Small 2003), a map upon which government bureaucrats or decision-makers of any R&D intensive institution could monitor a large display of the scientific terrain, planning the next manoeuvre.

References

Börner K, Chen C, Boyack KW (2003) Visualizing knowledge domains. Ann Rev Info Sci Technol 37:179–255

Cambrosio A, Keating P, Mercier S et al (2006) Mapping the emergence and development of translational cancer research. Eur J Cancer 42:3140–3148

Garfield E, Sher IH, Torpie RJ et al (1964) The use of citation data in writing the history of science. Institute for Scientific Information Inc, Philadelphia

---

[3] ANEP stands for *Agencia Nacional de Evaluación y Prospectiva*, the Spanish National Agency for Evaluation and Prospection of the Spanish Ministry of Science and Innovation.

Leydesdorff L, Rafols I (2009) A global map of science based on the ISI subject categories. J Am Soc Info Sci Technol 60:348–362

McCain KW (1998) Neural networks research in context: a longitudinal journal co-citation analysis of an emerging interdisciplinary field. Scientometrics 41:389–410

Moya-Anegón et al. (2004) A new technique for building maps of large scientific domains based on the co-citation of classes and categories. Scientometrics 61:129–145

Noyons C (2005) Science maps within a science policy context, Handbook of quantitative science and technology research, pp 237–255

Porter AL, Cunningham SW (2005) Tech mining: exploiting new technologies for competitive advantage, Wiley-Interscience, London

Price DJ (1965) Networks of Scientific Papers, Science New York. vol.149 pp 510–515

Rafols I, Porter A, Leydesdorff L (2010) Science overlay maps: a new tool for research policy and library management. J Am Soc Info Sci Technol 61:1871–1887

Rostaing H (1996) La bibliométrie et ses techniques. Éditions Sciences de la Société

Small H (1973) Co-citation in the scientific literature: a new measure of the relationship between two documents. J Am Soc Info Sci 24:265–269

Small H (1999) Visualizing science by citation mapping. J Am Soc Info Sci 50:799–813

Small H (2003) Paradigms, citations, and maps of science: a personal history. J Am Soc Info Sci 54:394–399

Small H, Griffith BC (1974) The structure of scientific literatures I: identifying and graphing specialties. Science studies pp 17–40

Spasser M (1997) Mapping the terrain of pharmacy: co-classification analysis of the international pharmaceutical abstracts database. Scientometrics 39:77–97

White HD, Griffith BC (1981) Author co-citation : a literature measure of intellectual structure. J Am Soc Info Sci 32:163–171

Chapter 13
Technology-Based Firms in Universities in the Basque Country-Aquitaine Euroregion

Enara Zarrabeitia, Aline Dupouy, Pablo Díaz de Basurto, Patxi Elissalde and Patxi Ruiz de Arbulo

Abstract In the quest for regional economic growth and sustainable development increasingly greater significance is being attributed to the need for a change of attitude towards what is referred to as the knowledge-based economy. Within this context, higher education plays a leading role through its support for the creation and development of technology-based firms (TBFs). Accordingly, in view of the importance that such firms have for the regions of the Basque Country and Aquitaine, the objective of this paper is to clarify the factors that underpin the success of university TBFs, and also to pool existing knowledge in each of these regions. An additional aim is to remove the barriers between these two territories in order to increase potential cross-border activity. The settings chosen for the study are the University of the Basque Country (UPV/EHU—Basque Country) and the École Supérieure des Technologies Industrielles Avancées (ESTIA—Aquitaine), where an analysis has been made of the case of 20 firms through personal in-depth interviews conducted with entrepreneurs in different university TBFs.

13.1 Background and Context

Today more than ever, due to an increasingly globalised and competitive market, the support provided for the creation of technology-based firms (TBFs) is a crucial matter for fostering regional development. It is essential to support the

E. Zarrabeitia (✉) · P. Díaz de Basurto · P. Ruiz de Arbulo
Department of Management Engineering School of Bilbao,
University of the Basque Country, Alameda Urquijo s/n, 48013 Bilbao, Spain
e-mail: enara.zarrabeitia@ehu.es

A. Dupouy · P. Elissalde
ESTIA Recherche et CREG UPPA, École Supérieure des Technologies Industrielles
Avancées, ESTIA, Technopôle Izarbel, 64210 Bidart, France

S. P. Sethi et al. (eds.), *Industrial Engineering: Innovative Networks*,
DOI: 10.1007/978-1-4471-2321-7_13, © Springer-Verlag London Limited 2012

Table 13.1 Extended role of universities

Teaching	Research	Spin-offs—tbfs
Preservation and dissemination of knowledge	First academic revolution	Second academic revolution
Twelth–early nineteenth centuries New missions generate a conflict of interests and discord Teaching University	End nineteenth–early twentieth centuries Two missions: Teaching and research Research University	End twentieth century Third mission: Economic and social development (in addition to the two prior missions) Enterprising University

Source Zarrabeitia et al. 2010; Etzkowitz 2004

incorporation of technology firms with a high growth potential, which contribute added value to the transformation of the economic system as a whole, introducing new businesses, new production techniques and new forms of management, while enriching the stock of technical and entrepreneurial qualifications; in short, firms that reinforce all those areas leading to integration in the knowledge society (Martínez and Tadeo 2006).

In turn, although teaching and research have traditionally been considered the overriding goal of higher education, the current prevailing opinion is that in addition to contributing to innovation through the advancement of basic science, a university should extend its remit to include a direct role in the economic development of the territory by promoting and supporting the creation and development of TBFs (Aceytuno and de Paz 2008; Gómez et al. 2007), Table 13.1.

Given the importance of TBFs, the appropriate strategies and measures of support need to be designed and implemented within the field of higher education in order to drive their creation and development. This means that it will be important to analyse the specific conditions of each university and its environment. Accordingly, there is a pressing need to understand the diversity of university TBFs, as well as the different requirements of these business ventures throughout the protracted period of generation, formation and development (Gómez et al. 2007). Accordingly, resources need to be earmarked for in-depth research into the characteristics, determinants and results of these firms that will lead to a better understanding of the factors that inform their creation and development.

Furthermore, although the relations between the border territories of the Basque Country and Aquitaine are extremely complex, the wish to create a Euroregion, and thereby gain more weight within the European Union, requires that both regions join forces and work together. This renders it essential to promote cooperation between these two regions and to deal with the challenges posed by the removal of frontiers, growing inter-dependence and global competition (Urteaga 2007).

Thus, although certain factors cannot be emulated and the bulk of the determinants of the success of the regions of the Basque Country and Aquitaine need to be contextualised within the regional sphere, what can be done is to pool the knowledge and experience available in each region. Likewise, the aim is to remove the barriers between the two territories with a view to increasing potential cross-border activity and favour the expansion and development of these ventures on both sides.

The research conducted here seeks to provide existing literature with new empirical evidence on the creation and development of university TBFs in the regions of the Basque Country and Aquitaine.

13.2 Research Goals

This research paper's overriding aim is to shed light on those factors with an emphasis on the creation and successful development of university TBFs in the regions of the Basque Country and Aquitaine.

In particular, the following specific goals are addressed:

- Understand the scope and main features of recent experiences in the creation and development of university TBFs in the Basque Country—Aquitaine.
- Analyse the factors that encourage people with advanced scientific-technological know-how to become involved in the creation of advanced technology firms in order to commercially exploit such expertise.
- Analyse specific experiences that have subsequently proven to be the most successful, as well as those that have failed.
- Extend the analysis to other key outside or scenario factors with an impact on the creation and development of university TBFs.

13.3 Research Methods and Scope of the Study

To date, the method of diagnosis has focused heavily focused on the use of the different resources available (e.g. databases) as the only focus of study on TBFs, whereas there has been hardly any exploration of the expectations and opinions of the entrepreneurs themselves. In order to change this, the method used here to analyse the success factors of university TBFs has involved investigating different cases by conducting several interviews with the actual entrepreneurs behind the ventures of interest on both sides of the border.

These interviews have led to an analysis of the determinants of the attitudes and skills of these entrepreneurs, as well as the most important factors in the different stages that university TBFs go through on the path to business success, including the main obstacles or hindrances for their growth and consolidation (Hlady-Rispal 2002; Moreau 2004; Molina et al. 2005; Espí et al. 2007; Dupouy 2008; Merino and Villar 2008; Ortín et al. 2008; Yin 2009).

Table 13.2 Technical details of the research

Sampling universe	TBFs set up under the auspices of the business creation schemes of UPV/EHU and ESTIA
Scope	– Basque Country (University of the Basque Country—UPV/EHU) – Aquitaine (École Supérieure des Technologies Industrielles Avancées—ESTIA)
Sampling procedure	Non-probability sampling. Quota sampling
Sample size	20 TBFs (3 of which have ceased trading) (10 in UPV/EHU and 10 in ESTIA)
Data gathering method	Personal interviews held over two sessions Each session lasted approximately 1 h
Field work	January 2010–July 2010
Data analysis	Qualitative and quantitative analysis (Microsoft Excel)

Regarding the stages of a TBF, the following aspects have been highlighted: inception and consolidation of the idea, drafting of the business plan, incorporation of the TBF and its subsequent expansion (Palacios et al. 2005 Merino and Villar 2008). Within each stage, therefore, an analysis has been made of the importance of different factors for the achievement of business success, grouped into: human capital, organisational capital, technological capital, tangible resources, social capital and scenario conditions.

Finally, in regard to the sample used for conducting the study, the firms analysed are TBFs set up under the auspices of the following business creation schemes: "EHU-Abiatu" and "Entreprenari" of UPV/EHU and "ESTIA Entreprendre" of ESTIA. The following are the technical details of the study Table 13.2.

13.4 Results and Conclusions

13.4.1 Scope and Nature of Recent Experiences in the Creation and Development of University TBFs in the Basque Country and Aquitaine

Although in terms of their trajectories and respective sizes the University of the Basque Country and ESTIA differ greatly,[1] both universities are nevertheless mindful of the importance of nurturing an entrepreneurial spirit and supporting the creation of technology-based firms, and have therefore implemented the necessary programmes and structures for fostering TBF's. The UPV/EHU has witnessed the

[1] The "Universidad del País Vasco" (UPV/EHU), created as such in 1980, as the culmination of a long process, is the public university of the Basque Autonomous Community, and has 31 faculties and schools and 195 degree courses. ESTIA, set up in 1996, is a private university that provides three specialist courses in engineering.

creation of 62 firms over the period 1997–2009, while ESTIA has seen the creation of 18 firms over the 2000–2010 period.

Accordingly, although ESTIA is a small university, the creation and development of university TBFs is being favoured by lesser expectations and a markedly innovative approach of the engineering school.

In regard to the factors that encourage people with advanced scientific and technical knowledge to involve themselves in setting up firms to commercially exploit their expertise, the sensing of a business opportunity (according to 80% of the entrepreneurs) and the chance to implement their own ideas (according to 90% of the entrepreneurs) are either very important or the main reasons for starting up a TBF.

13.4.2 Success Factors in the Processes for Creating and Developing University TBFs

The information garnered from the personal interviews with the various entrepreneurs has revealed that the real circumstances of university TBFs are extremely complex. So although this project has sought to further describe the so-called factors of success for the consolidation of these ventures, there has been no attempt to provide a single answer through a single outcome.

Based on this premise, there follows an overview of what these business people on both sides of the border consider to be the main factors of success in the entrepreneurial process (see Fig. 13.1).

In addition to the performance skills and abilities of the entrepreneurs themselves, such as the capacity for work, the desire to undertake new projects and the awareness to adapt to the environment, human resources are a crucial factor in university TBFs. A creative and motivated human component is essential from the very moment an idea is first conceived. Moreover, the reliance on qualified staff with the right profile in terms of scientific and technological know-how and business management skills, on the one hand, and on the other an executive team that is wholly familiar with the target market, are two aspects that become critical factors once the idea has been developed, above all during the stages of creating and expanding TBFs. Although deemed important throughout the firm's development stages, undivided dedication by the entrepreneurs becomes a crucial factor, in the opinion of those interviewed, during the final stages of a TBF's implementation.

The business plan's clarity and credibility and the preparation of sound strategic planning are key variables for ensuring that the venture has a smooth commencement. Regarding the product or service provided, key aspects at all stages of a TBF are, in addition to filling a gap in the market, innovation, quality and the degree of differentiation. Relations within the business structure and, above all, dealings with potential customers are decisive during the stages of creating and expanding a TBF.

	Idea	Business Plan	Creation	Expansion	Average
HUMAN CAPITAL					
Prior experience of the executive team in the business sector	3.95	4.00	4.00	4.40	4.0875
Staff highly skilled in science and technology	3.10	2.70	3.15	3.80	3.1875
Staff highly skilled in business management	3.00	3.80	3.85	4.10	3.6875
Creative and motivated staff	4.50	3.95	4.30	4.30	4.2625
Attraction of partners with backgrounds in the sector	2.65	2.65	3.15	3.90	3.0875
Attraction of risk-taking partners	2.70	2.75	3.25	3.95	3.1625
Prestige of the executive team	2.70	2.50	3.20	3.40	2.95
Entrepreneurs' sole dedication to the venture	3.40	3.70	3.90	4.40	3.85
ORGANISATIONAL CAPITAL					
Clarity and credibility of the business plan	3.30	4.00	3.70	4.00	3.75
Strategic planning	3.00	3.75	3.80	4.45	3.75
Adaptability	4.05	4.05	4.05	4.45	4.15
Brand image	2.55	3.00	3.50	4.25	3.325
TECHNOLOGICAL CAPITAL					
Degree of product differentiation	4.20	4.10	4.20	4.30	4.2
Value added and product quality	4.00	3.90	4.30	4.55	4.1875
Time allocated to the product's technological development	3.06	2.83	3.89	3.89	3.4175
Availability of a suitable appraisal of the technology	3.05	3.25	3.85	3.95	3.525
Legal protection of innovations	2.75	2.50	3.05	3.60	2.975
TANGIBLE RESOURCES					
Founders' own financial resources	2.90	3.20	3.50	3.40	3.25
Own technology equipment	2.60	2.60	3.50	2.95	2.9125
Incubation services	3.50	3.70	4.20	2.95	3.5875
SOCIAL CAPITAL					
Dealings within the business fabric	3.10	3.20	4.00	4.15	3.6125
Dealings with potential customers	3.70	3.85	4.50	4.60	4.1625
Dealings with R&D centres and universities	3.70	3.20	3.60	3.60	3.525
SCENARIO CONDITIONS					
Availability of qualified staff	3.75	3.50	4.00	4.70	3.9875
Financial market offer	2.50	3.00	3.80	4.00	3.325
Public subsidies overall	3.00	3.50	4.15	3.35	3.5
Business support programmes	3.30	4.00	4.00	3.35	3.6625
Presence of science and technology parks	2.90	2.80	2.80	2.80	2.825
University's prestige	2.83	2.67	2.89	2.78	2.7925
Prevailing business culture	4.10	4.10	4.30	4.40	4.225

Fig. 13.1 Rating of the degree of importance of different factors underpinning business success in the opinion of the entrepreneurs (average score of the responses provided by the 20 entrepreneurs on a five-point Likert-type scale from "hardly important" to "very important")

Another vital factor throughout the entire process of developing a TBF is the prevailing business culture, within both society at large and the university itself. Furthermore, business support schemes and public subsidies are essential for drafting the business plan and setting up the TBF.

Therefore, given the characteristics of university TBFs, within all the groups of factors analysed there are determinants that are crucial to the creation and/or development of these ventures. Accordingly, in order to pursue their successful implementation, any support that they receive should focus on their human capital, organisational capital, technological capital, tangible resources, social capital and the scenario conditions; and, moreover, this support will depend on the venture's sequential stage of development.

Furthermore, it is apparent that as a TBF progresses through the venture's sequential development, the factors with a bearing on its business success increase in number. This is confirmed by an analysis of the influence that different groups of factors have on the various stages that a TBF passes through on the path to success (see Fig. 13.2). The importance of all these groups follows a generally upward trend with the exception of the group of tangible resources. The latter is because once the venture has been consolidated, and as is to be expected, factors such as the founders' own financial resources and incubation service have diminished importance.

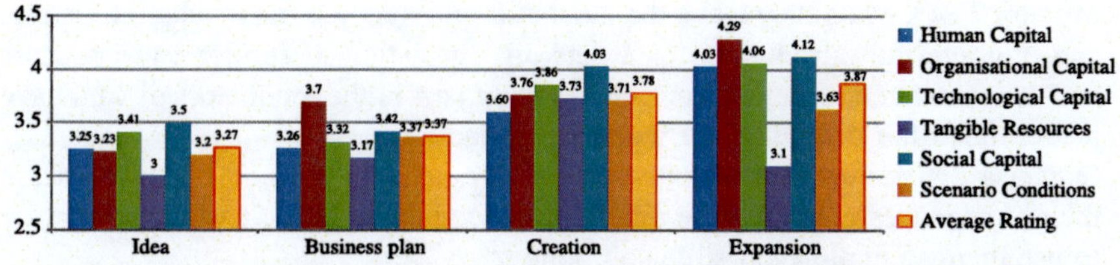

Fig. 13.2 Influence of the various groups during the stages on a TBF's path to success

Within this context, despite the fact that diverse circles in both the Basque Country and Aquitaine are backing the creation and development of TBFs through the design and implementation of different support measures and strategies, the schemes and aid analysed are geared towards the process of adapting to the new business paradigm in higher education, focusing on aspects that boost the development of an entrepreneurial culture and support entrepreneurs during the stages involved in consolidating the idea and starting up a TBF.

All these measures are required to ensure that a university can undertake the new mission that it has been entrusted, namely, to contribute directly to a territory's economic development. Nevertheless, once knowledge transfer has been achieved through the creation of a TBF, it should be remembered that although ventures of this nature may provide considerable added value, their small size makes them fragile and they will continue to need a helping hand or receive mentoring once they have left the incubator. This means that if the aim is to create businesses through a university that will play a part in territorial development, their monitoring or support should not be terminated when they are created.

13.5 Recommendations

There follows details of some of the proposals for actions to be undertaken by universities, as deemed convenient to favour the successful proliferation of university technology-based firms and limit their risks:

- Further develop the way of fostering business acumen, even though there are currently actions designed for this purpose. The aim should be, on the one hand, to intensify the present awareness schemes in the field of entrepreneurship, and on the other, to organise new actions such as awareness sessions for dynamising the enterprising spirit that target specific segments (students, researchers, teaching staff, etc.) and visits to the UPV/EHU's business incubators, amongst others.
- Provide TBFs with tailor-made professional support throughout the period of pre-incubation and incubation, which will have a constant impact on the key factors for the development of the ventures depending on the phase or stage that they are undergoing.

- Monitor TBFs when they leave the incubator or overcome the incubation period, and promote the introduction of a network consisting of former entrepreneurs and the present entrepreneurs currently taking part in the incubators or who have just climbed the first rungs on the business ladder.
- Organise discussion meetings involving entrepreneurs to enable them to share their experiences regarding the difficulties encountered in their ventures in order to use them to explore possible solutions.
- Share experiences with established entrepreneurs in the Basque Country and Aquitaine to understand the reality of the business environment in the two regions and learn about the best practices they apply.

References

Aceytuno MT, de Paz MA (2008) La creación de spin-off universitarias. El caso de la universidad de Huelva. Economía Industrial 368:97–111

Dupouy A (2008) L'accompagnement du porteur de projet innovant en incubateur: une question d'apprentissage. A partir du cas de l'incubateur de projets innovants de l'ESTIA. Thèse de doctorat en Sciences de Gestion, Université de Pau et des Pays de l'Adour

Espí MT, Arana G, Heras I, Díaz de Junguitu A (2007) Perfil emprendedor del campus de Gipuzkoa de la UPV/EHU. Revista de Dirección y Administración de Empresas 14:83–110

Etzkowitz H (2004) The evolution of the entrepreneurial university. Inter J Technol Global 1:64–77

Gómez JM, Mira I, Verdú AJ, Sancho J (2007) Las spin-offs académicas como vía de transferencia tecnológica. Economía Industrial 366:61–72

Hlady-Rispal M (2002) La méthode des cas, application à la recherche en gestion, Col. Perspectives Marketing. De Boeck University, Leuven

Martínez A, Tadeo J (2006) Las políticas de desarrollo local. El soporte público a las políticas de creación de empresas en las universidades como contribución al desarrollo. In: Fundación CYD (ed) I jornada técnica fundación CYD. Creación de empresas, desarrollo territorial y el papel de la universidad, Mexico

Merino C, Villar L (2008) Factores de éxito en los procesos de creación de empresas de base tecnológica. Economía Industrial 366:147–167

Molina M, Tobar M, Jara V (2005) Diseño de un modelo de evaluación técnico-económica y de gestión de empresas tecnológicas universitarias. Proyecto Alfa (2000-2005), n° II-0210-A

Moreau R (2004) L'émergence organisationnelle: Le cas des Entreprises de Nouvelle Technologie. Thèse pour le Doctorat de Sociologie, Université de Nantes, Nantes

Ortín P, Salas V, Trujillo MV, Vendrell F (2008) El spin-off universitario en España como modelo de creación de empresas intensivas en conocimiento. Ministerio de Industria, Turismo y Comercio, Madrid

Palacios M, Del Val T, Casanueva C (2005) Nuevas empresas de base tecnológica y business angels. Revista Madri + d 31

Urteaga E (2007) La coopération transfrontalière en Pays Basque. Harmattan, Paris

Yin R (2009) Case Study Research: Design and Methods. Sage, Los Angeles

Zarrabeitia E, Díaz de Basurto P, Ruiz de Arbulo P, Álvarez I (2010) Análisis de las aportaciones teóricas y regularidades empíricas en la literatura nacional reciente sobre el fenómeno de la universidad emprendedora y las spin-offs universitarias. In: Departamento de Organización de Empresas (ed) Industrial Engineering as University Third Mission agent

Chapter 14
Innovation in Cost Management. A Comparison Between Time-Driven Activity-Based Costing (TDABC) and Value Stream Costing (VSC) in an Auto-Parts Factory

Patxi Ruiz de Arbulo, Jordi Fortuny, Javier García, Pablo Díaz de Basurto and Enara Zarrabeitia

Abstract This paper deals with two new approaches to cost accounting: Time-Driven Activity-Based Costing (TDABC) and Value Stream Costing (VSC). First, the paper introduces both methods, focusing on TDABC. Next, we analyse the experience of an auto parts manufacturer who shifted from ABC to TDABC. Finally, we test how both methods work in a multiproduct/multi-process manufacturing environment adapted from reality, and we compare them. Although their aims are very different, they yield similar results.

14.1 The Importance of Costing

The importance of costing in the auto parts industry is out of question. Knowing the true cost is necessary for the manufacturers of auto parts because the selling price is imposed by the car manufacturers and therefore the manufacturers of auto parts cannot set the price on the basis of the cost of the product. The only variable that they can act upon is the manufacturing cost. In consequence, the only way to

P. Ruiz de Arbulo (✉) · P. Díaz de Basurto · E. Zarrabeitia
Dpto. de Organización de Empresas. Escuela Técnica Superior de Ingeniería de Bilbao, Universidad del País Vasco, Alameda Urquijo s/n, 48013 Bilbao, Spain
e-mail: patxi.ruizdearbulo@ehu.es

J. Fortuny
Dpto. de Organización de Empresas. Escuela Politécnica Superior de Ingeniería de Manresa, Universidad Politécnica de Cataluña, Av. Bases de Manresa, 61–73, 08240 Manresa, Spain

J. García
Dpto. de Organización de Empresas. Escuela Técnica Superior de Ingeniería de Bilbao, Guardian Industries, Alameda Urquijo s/n, 48013 Bilbao, Spain

S. P. Sethi et al. (eds.), *Industrial Engineering: Innovative Networks*,
DOI: 10.1007/978-1-4471-2321-7_14,

achieve a certain profit is by cutting costs, especially in an industry where competition is fierce.

A company needs to know whether a product is profitable or unprofitable. This would be easier if each factory produced just one item (let us say a single model of a single type of product), but this does not usually happen. In fact, auto parts companies tend to manufacture many different items by means of the same processes: each product goes through a number of different processes and each process may be necessary for different products. In running a certain operation, the consumption of resources (and therefore the cost) is different for each model. Sometimes these items are for the same car manufacturer but sometimes they are not. Furthermore, products are usually manufactured on a just-in-time basis, and assembled on a cellular layout.

Companies have long realized that traditional cost accounting systems, which were created for an environment where the most important costs were direct costs such as labour, and which were driven by mass manufacturing strategies, fail to properly describe these current situations.

In our research, we have tried to understand the reasons that make a company move from traditional cost-accounting methods to Time-Driven Activity Based Costing (TDABC), and how this new approach can cope with the existing difficulties. Besides, we compare TDABC with another recent approach termed Value Stream Costing (VSC), which is intended for lean manufacturing.

14.2 An Introduction to TDABC

Activity-Based Costing (ABC) was created in the 1980 s to overcome the problems of traditional full costing systems (such as the allocation of indirect costs on the basis of direct labour). While traditional costing systems allocate costs to products, ABC allocates costs to activities (because activities consume resources) and then, it relates activities to products. However, ABC has not achieved much success (Rigby 2003). According to Kaplan and Anderson (2007) the main reasons for the low adoption of ABC are: (1) Starting an ABC system needs a lot of time because it is based on surveys and interviews; (2) Since the time devoted to each activity is based on subjective methods, poor precision can be expected; and (3) ABC cannot capture the complexity of processes in real companies.

To overcome the problems of ABC, Robert Kaplan (creator of ABC) and Steven Anderson developed a new approach termed Time Driven ABC (TDABC) (Kaplan and Anderson 2007).

ABC has a sole cost driver for each activity. When the activities are slightly different for one product or another (this is particularly true in the service sector and in warehousing), ABC can only report an average cost for all products. A possible solution is to split the activity in two or more actions but then it is necessary to study each of them, and determine the time that an operator spends performing each task. The ABC system becomes more and more complex.

The advantage of TDABC lies in the way in which this method estimates time. In TDABC, the main cost driver is time. Therefore, it is not necessary to allocate the costs of the resources to activities before relating activities to products or other cost objects.

Besides, when the time to perform a task depends on several factors, TDABC uses time equations. A time equation is an equation that describes the time to perform an activity as a function of several drivers, as shown in Eq. 14.1:

$$\text{Process time}_{\text{(per unit)}} = \beta_0 + \sum \beta_i \cdot X_i \tag{14.1}$$

Where β_0 is the standard time to perform the basic activity, β_i is the time for each incremental activity, Xi is the quantity of each incremental activity.

There are not many papers available in literature that deal with TDABC (Villarmois and Levant 2007). Kaplan and Anderson (2007) refer to more than 100 real cases but there is only brief information on two examples. Everaert et al. (2008) describe in detail the case of a wholesale company. In Spain, García and Ruiz de Arbulo (2010) studied how TDABC was implemented in an auto parts company.

14.3 Value Stream Costing

Maskell & Baggaley developed a tool named Value Stream Costing (VSC) in order to simplify cost management in companies that were achieving a certain degree of maturity in the implementation of lean manufacturing (Maskell and Baggaley 2003; Ruiz de Arbulo 2007). VSC is based on the differences between lean manufacturing (one-piece flow and cellular manufacturing) and mass production (batch-and-queue scheduling in process-oriented factories). Usually lean companies group similar products and manufacture them together in the same cells or assembly lines. According to Maskell and Baggaley (2003), this grouping greatly simplifies costs calculations because it is easier to calculate the cost of the value stream as a whole, rather than working out the cost of isolated products.

The different cost elements that are taken into account by VSC are shown in Fig. 14.1. There is no difference between direct costs and overheads. All costs are considered to be direct costs and there is no need to allocate overhead costs to individual products. All labour costs are part of the VSC: there is no distinction between people who assemble the product, design it, move materials or do maintenance.

VSC is simpler than traditional costing systems because it does not need to collect information on the costs of each task or product. Costs are captured for the whole value stream and summarized for the chosen period (a week, a fortnight or a month). Traditional cost accounting systems need an expensive system to collect data because they want to know all the costs involved in each step of the production process. That needs a complex process that, most of the time is useless and slow, and leads to unclear, and sometimes unnecessary, information.

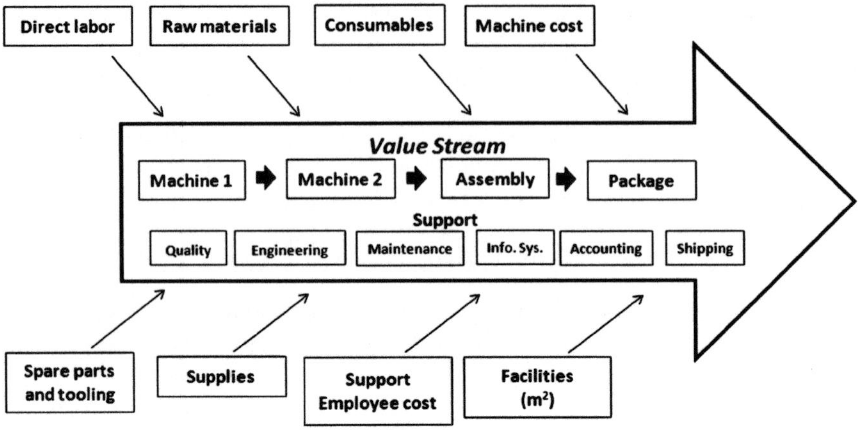

Fig. 14.1 Types of cost considered in value stream costing (Maskell and Baggaley 2004)

The philosophy of lean manufacturing lies on the elimination of all types of non-value-added activities and therefore is contrary to such wasteful costing systems.

14.4 Case Study. SRG Europe: Supplier of Plastic Parts for the Automotive Industry

SRG is a part of the multinational group Guardian Industries, headquartered in Detroit (Michigan USA). The company has 3 divisions: glass, building materials and plastic parts. In its plant in Valencia (Spain), SRG manufactures plastic parts for the automotive industry as a Tier 1 supplier (as an Original Equipment Manufacturer, it provides parts to the car assembler) or Tier 2 supplier (it provides parts for subassemblies to other Tier 1 manufacturers).

At SRG Valencia we can find two main processes:

- Pre-fabrication: made up of sales, offers and development engineering tasks.
- Fabrication: made up of

a. Direct processes: injection moulding, paint, chrome plating and assembly.
b. Auxiliary processes: material handling, final quality, warehouse management, supplies, office processes (human resources, finance).

Final products can be made with one process (just injection moulding) or they may need several processes (injection moulding + paint or injection moulding + paint + chrome plating according to the terms used in the company). All of them need a final assembly.

Table 14.1 Assembly time for each boot locking knob

Type of boot locking knob	Assembly time (s)
Model A	140
Model B	100
Model C	80

SRG Valencia manufactures more than 10 different product families, with different requirements, for different customers. This situation causes a neither homogeneous nor constant consumption of resources in the different processes of direct fabrication, assembly, auxiliary tasks and tasks before fabrication.

Because ABC was not able to track the complexity of such a multi-process, multi-product structure, SRG Valencia has been using TDABC since 2009. The current production mix has led to the development of new processes that support manufacturing, development and sales (such as final product audit, customer support, offer management, etc.). Within manufacturing processes, the number of indirect tasks has also risen.

TDABC has been useful for the company because it is able to provide accurate cost information. TDABC has allowed the plant managers:

1. To appreciate the importance of some indirect processes (for example, material handling was as costly as the assembly process).
2. To track efficient use of resources (for example, for indirect labour, hours really used can be compared to paid hours).
3. To have an idea of the profitability of each part as a function of their routing (manufacturing process), leading to consider value-added operations and non-value-added operations.

14.5 TDABC and VSC in the Assembly Process

To compare TDABC with VSC, both methods will be used to compute the costs of the assembly process of a boot locking knob. In the assembly process, several parts (switch, foam, clip, fastener, screws, etc.) are mounted on the plastic parts coming from the paint section. The switch is used to open the boot of the car. It is placed in the centre of the plastic part and fixed with screws. The foam seal is a long plastic strip that is stuck along the edge of the part; its role is to seal the part against the various elements (water, humidity and dust) so that they cannot get inside the part. The plastic fastener and the screws allow the sub-assembly to be mounted on the car at the car assembly line. The number of screws depends on the type of boot locking knob. Table 14.1 shows the work time to assemble each knob.

Table 14.2 Steps to calculate cost per minute

1.	Design capacity (per week)	40 h × 0.46875 people =	18.75 h
2.	Effective capacity(per week)	80% × 18.75 h =	15 h
3.	Total cost(per week)		4,000 €
4.	Cost per minute	4,000 €/(15 h × 60 min/h) =	**4.44 €/min**

We consider a production mix of 2,000 units of A; 2,000 units of B and 1,000 units of C. After the assembly, a final quality control is performed.

14.5.1 Computing the Cost of the Product Using TDABC

The process includes several steps. The first one is to develop time equations; the second one is to calculate the cost per unit of time for every group of resources and the third step is to estimate the cost for each activity.[1]

Development of time equations for quality control:

The time for performing the quality control of each knob depends on the number of points that have to be checked. There are 4 check points in type 'A' locks; 4 in type 'B' and 6 in type 'C'. Equation 1.2 gathers all this information.

$$\text{Timeforqualitycontrol(inminutes)} = 0.05X \tag{14.2}$$

Where $X =$ number of checkpoints (2 in boot locking knob A, 4 in B and 6 C)

Cost per unit of time for groups of resources

The cost per unit of time (minute) for each group of resources can be computed as shown in Table 14.2. Design capacity is based on a 40-hour schedule per employee per week. Effective capacity is 80% of the theoretical capacity. We consider that a worker only spends a part of the day on quality control.

Computing the cost of the final products

Table 14.3 shows the cost of quality control, according to Table 14.2 (cost per minute) and Eq. 14.2 (time equation).

Finally, we add the cost of quality control and the cost of assembly[2] in order to reveal the total cost of production (Table 14.4).

14.5.2 Computing Cost by Value Stream Costing

We have monitored VSC on a monthly basis. Table 14.5 shows the cost centres that are going to be used in this example. Materials, labour, depreciation and other conversion costs are considered.

---

[1] In this example, we only develop equations for the activities that would need more than one cost driver because of their variability.

[2] The cost of assembly is 24,888.88 €. Therefore, the cost per labour hour (which is the cost driver) is 160 €/hour.

Table 14.3 Computing the cost of quality control using TDABC

Product	Time consumed by inspection (m)	Cost per unit (€)
Model A	0.10 × 2,000 = 200	200 × 4.44/2,000 = 0.444
Model B	0.20 × 2,000 = 400	400 × 4.44/2,000 = 0.888
Model C	0.30 × 1,000 = 300	300 × 4.44/1,000 = 1.333

Table 14.4 Cost of assembly and quality control (according to TDABC)

Product	Assembly (€/unit) According to Table 1	Quality (€/unit) in Table 3	Total (€/unit)
Model A	(140/3,600) × 160 = 6.22	0.444	6.666
Model B	(100/3,600) × 160 = 4.44	0.888	5.332
Model C	(80/3,600) × 160 = 3.55	1.333	4.888

Table 14.5 Value stream costing

Centre	Material (€)	Labour (€)	Depreciation (€)	Other costs (€)	TOTAL COST (€)
Assembly	100	18,900	5,060	928.88	24,988.88
Quality	–	3,000	–	1,000	4,000
Total	100	21,900	5,060	1928.88	28,988.88

When we get the value stream average cost (which is the main goal of VSC), we can calculate a product cost by means of the product features and characteristics that have an impact on the flow of the product through the value stream.

In our example, we have chosen the cycle time for each knob. In a month (155.5 h), it is possible to assemble 4,000 type 'A' knobs (because the cycle time is 140 s) according to Table 14.1. Likewise, the maximum production is 5,600 knobs B and 7000 knobs C. Then, the cost of a knob A would be 7.25 Euros (28,988.88/4,000); the cost of a knob B would be 5.18 Euros, and the cost of a knob C would be 4.14 Euros.

14.6 Comparison and Conclusions

In this paper we have compared TDABC with VSC in a lean manufacturing environment. We have found no other comparisons in literature between both methods. There are not many companies using TDABC, but the method seems to be especially useful in plants with different products and when processes may depend on the type of product. We have not been able to find a company that has implemented VSC in Spain. Existing literature on this subject has been produced mostly in the USA. In our example:

TDABC is easier to use than ABC and traditional full costing in multiproduct/multi-process companies. Besides, TDABC can capture the costs of the different products in the product mix (which changes very often), fulfilling the needs of the company.

While TDABC accurately calculates the cost of a product, VSC gives just an average value, because the latter looks at the profitability of the value chain, not the isolated product.

VSC should be used in lean environments with cell manufacturing where data collection can be greatly simplified. It is used to drive improvements in the chain, not for product costing. In our case, where assembly and quality control processes coexist, TDABC is able to calculate costs more accurately.

It may seem that data collection for TDABC is complex, but it can easily model the activities of the company with not many time equations.

In VSC, we could get the same cost as in TDABC if the cost per product were calculated on the basis of time attributes for each process (assembly and quality).

Both methods can help analyse how capacity is used (overused or underused).

References

Everaert P et al (2008) Cost modelling in logistics using time-driven ABC. Experiences from a wholesaler. Inter J Phys Distrib Logist Manag 38(3):172–191

García J, Ruiz de Arbulo P (2010) El uso del Time Driven Activity Based Costing en la industria de componentes plásticos para automóvil. 4th International Conference on Industrial Engineering and Industrial Management. Donostia, San Sebastian

Kaplan RS, Anderson S (2007) Time-Driven Activity-Based Costing. A simpler and more powerful path to higher profits. Harvard Business School Press, Boston

La Villarmois (de) O, Levant Y (2007) Le time-driven ABC: la simplification de l'évaluation des coûts par le recours aux équivalents-un essai de positionnement. Finance Contrôle Stratégi 10(1):149–182

Maskell B, Baggaley B (2003) Practical Lean Acoounting. Productivity Press

Rigby D (2003) Management Tools 2003. Bain and Company

Ruiz de Arbulo P (2007) La gestión de costes en lean manufacturing. Cómo evaluar las mejoras en costes en un sistema lean, NetBiblo. La Coruña

Chapter 15
The Link Between Following a Project Management Standard and the Technology Transfer Results in Basque R&D Centres

Noemi Zabaleta, Juan Ignacio Igartua and Nekane Errasti

Abstract The relocation of industry to countries with lower labour costs means that business approaches which have been valid so far are now becoming obsolete. With this in mind, the search and exploitation of opportunities constitute one of the current ways for companies to get ahead, and the project's success becomes even more critical due to its performance. On the other hand, technology transfer, which is not a new concept, has taken a special relevance in the last two decades. This paper examines the link between project management standards and the results of technology transfer and aims to find the way to work at the lowest level of technology transfer, through a case study analysing 44 projects in R&D Centres in the Basque Country in Northern Spain. The most important findings are connected with the interactions between the project management variables and their influence on project success.

15.1 Introduction

Interest in technology transfer has increased dramatically in recent years, both from an academic and a business point of view. That growing commitment to technology transfer is intended to improve innovation and economic performance of the organization, which is very necessary at the beginning of the twenty-first century. However, it can be said that in most cases today's technology transfer takes place at a domestic level, and is firmly rooted among the leaders of a country

N. Zabaleta (✉) · J. I. Igartua · N. Errasti
Mechanical and Manufacturing Department, Mondragon University, Loramendi 4, 20500 Arrasate-Mondragon, Spain
e-mail: nzabaleta@eps.mondragon.edu

S. P. Sethi et al. (eds.), *Industrial Engineering: Innovative Networks*,
DOI: 10.1007/978-1-4471-2321-7_15, © Springer-Verlag London Limited 2012

with a view to developing their innovation policies. It is contradictory, but the same is not true for technology transfer carried out at the enterprise level, where the day-to-day problems are very different from national decisions on the topic.

As a central element in the process of innovation, project management has also become a key activity in most industrial organizations and across many industries (Shenhar and Dvir 1996). Although projects have existed since the beginning of civilization, project management standards emerged in the 1960s with the establishment of the Project Management Institute (PMI) in the United States. After that, other project management standards appeared such as the ICB standard in Europe (developed by the International Project Management Association, IPMA) or the P2 M standard in Japan (established by the Project Management Association of Japan).

Based on a sample of 44 projects, the interactions between the project management standards and the technology transfer results are analysed in this study.

15.2 Technology Transfer and Project Management

Throughout history, the evolution of knowledge and its transformation into constructive use by the company has been provided by the interaction between universities, industry, government and other organizations. Technology transfer is the conceptual framework for this interaction, and is related to the relationships that facilitate the transformation of research results into knowledge, products and services (Murrell 1996).

Technology transfer is not a static concept, but has constantly adapted to the needs of the environment. The traditional and simplest approach has focused primarily on international transfers of technology between rich and poor countries (Hameri 1996).

This trend of study between countries has been losing potency in recent years, and the organizational level of study has become increasingly important. However, most of the work around the theme focuses more on the strategic or corporate level of analysis (focusing more on the legal issues, contractual and industrial property) than in a co-operational or project level (Stock and Tatikonda 2000). This article aims to go one step further, and advocates the concept of technology transfer to a lower level (see Fig. 15.1), in order to assist in the daily life of the organizations involved in a process of this type with the ultimate aim of transferring Technology. Stock and Tatikonda (2000) are among the few authors who argue for technology transfer at the project level.

There are various ways to highlight technology transfer as a fundamental process within the area of innovation in a company (Jimenez 2002; Arvanitis and Woerter 2006). Improving the interface between R & D and production is essential if we are to improve environments in constant state of change and becoming increasingly competitive (Ginn and Rubenstein 1986).

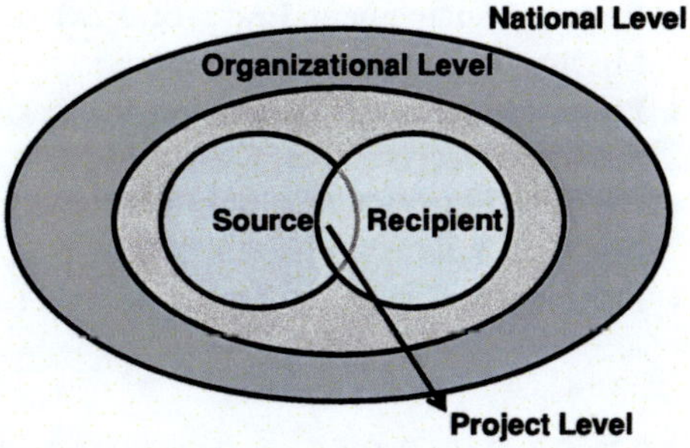

Fig. 15.1 Different levels of performance in technology transfer

Project Management has been practiced since early civilization but it is an activity that emerges as a discipline in the twentieth century and it has taken a more relevant importance in the last few years. The direction now taken by global enterprise activities justifies the relevance of projects. It is also expected that the trend to work for projects will continue to grow in the future (Shenhar and Dvir 1996; Stoneburner, 1999; O'Neal et al. 2006). In these circumstances, the ability of organizations to develop their key project activities more effectively and efficiently has become an important competitive factor (Pinto 2002; Mathur et al. 2007).

There have been several attempts to develop Project Management standards, such as:

IPMA Competence Baseline (ICB): developed by the European-based International Project Management Association (the oldest project management association, founded in Switzerland in 1965). Professional Project Management is broken down into 46 competence elements that cover the following: technical competencies for project management (20 elements); behavioural competences of project personnel (15 elements); and the contextual competences of projects, programmes and portfolios (11 elements). The Eye of Competence represents the integration of all the elements or project management as seen through the eyes of the project manager when evaluating a specific situation. The eye represents clarity and vision (International Project Management Association 2006).

The PMBOK (Project Management Body of Knowledge): It is the Standard developed by the Project Management Institute. This organization is the world's leading not-for-profit membership association for the Project Management profession, with more than half a million members and credential holders in 185 countries.

According the PMBOK, Project Management is the application of knowledge, skills, tools, and techniques to project activities to meet the project requirements. It identifies 9 areas of knowledge, 42 processes and 5 process groups (Project Management Institute 2008).

Project & Program Management for Enterprise Innovation (P2M): It is a standard issued by the Project Management Professionals Certification Centre (PMCC) of Japan, now renamed as Project Management Association of Japan (PMAJ 2005). The "Project Management Tower" shows the overview of P2M:

- Project Management Entry of P2M describes how to make a first step as a Mission-achievement professional.
- Project Management explains the basic definition and framework of project management.
- Program Management introduces program management that organically combines multiple projects.
- Project Domain Management offers 11 domains of project management.

15.3 Study of Research and Methodology

Taking into account the research objectives, the methodology was based on a questionnaire comprising 20 questions covering 2 basic aspects of the technology transfer projects: the details of following a project management standard and the technology transfer activities pursued by the R&D Centres during the project's implementation.

The research work was based on a questionnaire filled in through personal interviews carried out with the 44 project managers of 22 R&D Centres in the Basque Country (Northern Spain). The questions were designed as closed questions. In order to obtain the most reliable results possible, it was ensured that the questions were properly understood and fully completed.

With the aim of simplifying the statistical analysis, a number of composite variables were built from the replies obtained. The key variables utilised for the analysis were as follows: the utilization of a project management standard in the technology transfer projects (V1) and the technology transfer activities pursued by the R&D Centres (V2) during project implementation. Each of them was optimised and calculated through a combination of single variables (α Cronbach = 0.705–0.875).

As regards the output variables, turnover per employee (V3), innovation intensity (V4) and project results satisfaction (V5) were selected. Innovation intensity was measured by the number of patents, new activities, new products, new spin off firms, and scientific publications read per person in the last three years.

The data analysis was based on two steps. A Cluster analysis permitted the classification of the different ways of working and an ANOVA test was formulated in order to validate it.

Ward's method was followed given its acceptable performance as compared with other clustering methods (Milligan and Cooper, 1987). The clustering variables used were: V1, V2, V3, V4 and V5. Three differentiated groups were found. Their main characteristics were as follows (see Fig. 15.2).

Cluster 1: This group comprises 14 projects (31.82% of the sample) characterised by following at least one project management standard, and by operating in fairly stable environments with a lower ratio of rotating technology. These projects

Fig. 15.2 Final clusters achieved in the study

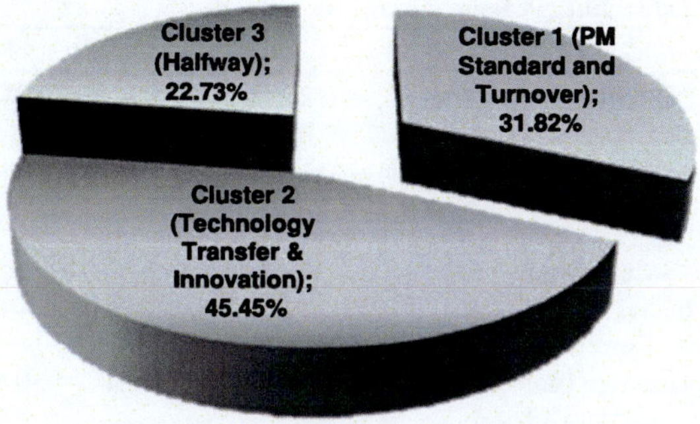

generally exhibit the highest turnover results and the lowest innovation performance. In these cases, their project managers are the least satisfied with the results.

Cluster 2: This dynamic environment drives 20 projects (45.45% of the sample) to take risks researching technologies that have not yet found a position in the market. They do not see the need for project management standard. With few exceptions, the vast majority of the most innovative R&D projects are in this group, but they have the worst turnover results.

Cluster 3: It is composed of 10 technology transfer projects (22.73% of the sample). Their environment is less dynamic and less uncertain than the second, but higher than the first one, with a medium technology rotation ratio. They are characterised by maintaining a good market position but without risking too much in order to be leaders in their field. Their project management standard is not as in Cluster 1 but they sometimes follow one standard to carry out technology transfer projects. Turnover and innovation results are not as good as in the other clusters, but these project managers are the most satisfied with them.

15.4 Results

Correlations of the model variables are shown in Table 15.1. The variable Project Management Standard (PM) exhibits a consistently high correlation with project turnover results. Technology transfer performances during project implementation are linked with project satisfaction and innovation results (this trend is positively correlated and significant) but not with turnover. Moreover, turnover results have a significant negative association with innovation.

Table 15.2 reports the means of the variables used. The ANOVA results indicate that the main differences between clusters lie in all the variables selected.

It should be pointed out that Cluster 2 has the highest Innovation performance; Cluster 1 is the one most concerned with the project management standard and has the highest turnover. Comparing these results through an ANOVA test, the mean difference is significant (Table 15.2, $p < 0.05$). Adopting a project management

Table 15.1 Pearson correlation results

	TT Activities	Satisfaction	PM Standard	Innovation	Turnover
TT Activities Pearson Correlation1	1	0.3533**	0.0993	0.3597**	−0,2764*
Satisfaction Pearson Correlation	0.3533**	1	−0.0386	0.4135**	−0.2130
PM Standard Pearson Correlation	0.0993	−0.0386	1	−0.2078	0.4653**
Innovation Pearson Correlation	0.3597**	0.4135**	−0.2078	1	−0.3916**
Turnover Pearson Correlation	−0.2764*	−0.2130	0,4653**	−0.3916**	1

** $p < 0.05$
* $p < 0.1$

Table 15.2 Technology Transfer Projects: cluster profile and analysis of the ANOVA test

ANOVA

	Cluster		Error		F	Sig.
PM Standard	6.03	2	1.24	35	4.8520	0.0138**
TT Activities	72.69	2	17.84	35	4.0738	0.0256**
Turnover	5120344957.54	2	32115821.14	35	159.4337	0.000**
Innovation	0.28	2	0.14	35	4.9621	0.015**
Satisfaction	43.28	2	16.16	35	2.6777	0.082*

** $p < 0.05$
* $p < 0.1$

standard helps in terms of performance in turnover, but projects with the highest turnover are not always the most innovative. Furthermore, the relationship (innovation performance versus turnover) is inversely proportional. This may be explained by the orientation of Cluster 1 projects, more focused on mature technologies, more involved in day-to-day innovation problems; therefore turnover is the main concern for their management.

15.5 Discussion and Conclusions

The main purpose of the article was to identify the link between following a project management standard and the technology transfer project results. Three project management standards have been cited and the need for managing technology transfer at project level has been shown. The approach used five variables addressing aspects of project management standards, technology transfer activities and performance results. All the data collected for this study were qualitative, originating from extensive interviews with 44 project managers.

The five variables emerging from the analysis form three clusters. The first cluster is associated with the project management standard and the turnover result. The second clusters deals with the technology transfer activities and innovation performance. Finally, the third cluster is halfway between Clusters 1 and 2. After that, a correlation analysis has been developed.

Following this clustering and relationship scheme, the findings suggest that the advantage of following any project management standard can help in the project turnover results as it helps in identifying critical factors that contribute to success, and not determining them can lead to a project failing. In this way the process is more controlled, but innovation has the risk of staying in the background as the project team is not allowed to work beyond the standard.

Special attention has been paid to technology transfer activities. The results show that the importance placed upon this variable promotes innovation performance and project management satisfaction. But working without a standard scheme, turnover is not as good as in Cluster 1. Therefore, it is necessary to maintain a balance between the use and application of a project management standard and the approach related with technology transfer activities. Thus, it is possible to make the most out of innovation and turnover.

The limitations of this paper result from the small number of variables used to create the project clusters. Further research and analysis would provide more detailed classifications. On the other hand, the contributions of this study must be interpreted with a degree of caution since it has been focused on the Basque context, which may have certain characteristics that can affect the final performance.

References

Arvanitis S, Woerter M (2006) Firms' Strategies for Knowledge and Technology Transfer with Public Research Organizations and their impact on Firms' Performance. Technology Transfer Society Conference Atlanta, Georgia, pp 1–32

Ginn ME, Rubenstein AH (1986) The R-And-D-Production Interface: A Case-Study of New Product Commercialization. J Prod Innov Manage 3(3):158–170

Hameri AP (1996) Technology transfer between basic research and industry. Technovation 16:51–57

International Project Management Association (2006) ICB-IPMA Competence Baseline Version 3.0

Jiménez JM (2002) Aspectos de la eficiencia en la Transferencia de Tecnología, La Transferencia de Tecnología en España. Monografía 5. Madrimasd

Mathur G, Jugdev K, Fung TS (2007) Intangible project management assets as determinants of competitive advantage. Management Research News 30(7):460–475

Milligan GW, Cooper MC (1987) Methodology review: clustering methods. Applied Psychological Measurement 11:1–27

Murrell KD (1996) Communications Technology transfer in the developed world. Vet Parasitol 64:107–120

O'Neal C, Tabatabaei M, Schrottner B (2006) A trend analysis of project management epistemology in information technology research, en: 2006 Southern Association for Information Systems Conference

Pinto JK (2002) Project management. Research Technology Management 45:22–37

Project Management Association of Japan (2005), A Guidebook of Project & Program Management for Enterprise Innovation

Project Management Institute (2008) A Guide to the Project Management Body of Knowledge

Shenhar A.J, Dvir D (1996) Toward a typological theory of project management Research Policy vol. 25, pp. 607–632

Stock GN, Tatikonda MV (2000) A typology of project-level Technology transfer processes. Journal of Operations Management 18:719–737

Stoneburner JD (1999) Project Management methods for accelerated product management

Chapter 16
External Sources of Knowledge and Innovation Performance: Evidence from Spanish Industrial Firms

M. Segarra Ciprés, S. Palomero Ródenas and V. Roca Puig

Abstract Previous studies indicate that the open innovation model is a central operating logic for innovative firms and leading countries in innovation which have developed public support for this open approach. This paper examines the basic premise of this model by analyzing the continuity in using institutional, vertical and horizontal alliances, and the recruitment of skilled R&D personnel over time, and their influence on innovation performance, in 541 Spanish industrial firms. The results show that each external knowledge source has a different impact on innovation, although the simultaneous use of external knowledge sources has a positive effect on innovation performance.

16.1 Introduction

Market dynamism, the increase in human mobility and rapid technological changes are some of the factors that have contributed to the evolution of the innovation models towards more open approaches (Chesbrough 2006). At present, innovation tends to be interactive, that is, relationships with producers, users and institutions are the centre of the process (von Hippel 1988; Chesbrough 2006; Leiponen and Helfat 2010; Fey and Birkinshaw 2005). Alliances with universities

M. Segarra Ciprés (✉) · S. Palomero Ródenas · V. Roca Puig
Department of Business Administration and Marketing,
University Jaume I, Campus Riu Sec, 12071 Castellón, Spain
e-mail: msegarra@emp.uji.es

S. Palomero Ródenas
e-mail: palomero@emp.uji.es

V. Roca Puig
e-mail: roca@emp.uji.es

S. P. Sethi et al. (eds.), *Industrial Engineering: Innovative Networks*,
DOI: 10.1007/978-1-4471-2321-7_16,

allow firms access to scientific knowledge which they are not usually able to produce alone. Cooperation with competitors provides complementary knowledge, which makes them attractive partners to the extent that costs and risks are reduced in both large projects and in dynamic environments (Miotti and Sachwald 2003). Recruitment of personnel is another mechanism to access the tacit and complex knowledge embedded in individuals (Song et al. 2003). Building on this perspective, using a continuous shared search for knowledge is an investment in abilities to create, use, and recombine knowledge that, together with internal knowledge, will be the basis of innovation.

On these grounds, this paper explores the basic premise of this model by analyzing the contribution of external knowledge sources to innovation performance in Spanish firms. Based on a statistical analysis of the Survey on Spanish Firm Strategies ("Encuesta sobre Estrategias Empresariales", henceforth referred to as ESEE), this paper examines how the contribution of technological alliances with universities, suppliers, clients and competitors, and the recruitment of skilled R&D personnel influences innovation performance over time.

This paper is arranged in four sections. The next section reviews the theoretical background of the study, describing the hypotheses that drive the analysis. The following section presents the sample and the methodology. Then, the most relevant results are presented, and the final section contains the discussion and conclusions.

16.2 Theoretical Background and Hypotheses

16.2.1 External Knowledge Sources and Innovation Performance

Because innovation occurs primarily through new combinations of resources, ideas, and technologies, a fertile R&D environment relies on a constant inflow of knowledge from other places (Fey and Birkinshaw 2005). External sources of knowledge become mechanisms to access valuable knowledge for innovation. On this basis, we propose analyzing the contribution of several external knowledge sources to innovation development. Specifically, we study the continuity in using vertical, horizontal and institutional alliances, and the recruitment of skilled R&D personnel.

- *Technological Alliances*

A common way to access external knowledge is through technological alliances with different partners such as universities, technological institutes, suppliers, clients, competitors (von Hippel 1988; Powell et al. 1996; Baum and Ingram 1998; Hagedoorn and Duysters 2002).

Collaboration with several actors facilitates access to different external knowledge (Song et al. 2003; Leiponen and Helfat 2010). Thus, universities are considered as especially useful for basic and long-term strategic research and as an inexpensive and low-risk source of specialist knowledge (Tether 2002).

Alliances between universities and firms represent a linkage between scientific and industrial knowledge (Godin and Gingras 2000) that constitutes a key input for obtaining innovation performance. In addition, vertical relationships can contribute to innovation results in as far as they provide valuable knowledge about technology, the market and the clients' needs. Another source of innovation for firms is knowledge from competitors. The main reason to cooperate with competitors is the need for solving common problems that are outside the competitors' area of influence (Tether 2002; Cassiman and Veugelers 2002), also when there are incentives for sharing innovator resources and for integrating into networks (Dussauge and Garrette 1998). From the above reasoning, we hypothesize the following:

Hypothesis 1 *Long-term alliances with universities positively affect innovation performance*

Hypothesis 2 *Long-term vertical alliances positively affect innovation performance*

Hypothesis 3 *Long-term horizontal alliances positively affect innovation performance*

- *Recruitment of skilled R&D personnel*

The recruitment of employees with knowledge and experience in R&D is another way for firms to access valuable external knowledge (Galende and De la Fuente 2003; Rosenkopf and Almeida 2003; Song et al. 2003; De Saá and Díaz 2007). Song et al. (2003) call this way of accessing the experts' knowledge and capabilities "learning-by-hiring".

According to Chesbrough (2003), the increase in the mobility of knowledge workers has been one of the factors that eroded the closed innovation model. This factor makes it increasingly difficult for firms to control their proprietary ideas and expertise. Thus, the increase in worker mobility has encouraged the recruitment of R&D people as a way to access this kind of external knowledge that can be beneficial for innovation. In particular, new skilled employees are a source of new ideas, since they promote the renovation of the internal knowledge base (Ireland 2002), and they have the opportunity to apply expertise knowledge to new contexts (Argote and Ingram 2000). Accordingly, the next hypothesis can be stated as:

Hypothesis 4 *Continuity in recruitment of skilled R&D personnel over time positively affects innovation performance*

16.2.2 Simultaneous Effect of External Knowledge Sources on Innovation Performance

Recent studies suggest that the basis of innovator success is to make use of a wide range of external knowledge sources. Leiponen and Helfat (2010) argue that the successful innovation is uncertain and firms may be able to increase its likelihood and value successful innovation by sampling a range of knowledge sources. Furthermore, knowledge sources can have complementarities and positive synergies in such a way that, collectively, they reinforce the effect of external knowledge acquisition on innovation performance. Similarly, Laursen and Salter (2006) refer to the external search breadth as the number of different search channels that a firm draws upon in its innovative activities.

Specifically, we explore the simultaneous effect of different kinds of knowledge sources that are connected to the different parts of the firms' value chain. So in the next hypothesis we propose that alliances with universities, vertical and horizontal alliances, and the recruitment of R&D personnel involve an interaction with so many diverse actors that, in turn, it facilitates access to different knowledge, promotes positive synergies, and avoids a knowledge overlap.

Hypothesis 5 *Using a diversity of external knowledge sources over time positively affects innovation performance*

16.3 Methodology

16.3.1 Data and Sample. Definition of Variables

The dataset used is the ESEE carried out by the SEPI Foundation in collaboration with the Spanish Ministry of Industry, Tourism and Trade. This survey aims to find out the characteristics of Spanish industrial firms and constitutes an unbalanced panel, as some firms fail to provide information on a regular basis, while others do so every year. Within this dataset, we selected the firms that have been in the dataset for at least 5 years and have reported positive R&D outputs over the last three years. This selection and the rejection of missing cases in the basic variables result in a sample of 541 firms covering the period 2002–2005.

Dependent Variable

We used innovation performance (IP) as the *dependent variable*, which is defined as the results of the innovation activity developed by a firm. This variable includes the sum of the new products and processes, patents and utility models developed by a firm in 2005.

Independent Variables

Given the difficulty to empirically demonstrate with accuracy when the use of external knowledge sources will produce benefits for innovation performance, we consider the lag with which innovation strategies are allowed to impact innovation performance.

External knowledge sources are represented by four variables: alliances with universities (ALU), vertical alliances (VA), horizontal alliances (HA), and recruitment of skilled R&D personnel (RRD). All 4 sources are coded as a binary variable, where 0 is no use and 1 is use of the given knowledge source.

The long-term effect of these variables is measured as follows: alliances with universities (ALU) takes a value of 0 when firms have made no use of this source in any of the last four years, and 4 is scored when a firm has used this source over the last four years. We use the same process to measure the rest of the sources. In order to test Hypothesis 5, we include variable BES, which represents the breadth of using external sources over time.

Control Variables

We distinguish three groups in relation to the sector's technological intensity (low, medium–low and high) and also include the following internal factors as control variables: firm size (LS), measured as the logarithm of total employees; firm age (LA), measured as the logarithm of the firm's age; R&D intensity (RDIn), measured as the firm's R&D expenditure divided by firm sales and the mean innovation performance in the last two years (MIP).

16.4 Results

Of the 541 firms sampled, 62% have used some kind of external knowledge source over the last four years (Table 16.1). Furthermore, those firms with a positive innovation performance in 2005 have used more external knowledge sources (37.7%) in the last four years than firms with no positive innovation results in 2005 (24.4%). Suppliers and clients were the most widely used cooperation partners (52.1%), but a significant proportion also engaged universities (45.3%) as partners in these arrangements. Recruitment of skilled R&D personnel (17.4%) and horizontal alliances (7.9%) were the least used sources in the last four years.

The hypotheses were tested by estimating a negative binomial model to consider the characteristics of the dependent variable. Table 16.2 shows the results of the estimation of the three models.

Model 1 shows the effect of the control variables on innovation performance. The results indicate that firm size and innovation performance do have a positive and significant effect on innovation performance. The results of Model 2 show the contribution of external knowledge sources to innovation performance. Alliances with universities positively affect innovation performance, although this effect is not significant. Therefore, Hypothesis 1 is not supported. This result confirms the limited contribution of universities to the innovation results of Spanish industries.

The estimation of Model 2 also confirms the positive and significant effect that vertical alliances ($p < 0.01$) and the recruitment of skilled R&D personnel ($p < 0.05$) have on innovation performance. Thus, Hypotheses 2 and 4 are strongly supported. These results demonstrate that long-term vertical alliances and the continuity to recruit skilled R&D personnel over time positively affect innovation performance. However, Hypothesis 3 was not supported because alliances with competitors have a negative and non-significant effect on innovation performance.

The results obtained in Model 3, which includes the simultaneous effect of external knowledge sources on innovation performance, support Hypothesis 5.

Table 16.1 External knowledge sources and innovation performance

Type of source (from 2002 to 2005)	$IPerformance_{2005} = 0$ (%)	$IPerformance_{2005} > 0$ (%)	Total (N = 541) (%)
Alliances with universities	16.3	29	45.3
Vertical alliances	19.8	32.3	52.1
Horizontal alliances	3.5	4.4	7.9
Recruitment of skilled R&D personnel	4.8	12.3	17.4
All external sources	24.4	37.7	62.1

Table 16.2 Results of the negative binomial model

Variables	Model 1	Model 2	Model 3
Intercept	−0.655**	−0.454	−0.369
Alliances with universities (ALU)		0.078	
Vertical alliances (VA)		0.086***	
Horizontal alliances (HA)		− 0.097	
Recruitment R&D personnel (RRD)		0.125**	
Breadth of using external sources (BES)			0.136***
R&D Intensity (RDIn)	0.018	0.003	0.004
Log Employees (LE)	0.407***	0.110	0.124
Log Age (LA)	0.023	− 0.025	− 0.085
Mean IPerformance (MIP)	0.118***	0.100***	0.107***
Tech. Intensity-Low	0.003	0.187	0.144
Tech. Intensity-Med	− 0.124	0.001	− 0.016
Omnibus test	389.966***	426.783***	420.643***

Note: Unstandardized regression coefficients *** $p < 0.01$; ** $p < 0.05$; * $p < 0.10$

The joint effect of external sources therefore reinforces the success of open strategies on innovation performance. However, as Laursen and Salter (2006) suggest, firms which are too open may encounter negative consequences for their innovation performance, which derive from the difficulty of managing many different sources. Consequently, the success of open strategies is based on using a small number of diverse external sources.

16.5 Discussion and Conclusions

In this paper we have empirically tested the basic premise of the open innovation model: after trying to ascertain whether knowledge searching processes are the centre of innovation in Spanish industrial firms, our results confirm that external knowledge is a useful resource for innovation, but not all external knowledge sources have the same importance for innovation. The empirical results support that long-term vertical relationships have a positive effect on innovation performance. Alliances with clients are beneficial for innovation in as much as they reduce the risk associated with introducing a new product into the market, facilitate rapid responses to the customers' needs, and allow the development of more novel and complex innovations (Tether 2002). In accordance with Miotti and Sachwald (2003), cooperation between competitors is particularly risky and should be limited to the existence of a strong common interest, and when cooperation concerns far-from-market research leading to generic results.

Despite the growing literature on cooperation pointing out that arrangements with universities are beneficial for firms, our analysis does not support this result. There are different motivations to collaborate with universities. One is that public research partners do not seek commercial applications and tend to focus on the most generic or basic ends of the R&D complex (Miotti and Sachwald 2003). Another motivation indicates that the financial pressure on R&D budgets encourages many firms to access the basic research in universities as it is far more cost-effective than them undertaking it themselves (Leonard-Barton 1995). However, the Spanish innovation system is characterized by the difficulties of transferring knowledge from universities to firms. Indeed, Spanish firms do not consider the public R&D system as a source of innovation because they ignore the universities' scientific capabilities.

Results also highlight that not all the open strategies support innovation, although the simultaneous use of external knowledge sources on innovation performance has a positive effect. In our analysis, we consider four diverse knowledge sources (alliances with universities, vertical and horizontal alliances, and recruitment of skilled R&D personnel) whose joint effect reinforces the contribution of external sources to innovation performance.

Finally, we suggest that our findings can be strongly motivated by the status of the Spanish innovation system and the characteristics of Spanish firms. According to Bayona et al. (2002), given the more limited demand for high-tech goods and services, Spanish firms have had to operate in a less stimulating environment than European or American firms. On the other hand, Spanish firms participate less than those of other European countries because Spain lacks a collaborative culture: R&D is not carried out internally, and in this respect, firms neither collaborate with other firms nor with the public R&D system (Bayona et al. 2002).

Acknowledgements This work was carried out as a part of a research project (ref. ECO2008-01674) subsidized by the Ministerio de Ciencia e Innovación.

References

Argote L, Ingram P (2000) Knowledge transfer: A basis for competitive advantage in firms. Organ Behav Hum Decis Process 82:150–169

Baum JAC, Ingram P (1998) Survival-enhancing learning in the Manhattan hotel industry, 1898–1980. Manage Sci 44:996–1016

Bayona C, García M, Huerta E (2002) Collaboration in R&D with universities and research centres: an empirical study of Spanish firms. R&D Manage 32:321–341

Cassiman B, Veugelers R (2002) Spillovers and R&D cooperation: some empirical evidence from Belgium. Am Econ Rev 92:1169–1184
Chesbrough H (2003) Open innovation. Harvard University Press, Cambridge
Chesbrough H (2006) Open innovation: A new paradigm for understanding industrial innovation, Open Innovation: Researching a New Paradigm, Chapter 1 Oxford University Press
De Saá Pérez P, Díaz Díaz NL (2007) Incidencia de los recursos humanos de I + D internos y contratados en la innovación. Cuadernos de Economía y Dirección de la Empresa 33:7–30
Dussauge P, Garrette B (1998) Anticipating the evolutions and outcomes of strategic alliances between rival firms. Int Stud Manage Organiz 27:104–126
Fey CF, Birkinshaw J (2005) External sources of knowledge, governance mode, and R&D performance. J Manage 31:597–621
Galende J, De la Fuente JM (2003) Internal factors determining a firm's innovative behaviour. Res Policy 32:715–736
Godin B, Gingras Y (2000) The place of universities in the system of knowledge. Res Policy 29:273–278
Hagedoorn J, Duysters G (2002) External sources of innovative capabilities: The preferences for strategic alliances or mergers and acquisitions. J Manag Stud 39:167–188
Ireland RD, Hitt MA, Vaidyanath D (2002) Alliance management as a source of competitive advantage. J Manage 28:413–446
Laursen K, Salter A (2006) Open for innovation: The role of openness in explaining innovation performance among UK manufacturing firms. Strateg Manag J 27:131–150
Leiponen A, Helfat CE (2010) Innovation objectives, knowledge sources, and the benefits of breadth. Strateg Manag J 31:224–236
Leonard-Barton D (1995) Wellsprings of knowledge: building and sustaining the sources of innovation. Harvard Business School Press, Boston
Miotti L, Sachwald F (2003) Co-operative R&D: why and with whom? An integrated framework of analysis. Res Policy 32:1481–1499
Powell WW, Koput KW, Smith-Doerr L (1996) Interorganizational collaboration and the locus of innovation: Networks of learning in biotechnology. Administrative Science Quarterly 41
Rosenkopf L, Almeida P (2003) Overcoming local search through alliances and mobility. Manage Sci 49:751–766
Song J, Almeida P, Wu G (2003) Learning-by-hiring: Mobility and knowledge transfer? Manage Sci 49:351–365
Tether BS (2002) Who co-operates for innovation, and why: An empirical analysis. Res Policy 31:947–967
von Hippel E (1988) The sources of innovation. Oxford University Press, New York

Part III
Management

Chapter 17
Eco-Design in the Furniture and Wood-Processing Sector in Spain: Benefits and Constraints

María Pilar Cordero and Raúl Poler

Abstract The objective of this study is to analyze the main benefits and constraints of the implementation of eco-design solutions in wooden products manufacturers and furniture companies in Spain, and whether this results in significant differences between both sectors. Although the level of eco-design implementation in both sectors is rather low, companies have considered some new eco-design strategies in product process development which are mainly related to energy and raw material optimization, use of non-hazardous substances and reduction of waste, obtaining some benefits but sometimes facing impediments that hinder its application. This study reveals that there are not statistical differences between both sectors in terms of their level of understanding of the benefits and constraints of eco-design. Most of the companies surveyed agreed that the main benefits derived from eco-design are not only related with reducing their environmental impact but also with other advantages such as improved companýs image, new market opportunities and meeting normative and legal requirements. On the other hand, limitations are mainly related to lack of knowledge and conflict with functional and quality product requirements.

17.1 Introduction

Environmental problems derived from current production and consumption patterns makes necessary changes towards products and production processes more in tune with the environment. With this in mind, many different approaches and

M. P. Cordero (✉) · R. Poler
Research Centre on Production Management and Engineering, Universidad Politécnica de Valencia, Camino de Vera s/n, 46022 Valencia, Spain
e-mail: pcordero@cigip.upv.es

S. P. Sethi et al. (eds.), *Industrial Engineering: Innovative Networks*,
DOI: 10.1007/978-1-4471-2321-7_17, © Springer-Verlag London Limited 2012

concepts have been developed over the last decades, among them: more environmentally sound production, cleaner technologies, waste minimisation and recycling approaches, eco-design and design for sustainability (Mont 2002). The eco-design concept, also known as Design for Environment (DFE) aims to provide methods to minimize the environmental impacts of a product during its lifecycle (Kurk and Eagan 2008). Different studies analyze the main driving factors to undertake eco-design in a company. According to these authors the main points which influence a firm's decision to take up eco-design are not only related with the reduction of their environmental burden, but also with the improvement in their competitiveness: opportunities for innovation, an expected increase in product quality and potential market opportunities (van Hemel and Cramer 2002).

However, although more and more companies are interested in implementing eco-design, sometimes they are faced with limitations that prevent this implementation. According to different studies existing in literature the main constraints that limit the implementation of eco-design are those connected with lack of knowledge (Knight 2009), the perception that eco-design has no clear environmental benefit (van Hemel and Cramer 2002), organizational constraints such as lack of cooperation between departments, unclear responsibilities, cultural differences between departments and lack of appropriate infrastructure (Boks 2006).

The objective of this study is to practically test these hypotheses on the main benefits and constraints in eco-design implementation in one of the most important economic sectors in Spain: furniture and wood-processing. This paper is organized as follows: in Sect. 17.2 the research approach used in this study is described. Section 17.3 shows an overview of the level of implementation of eco-design and the eco-design strategies most commonly used in these sectors. Finally, Sect. 17.4 shows the results obtained in the statistical analysis of the main benefits and constraints obtained from survey data.

17.2 Research Approach

The results of the empirical analysis presented in this paper, which was carried out in spring/summer 2010, are based on the data collected from a questionnaire which was sent to a list of 3,500 furniture and wood-processing companies in Spain, selected on the basis of a simple random sampling procedure. The number of responses obtained from the questionnaire was 107, which exceeds the number of responses required to obtain a representative sample ($n = 96$) according to the formula to estimate sample size proposed by Miquel et al. (1997), with a confidence level of 95% and a standard error of 10%.

The survey included queries about the eco-design strategies implemented in the companies in addition to questions about main benefits and limitations encountered by them in these implementations. The questions about eco-design strategies applied by companies were designed by taking into consideration the eco-design

Table 17.1 Benefits and constraints to implement eco-design analyzed in the study (Van Hemel and Cramer (2002), Knight 2009 and Boks 2006)

Benefits	Constrains
Design of more innovative products	No clear environmental benefit
Meet customers' environmental requirements	Insufficient time period
New market opportunities	Insufficient knowledge
Improved company image	No innovation opportunities
Meet normative and legal requirements	Conflict with functional product requirements
Cost savings	Conflict with quality product requirements
Increase in product quality	Conflict with customer requirements
Increased efficiency in production, logistics, etc	
Reduction of environmental impact	

strategies proposed by van Hemel and Cramer (2002), who classify the strategies into the following categories related by the use of more sustainable materials and/ or the life cycle phase of the product where the eco-design strategy is applied:

- Selection of low-impact materials
- Reduction of raw materials consumption
- Optimization of production processes
- Optimization of distribution systems
- Reduction of impact during use
- Optimization of end-of-life systems

Additionally, respondents were asked about the level of agreement or disagreement concerning the benefits and constraints derived from eco-design implementation as analyzed in the reference material (van Hemel and Cramer, 2002; Mont 2002; Boks 2006; Kurk and Eagan 2008; Houe and Grabot 2009; Knight 2009) by using the Likert scale, where the response options ranged from 1 for "strongly disagree" to 5 for "strongly agree" (Table 17.1).

17.3 Results

According to the results obtained in this study, 33% of the 103 companies surveyed have implemented eco-design strategies in the design and development of their products. Comparing these results by sectors, the wood-processing sector has a greater percentage of companies that have implemented eco-design strategies (45%) than the furniture sector (28%). However, the results show that there are a greater percentage of furniture companies that would be interested in improving the environmental performance of their products in the future (60%) by using this methodology.

Figure 17.1 shows the most common eco-design strategies implemented in the furniture and wood-processing sector in Spain. These results tally with those

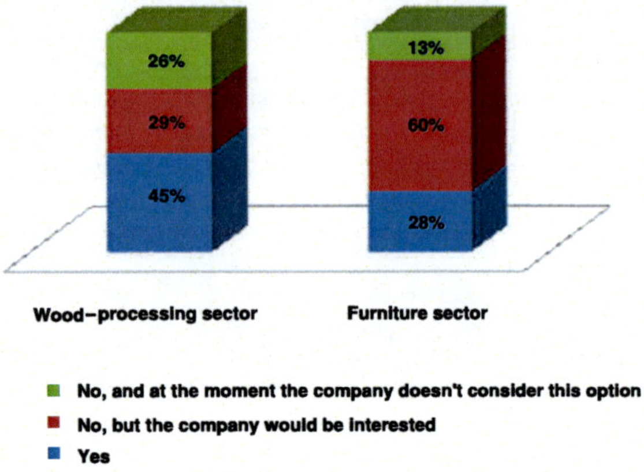

Fig. 17.1 Level of implementation of eco-design in the furniture and wood-processing sector in Spain

studies available in the literature that analyze the main environmental impacts in the furniture and wood-processing sector which, although depending on the type of products and the raw materials used in manufacturing, are related with raw material and energy consumption, emission of chemical substances and waste generation (Parikka-Alhola 2008).

Thus, the eco-design strategies most commonly used by the surveyed companies are related with:

1. the use of cleaner materials
   a. with or without a low content of toxic and hazardous substances such as Volatile Organic Compounds (VOCs) or formaldehyde, mainly used in finishing processes
   b. optimization of raw material consumption
   c. cleaner or reusable packaging used in distribution systems
2. reducing waste generation
   a. the design of products with high reliability and durability to increase product's life span and
   b. the use of recycling materials.

17.3.1 Benefits and Constraints in Eco-Design Implementation in the Furniture and Wood-Processing Sector in Spain

Figures 17.2 and 17.3 show the average level of agreement or disagreement of furniture and wood-processing companies with the main benefits and constraints

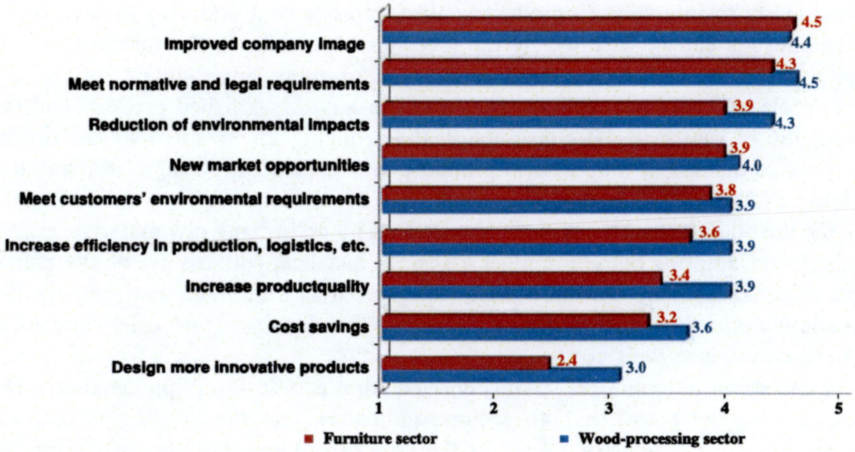

Fig. 17.2 Main benefits in the implementation of eco-design strategies in the furniture and wood-processing sector in Spain

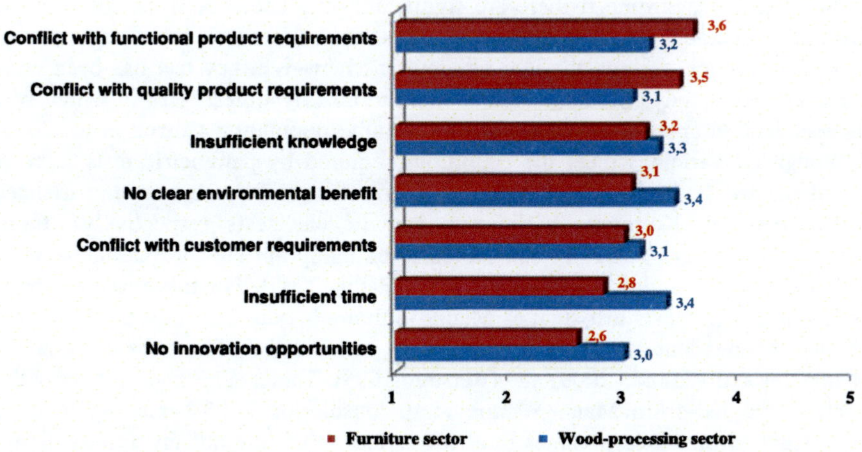

Fig. 17.3 Main constraints in the implementation of eco-design strategies in furniture and wood-processing sector in Spain

derived from eco-design implementation, considering a Likert scale from 1 to 5. Similar results have been obtained in both sectors and companies consider that the main benefits derived from eco-design implementation are: improved company image, meeting normative and legal requirements, the reduction of environmental impact and new market opportunities.

These results contrast partially with those identified in the literature review due to the fact that most of the companies surveyed did not consider eco-design as an opportunity to innovate, to product quality improvement and also to save costs. These results can be justified by the fact that, especially in the furniture sector, some eco-design strategies such as the use of more environmentally-friendly chemicals in finishing processes may result in a poorly-finished product, and the reduction of raw material consumption (by reducing thickness or by the use of alternative materials) may lead to products with inadequate strength or durability (Table 17.2).

On the other hand, cost savings are possible by optimizing raw materials and/or energy consumption but not in other cases; for instance, with the use of alternative raw materials that are normally more expensive than traditional ones, mainly due to a low production volume for these products which may not yet offer economies of scale to reduce their costs.

Concerning the main constraints derived from eco-design implementation, the results are very broad in both sectors, indicating that the limiting factors are dependent on characteristics such as the type of product, the company's size or having environmental expertise in product development, among others. Nevertheless, for most companies these limitations are related with lack of knowledge and conflict with functional and quality product requirements, especially in the furniture sector, where products typically have more strict quality requirements.

In order to determine if there are significant differences between the average level of agreement about benefits and constraints related to eco-design implementation among sectors, the non parametric Mann–Whitney test has been considered in this study, which is used to test the null hypothesis H0, in which two independent samples were drawn from the same population Currell et al. 2009. Although all variables meet the conditions required by parametric tests such as *t* student, that is, they also have normal distribution and homogeneous variances (determined by Kolmorogov–Smirnov and Levene tests respectively), these hypothesis tests are generally considered to be inappropriate for qualitative variables and small random samples ($n < 50$) (Panik 2005). For this reason, a non-parametric hypothesis test such as Mann–Whitney is considered as an alternative to the independent measures *t* hypothesis and suitable for ordinal variables (Gravetter and Wallnan 2009) and (Berstein 1999). Tables 17.3 and 17.4 show the results obtained from Mann–Whitney test considering a 95% reading to be a significant level. Since p-values in all cases are > 0.5, the null hypothesis is not rejected, concluding that the two samples come from the same population or there are not significant differences in the average levels of agreement between sectors.

17.4 Conclusions

Eco-design is a useful tool for furniture and wood-processing companies, not only helping them to reduce their environmental impact, but also assisting them in obtaining a competitive advantage. Although the level of eco-design implementation in both sectors in Spain is low, there are companies that have improved the

Table 17.2 Eco-design strategies in furniture and wood-processing sector in Spain

Eco-design strategy	Wood-processing sector (%)	Furniture sector (%)
Selection of low-impact materials		
Cleaner materials	**26**	**50**
Renewable materials	12	18
Recycled materials	18	21
Recyclable materials	21	29
Reduction of raw material consumption		
Reduction in weight	**18**	**41**
Reduction in volume	9	26
Optimization of production processes		
Alternative production techniques	6	12
Fewer production steps	18	29
Lower/cleaner energy consumption	18	32
Less production waste	29	35
Fewer/cleaner production consumables	**32**	**41**
Optimization of distribution systems		
Less/cleaner/reusable packaging	**26**	**47**
Energy-efficient transport mode	21	15
Use of cleaner fuels	6	0
Energy-efficient logistic	21	29
Reduction of impact during use		
Easy- maintenance product	18	21
Product with high reliability/durability	**26**	**44**
Optimization of end-of-life system		
Product recycling	9	9
Remanufacturing	3	0
Recycling of materials	**18**	**32**

environmental performance of their products through the implementation of eco-design strategies focused primarily on raw material and energy optimization, use of cleaner materials and waste reduction. The implementation of eco-design strategies has benefits for companies, but they are sometimes constrained. The Mann–Whitney test showed that there are no relevant differences between sectors regarding benefits and constraints of eco-design implementation. Thus, in both sectors the main benefits obtained by companies are not only related to the environmental improvement of products but also help to improve their image, meet legal and normative requirements and help them find new market opportunities. As for the main restrictions faced by companies, these are mainly related to lack of knowledge and conflicts with functional and quality product requirements that must be considered beforehand to achieve the expected benefits. New information, training campaigns on eco-design and the development of innovative and sustainable materials will overcome these barriers and will also seek to promote the production and consumption of sustainable products in the furniture and wood-processing sector.

Table 17.3 Mann-Whitney U test results

Benefits	Mean score		U	Z	p-value
	Wood-processing	Furniture			
Design more innovative products	3.00	2.41	83.0	−1.484	0.138
Meet environmental customer's requirements	3.93	3.76	108.0	−0.461	0.645
New market opportunities	4.00	3.88	112.0	−0.303	0.762
Improved company's image	4.43	4.47	116.0	−0.134	0.893
Meet normative and legislative requirements	4.50	4.29	89.5	−1.313	0.189
(continued)					
Cost savings	3.57	3.24	100.0	−0.775	0.438
Increase product quality	3.93	3.35	80.5	−1.598	0.110
Increase efficiency in production, logistics, etc.	3.93	3.59	96.0	−0.962	0.336
Reduction of environmental impact	4.29	3.88	85.0	−1.442	0.149

Main benefits in the implementation of eco-design strategies in the furniture and wood-processing sector in Spain

Table 17.4 Mann-Whitney U test results

Constraint	Mean score		U	Z	p-value
	Wood-processing	Furniture			
No clear environmental benefit	3.43	3.06	89.5	−1.261	0.207
Insufficient time	3.36	2.82	82.0	−1.553	0.120
Insufficient knowledge	3.29	3.18	113.5	−0.239	0.811
No innovation opportunities	3.00	2.59	89.5	−1.244	0.214
Conflict with functional product requirements	3.21	3.59	97.0	−0.935	0.350
Conflict with quality product requirements	3.07	3.47	93.5	−1.059	0.290
Conflict with customer requirements	3.14	3.00	106.0	−0.545	0.586

Main constraints in the implementation of eco-design strategies in the furniture and wood-processing sector in Spain

References

Boks C (2006) The soft side of eco-design. J Clean Prod 14:1346–1356

Currell G, Dowman A (2009) Essential mathematics and statistics for science. Wiley-Blackwell, Oxford

Gravetter FJ, Wallnan LB (2009) Statistics for the behavioural sciences. Wadsworth Cengage Learning

Houe R, Grabot B (2009) Assessing the compliance of a product with an eco-label: From standards to constraints. J Clean Prod 121:21–38

Knight P, Jenkins JO (2009) Adopting and applying ecodesign techniques: a practitioners perspective. J Clean Prod 17:549–558

Kurk F, Eagan M (2008) The value of adding design for the environment to pollution prevention assistance options. J Clean Prod 16:722–726
Miquel S et al (1997) Investigción mercados. Mc Graw Hill, Madrid
Mont OK (2002) Clarifying the concept of product–service system. J Clean Prod 10:237–245
Panik MJ (2005) Advanced statistics from an elementary point of view. Elsevier
Parikka-Alhola K (2008) Promoting environmentally sound furniture by green public procurement. J Clean Prod 68:472–485
van Hemel C, Cramer J (2002) Barriers and stimuli for eco-design in SMES. J Clean Prod 10:439–453

Chapter 18
Aligning Hotel Data Management for Customer-Related Resource Assignment Decisions

Rajesh Natarajan and Alfonso Duran

Abstract The hospitality industry is extremely competitive due to factors such as overcapacity, low entry barriers, perishable nature of services and long lead times, in addition to rising customer expectations. One possible way to address this challenge is for hotels to use their large customer database to adapt their products and services to specific customer requirements. However, efficient and effective utilization of customer data requires that hotels align their data management and infrastructure with decision-making processes, thus ensuring that the right data are available at the right moment. Within a research on hotel decision making, this paper discusses the various issues that hotels need to deal with in order to make a more effective use of their customer data. First, we provide a classification of hotel decision-related customer data. Then we analyze the availability of customer data for decision-making during the various stages of the customer life cycle. Finally, we discuss how hotels can prioritize their data management efforts in accordance with the criticality of data-supported decision making, and conclude the paper with a summary.

R. Natarajan (✉) · A. Duran
Área de Ingeniería de Organización, Universidad Carlos III de Madrid,
Avda. de la Universidad, 30, C.P. 28911 Leganés, Madrid, Spain
e-mail: rajesh.natarajan@uc3m.es

A. Duran
e-mail: alfonso.duran@uc3m.es

S. P. Sethi et al. (eds.), *Industrial Engineering: Innovative Networks*,
DOI: 10.1007/978-1-4471-2321-7_18,

18.1 Introduction and Related Work

The intense competition in the hospitality industry is a consequence of the very nature of this type of business, coupled with external factors such as reduction in customer search costs, new forms of intermediation, increasingly discerning and demanding customers and the current economic downturn (Rutherford and O'Fallon 2007). The industry pressures include overcapacity, perishable nature of the product, low barriers to entry, supplier fragmentation and long lead times for capacity change. Hotels have to be extremely pro-active and flexible in order to anticipate customer requirements and customize their products and services accordingly.

Hotels have responded to these challenges by strengthening their operations while simultaneously seeking a better understanding of their customers for effective personalization (Minghetti 2003; Noone et al. 2003). Extensive use of information technology in the form of systems for centralized reservation, property management, yield management systems and the like have granted hotels better operational efficiency and effectiveness, along with reduced operation and missed opportunity costs (Sigala et al. 2001; Buhalis and Law 2008; Luck and Stephenson 2009). These systems also capture large amounts of customer data. Hotels need to leverage this resource by ensuring that their products and services are tailored to specific customer requirements (Magnini et al. 2003; Minghetti 2003; Noone et al. 2003). However, efficient and effective utilization of customer data requires that hotels data management processes and infrastructure be in line with decision-making. This is because the value of captured data is realized only when it supports hotel decision-making. In particular, hotels need to ensure that the right data are made available at the opportune time. The study by Minghetti (2003) has categorized the types of hotel data that need to be collected and has developed a methodology for designing a customer information system by using a guest information matrix. Our study can be considered complementary to it, in so much as we consider a different categorization of customer data and link them to their availability at the appropriate moment to aid decision-making.

Within the framework of a research project on simulation-based enhancement of hotel decision making, this paper focuses on one of its most critical components, namely the customer data management process, and its design to maximize its value for decision support. The next section briefly discusses the life cycle of a hotel customer. We then provide a classification of customer data based on whether they are directly collected from the customer, derived from customer's past history or inferred from data that pertains to other customers. We then present a framework that analyzes the availability of customer data for decision-making at the various stages of the customer life cycle. Derived and inferred customer data are needed because sometimes direct decision-supporting customer data are available only after the decision has been made. Hotels need to prioritize their data management efforts in accordance with the time critical nature of decisions that the data support. We examine some relevant issues for such prioritization. We conclude the paper with a summary and provide a scope for future research.

18.2 Life Cycle of a Hotel Customer

Customer's personal data such as likes, dislikes and preferences can be used to influence his/her consumption decisions by tailoring products and services to suit specific customer requirements. Primary sources of hotel data may include property management, hotel reservation, customer relationship management systems, customer surveys etc. Each one of these systems caters to a different aspect of a hotel's operations and consequently only has a partial view of the customer. Therefore, to obtain an integrated view, it becomes important that the hotel's perspective is one that places the customer at the centre of its data requirements. One possible orientation is to use the *customer life cycle* as the basis for defining customer data requirements.

The customer life cycle depicts the various stages that the customers experience when served by a hotel. Each stage is characterized by possible multiple interactions with the firm. Each interaction presents the hotel with an opportunity to influence the guest by means of a customer-related decision. Most of the decisions that directly affect the customer take place during the customer life cycle. Quality of decisions is substantially enhanced if relevant customer information is available at appropriate time. Thus, by defining its data requirements around the customer life cycle, a hotel can concentrate on its most critical data requirements.

The customer life cycle (Fig. 18.1) begins with a search process that may be undertaken either with the help of travel agencies or by an Internet-based search. Various hotels and information websites help in this process by having online search menus (Buhalis and Law 2008). During this stage, the customer is primarily interested in information regarding room availability, current rates, ancillary facilities, etc. that would help him/her evaluate the firm's offers. If the potential client's needs are met, then he/she might make a direct enquiry with the hotel. If the hotel has sufficient information, then it can influence the customer to book by tailoring its package to specific customer requirements. The next interaction occurs when the customer checks into the hotel for the stay. During the stay, the customer may also avail of various ancillary services provided by the firm. The period before checking-out presents numerous opportunities for the firm to both offer its services and also to collect data about the customer and use it to influence his/her decisions. The checking-out process is the last customer interaction opportunity for the hotel during the current stay. Feedback elicited during this stage or later can help a hotel further refine its services to customer tastes. Customer feedback presented in various Internet-based opinion forums can influence the decisions of new customers during their search process. Sometimes the customer may cancel the room booking and later initiate a new search process.

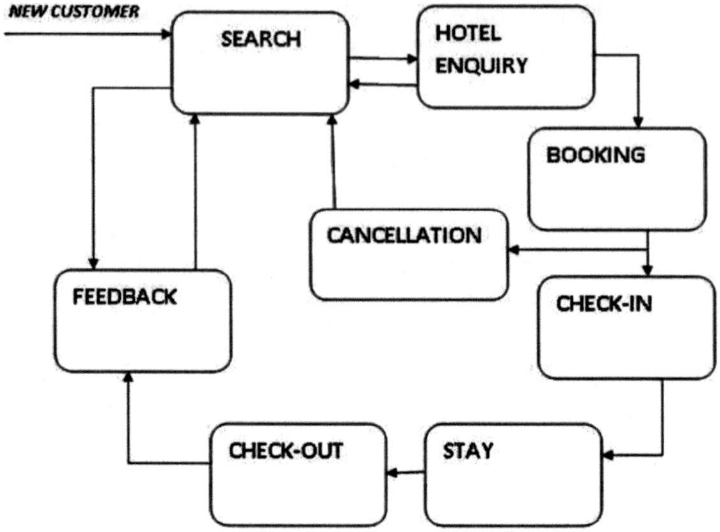

Fig. 18.1 Typical life cycle of a hotel customer

18.3 Hotel Decision Making and Availability of Customer Data

Data, collected by a hotel, is valuable only if the hotel is able to utilize the data to enhance the quality of its decisions. Then, the decision-maker should have access to relevant data at the time of need. This implies that the hotel should not only have identified its customer decision-related data requirements but should also have well-organized data collection and dissemination mechanisms. One possible approach for a hotel firm is to begin by analyzing its data requirements with respect to the decisions made during the customer life cycle and then align its data collection, analysis and dissemination mechanisms with it.

Any customer-related hotel decision would require data that can be classified into three main categories. The first category, *direct customer data*, deals with customer details which can be directly collected from customers themselves by information systems. For example, address, age, period of stay, etc. Some of these data are mandatory, as they are required by the hotel for its operations. For example, credit card information, date & time of arrival, etc. On the other hand, some data, especially of the attitudinal and behavioural type, though not compulsory might enhance the hotel's services. For example, favourite beverage, etc. Sometimes, customers' preferences may be revealed by behavioural data that are automatically collected when the customer uses varied services that the firm offers.

The second category, i.e. *Derived Customer Data*, includes data derived from related data pertaining to the *same* customer. The usefulness of derived data depends on the exactness of relationships used by the hotel in its derivation mechanism, in addition to accurate data collection. For example, a hotel might be able to estimate the supplies needed to serve a customer from his/her consumption in previous stays at the hotel or in the initial days of a longer stay. Finally, consider a variable such as a customer's *Propensity to Spend* on other hotel services, i.e. dining, health facilities, etc. The hotel would know whether a customer is a *High Spender* or *Low Spender* only after the actual utilization of its services. However, by using clustering and classification techniques (Magnini et al. 2003), the hotel may be able to predict with some degree of certainty a customer's behaviour from past behaviours of other customers who are similar to the current customer. Thus, the particular customer's behavioural data, inferred from the behaviour of other 'similar' customers can be referred to as *Inferred Customer Data*. We note that *Derived Customer Data* is different from *Direct Customer Data*, as it is extrapolated from related customer variables of the same customer. *Inferred Customer Data* is further removed as it comes from data pertaining to other customers in addition to the data of the same customer.

Figure 18.2 analyzes a hotel's temporal dimension of decision-making with respect to the data collection processes and availability of data during the life cycle of a typical customer. Data are valuable when they are at hand and can be used during the decision-making process by the hotel employee to influence customer's decisions. Figure 18.2 employs the customer life cycle of a typical customer as the underlying temporal dimension. The X-axis represents the data collection timeline that is further subdivided by each customer's life cycle. Each customer has one horizontal axis that is further subdivided by the life cycles of all of his/her visits. Note that each visit's life cycle can be further sub-divided by the various stages shown in Fig. 18.1. The gap between two life cycles represents the fact that there might be time gaps between two successive hotel visits by the same customer. Further, differences in the lengths of hotel stay and other stages of each life cycle might result in differences in the total length of the life cycles.

Consider a typical customer *X*. This customer's current and past life cycles are depicted by the top horizontal axis in Fig. 18.2. Time $t = 0$ represents the start of *X*'s current life cycle. Let us assume that the hotel has to make a decision with regards to this customer at time $t = T$. Here we assume that a decision made during time T_i of the customer life cycle will have all data collected until point T_i. Thus, data available for enabling this decision are all *direct data* pertaining to *X* collected till this point including data from previous life cycles.

Let point $t = T$ represents a moment during the check-in period of *X*'s current life cycle when the customer arrives at the hotel. Let us suppose that the hotel has a policy of offering discounts for its various ancillary services, like restaurants, gymnasium, spas, gaming facilities, etc. The hotel's receptionist may emphasize and offer discounts for some of its services. Due to limited time, the receptionist should highlight only those services that the customer is likely to make use of during the current visit. In other words, the receptionist should know the customer's propensity to spend on each service at the start of the visit. *Direct data* regarding this will be available only during the period between $t = T_1$ and $t = T_2$, when the customer uses the hotel's services during his/her stay. Availability of such data at the point of decision would enhance the quality of decisions. But as

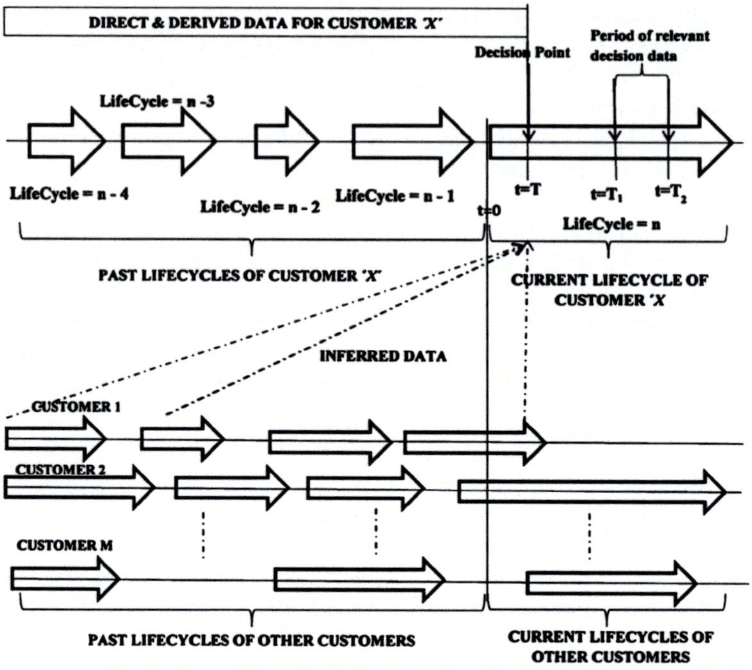

Fig. 18.2 Customer decisions and data availability

we can see, these data are not available. Thus the segment from period $t = T_1$ to $t = T_2$ represents a *gap* which can be termed as data essential for a decision that have not been collected yet due to the fact that they would be available only at a later stage in the customer life cycle. Note that the *gap* does not refer to the inadequacy of the hotel's data collection mechanisms. On the contrary, it just represents a fundamental constraint caused by the fact that relevant decision-enhancing data will only be available after the decision has been made. Decisions related to customer behaviour typically require this data. The best data collection mechanisms of a hotel cannot overcome this difficulty. One way to bridge this gap is to base the current decision on *derived* data from the previous visits of the same customer, or *inferred* data from other customers. The firm can use Data Mining, Statistics and other predictive techniques (Magnini et al. 2003) to gather the relevant data from other related data.

18.4 Streamlining Data Collection for Effective Decisions

The framework discussed in the previous section provides a useful starting point for analyzing data requirements for making hotel decisions. For supporting a particular decision, the hotel can gauge the competence of available data. This in turn helps it identify essential data that is unavailable at the point of decision making and possible ways of filling this limitation by means of derived or inferred data.

The data management infrastructure of hotels needs to be evaluated on the basis of its effectiveness to influence decision-making rather than by the quantum of data they capture and disseminate. Typical hotel firms have diverse information systems that enable their day-to-day operations. These include property management, hotel reservation, Customer Relationship Management (CRM), knowledge management, business intelligence and other related systems. Some systems act as data sources for others, for example the property management systems might get their data from reservation systems. Streamlined integration of diverse stand-alone systems will ensure that the full decision-supporting potential of customer data is realized.

Integration of operational and decision-support systems such as business intelligence, CRM and related systems should ensure timely availability of relevant derived and inferred data. The usefulness of derived and inferred data depends not only on the correctness of their generation mechanisms but also on the accuracy and relevance of the associated direct data used to create them. For example, data pertaining to a very old customer visit may not be useful in predicting current behaviour. Therefore, for increased effectiveness, the derived and inferred data creation mechanisms need regular updating and their dissemination must be synchronized with hotel decision cycles.

Another aspect concerns the rationalization of data management effort with respect to decision-making goals. The selection of decision processes that need to be supported must be based on the real value to the hotel. All decisions are not equal in terms of importance, criticality or value. In particular, decisions differ in their scope, duration of influence, risk and benefit. For example, a hotel may gain extra revenue if it offers discounts for its services that are of interest to the customer. On the other hand, the hotel may not gain much if services that are of little interest to the customer are highlighted. Alignment of data management with decision-making ensures that such revenue generation opportunities are not lost. Consider the scheduling of cleaning services. Hotel rooms are expected to be clean and making the right decision may be financially neutral. However, reducing cleaning services below an acceptable level might lead to dissatisfied customers and adversely affect future earnings. Thus, both benefits and risks of a decision should influence its inclusion in the list of important decision-making processes that need to be adequately supported by a hotel's data management infrastructure.

In addition to resources utilized in data collection, derived/inferred data may entail extra expenditure for additional software, manpower and associated costs. Given limited resources, it is important that the hotel prioritizes its data management efforts with decision support as the main objective. A data element is valuable if it helps many important hotel decisions. Thus, the decision to collect and maintain a particular customer data element should be based on a cost-benefit analysis of the different decisions that the data element supports. For example, certain ad-hoc customer requests can be fulfilled by the hotel without having information about related customer preferences. Such customer preference data need not be collected a priori.

18.5 Summary

The hospitality industry is addressing some of the issues faced in its challenging environment by leveraging its customer data. However, efficient and effective utilization of customer data requires that hotels align their data management processes and infrastructure with decision-making, in order to ensure that the right data are available at the relevant moment. This paper discussed some of the issues that hotels need to address in order to make a more effective use of their customer data. We first classified customer data typically needed for hotel decision making. Subsequently, we analyzed the availability of decision-supporting customer data during the various stages of the customer life cycle. Given limited resources, hotels need to prioritize their data management efforts in accordance with the criticality of decision-making that the data supports. As a line for future research, we will study the problem of evaluation and quantification of the extent of decision support provided by a particular hotel's data management infrastructure.

Acknowledgments This work has been supported by the research project funded by the Spanish National Research Plan, reference DPI2008-04872, titled "Optimización de la asignación de infraestructuras de servicios mediante simulación - sectores hotelero y sanitario."

References

Buhalis D, Law R (2008) Progress in information technology and tourism management: 20 years on and 10 years after the Internet–The state of eTourism research. Tour Manag 29:609–623
Luck D, Stephenson ML (2009) An evaluation of the significant elements of customer relationship management within the hotel industry. Tour today 9:7–26
Magnini VP, Honeycutt ED Jr, Hodge SK (2003) Data mining for hotel firms: use and limitations. Cornell Hotel Restaur Adm Q 44(2):99–105
Minghetti V (2003) Building customer value in the hospitality industry: towards the definition of a customer-centric information system. Inf Technol Tour 6:141–152
Noone BM, Kimes ES, Renahan LMI (2003) Integrating customer relationship management and revenue management: a hotel perspective. J Revenue Pricing Manag 2(1):7–21
Rutherford DG, O'Fallon MJ (eds) (2007) Hotel Manag Oper, 4th edn. Wiley, New Jersey
Sigala M, Lockwood A, Jones P (2001) Strategic implementation and IT: gaining competitive advantage from the hotel reservations process. Intern J Contemp Hosp Manag 13(7):363–371

Chapter 19
Environmental Management in Industrial Enterprises: A Multiple Case Study

M. Ormazábal and J. M. Sarriegi

Abstract This paper consists of an analysis of environmental management in different enterprises. A classification of the factors that affect the environmental management in a company has been made through a multiple case study of nineteen different companies. Some of the main factors that affect all businesses have been identified, such as customers, the law and competitors. Every company has to comply with the law, although the law affects each sector differently. Depending on the markets in which they operate, customers are more or less aware of environmental aspects. Moreover, it is important to distinguish public from private clients. On the other hand, competitors also need to be taken into account when company decisions are made. All these aspects can be found in different companies with different activities and different characteristics, so enterprises should be aware of them when taking environmental decisions.

19.1 Introduction

In recent years, environmental issues have become a very important aspect of any enterprise. The intensive use of natural resources and increasing waste generation are causing serious problems for the environment. This has led to a desire to improve environmental management, and an ever increasing social concerned over environmental problems (Ludevid 2004). For this reason, society is beginning to demand that companies take more and more environmental measures. Consumers understand that buying goods or services brings not only an economic cost,

---

M. Ormazábal (✉) · J. M. Sarriegi
Tecnun, University of Navarra, P/Manuel Lardizábal, 13, 20018 San Sebastián, Spain
e-mail: mogoenaga@tecnun.es

S. P. Sethi et al. (eds.), *Industrial Engineering: Innovative Networks*,
DOI: 10.1007/978-1-4471-2321-7_19,

but also an environmental one. As such, there tends to be an increase in market demand for products and services that minimize environmental impact (Abarca and Sepúlveda 2001). As a result of this, an essential objective of environmental management is to reduce the risk of a damaging environmental incident.

For years, environmental consulting firms have offered to help in the implementation of different certifications to improve environmental management. Certifications include the ISO 14000 international series and the European regulation EMAS. There are grants and procedures to help companies to obtain certification. Some of these certification schemes assist law enforcement; others improve the environmental aspects of processes and others are designed to improve the environmental aspects of products.

Nowadays, the fulfilment of environmental requirements, demanded by both customers and legislations, is carried out in the majority of companies. However, it is not easy to go even further than the current rules demand and thereby achieve corporate environmental excellence. The first thing that any company should do is to identify all the factors that may affect the environmental management in a business so improvements in their own activities can be made. These factors have been analyses through a multiple case study in different companies.

19.2 Methodology

This research is carried out through different interviews of environmental managers employed in different companies in the Basque Country. The study contrasts firms from different sectors such as chemical, automotive, railway, elevators and electrical devices, which differ in terms of environmental management. The study includes ten enterprises from different sectors. Apart from interviews, there has also been access to external sources, such as the press and web pages, to compare and complete information given by the environmental managers. Collecting information from multiple sources helps verify events or information more than a single source of evidence (Yin 2009).

The interviews with each environmental manager were carried out face to face. The main objective of each interview was to understand how environmental management has been deployed in each enterprise. The first aspect that has to be analysed is the characteristics of each business and the kind of activities carried out in each organisation. The second aspect that is discussed is what the current situation of the organisation is, what types of certificates and tools they have and why they have them. In regard to this, it is important to know why companies are introducing environmental aspects into their activities and why they think it is important. Through this multiple case study all the factors that could affect the environmental management in an enterprise have been analysed.

19.3 Results

After gathering all the information among different companies, a classification of the factors that may affect the environmental management in a business has been carried out.

19.3.1 Incentives or Constraints that have Driven Environmental Management

There have been several differing reasons that have prompted companies to take environmental management into account in their daily activities. Each enterprise belongs to a different sector. Consequently, the following reasons have affected enterprises in different ways.

Legislation imposes certain norms that force companies to follow the specific rules. Rules are becoming stricter so companies are being forced to reduce the environmental impact their products and processes have. All companies are trying to abide by current legislation and are making great efforts to implement certain laws. In this sense, many companies prefer to exceed the controls of present legislation in order to be prepared if a new law is issued.

Market One of the main factors which have driven companies towards environmental improvements is the requirements demanded by customers. Nowadays clients are more conscious of environmental problems and as such are beginning to demand sound environmental practices. Nevertheless, there is a distinction between public customers and private customers. When customers are public, enterprises have to participate in public tenders. In these public tenders there is usually a section which corresponds to environmental aspects, so enterprises have to justify how their products or services take into account the environment. Private customers are not so demanding but they are beginning to give more importance to the impact products or services have on the environment, companies must therefore, be aware of these demands.

Green brands Several enterprises have seen environmental management as an opportunity to gain advantage over their competitors. If companies do not limit themselves to simply complying with legislation, but instead try to optimise environmental management in their daily activities it can lead to greater product prestige. In this sense, some enterprises have not seen environmental management as a threat but rather use this environmental awareness in order to take advantage of the situation.

Horizontal Pressure In some enterprises, environmental management can be reinforced by pressure imposed by competitors. In sectors where environmental practices are widespread, environmental management can be considered an opportunity rather than a necessity.

Environmental Awareness Some decades ago, companies implemented environmental practices in their enterprises because of purely legal obligations. However, nowadays, more and more companies are improving the environmental credentials of their companies because of environmental awareness. In other words they take these steps because they believe it is important to protect the environment.

19.3.2 Certification Schemes and Tools

There are different certification schemes and tools that help companies reduce their environmental impact.

Ekoscan This is promoted by the Public Agency of Environmental Management of the Basque Country. Some of the companies that were interviewed had this certification before having the ISO 14001. Ekoscan helps companies in the first step of Environmental Management. This certification does not require authorisation so companies can gain certification relatively quickly. All enterprises think that as a first step, these certifications help to document and organise the environmental management in the company.

ISO 14001 Most of the companies interviewed have the ISO 14001. The main purpose of this international certification scheme is to provide organisations with the pillars of an Environmental Management System so they can reach environmental and economic goals. Therefore ISO 14001 is a way to systematize environmental management in enterprises. In some markets, it is unthinkable not to be certified with this norm. It is a minimum requirement to enter in most public tenders; so many businesses need this certification as a prerequisite to participate. A great number of organisations have ISO 14001 very well assimilated, so much so that they go further and look for new environmental improvements through different tools. For some companies the ISO 14001 certification is taken for granted and does not give any added value.

EMAS Not many companies are certified with the European certification EMAS. Despite not being an international norm, the requirement is much higher than ISO 14001; hence enterprises prefer to be certified with ISO 14001 which is known around the world. One of the differences between these certifications is that with EMAS all the environmental impact of the enterprise has to be made public. Some of the enterprises that have been certified with EMAS, have decided not to continue with this norm as they did not find it beneficial to make all the indicators public, it also proved to be costly. Furthermore, all those companies that are certified with EMAS usually continue with ISO 14001 so they can compete in international markets. The cost of maintaining two certifications is quite high.

Recycling Another step, that some companies are trying to follow, is to ensure the recycling of their products, to meet market demand and a possible legal requirement. For these companies, it is very important that all parts of the product are recyclable; to this end they are improving their products and processes to achieve this purpose. Furthermore, they have to consider how the product is constructed, as at the end of the useful life of the product it will have to be separated into distinct parts for recycling. In addition, several enterprises are reusing their packaging. They have started to use returnable packaging thus the product is sent to the customer and the packaging returns to the company to be used again.

Indicators All enterprises monitor different environmental indicators and take measures to improve different environmental aspects of the company. By monitoring the indicators, the effectiveness of the measures carried out can be checked. It is also advantageous to set some goals and track them monthly, quarterly, every six months or annually to move towards continuous improvement (Epstein 2008). In the beginning, it is quite easy to improve the indicators as there are many things that are easily changed for the better. Nevertheless, as time passes it is more difficult since improvements are not so readily detected and are not so easily implemented. On the other hand, quite a lot of businesses agree that there are difficulties when measuring these indicators. As previously mentioned, indicators are mainly used to control the evolution of certain parameters. The problem is that in some cases what it is being compared is different from one year to another since the production has changed or the project required by the customer is different. For this reason, occasionally it is difficult to devise a formula that compares effectively the different parameters of environmental business. For example, if the activity of an enterprise depends on the requirements of the customers and each year they demand a different product or service, it is very difficult to compare some indicators from one year to another as the decrease or increase of these variables will mean nothing.

Eco-design Those enterprises, which are quite advanced in their environmental management system, are beginning to improve environmental aspects from the design point of view. They analyse the entire life cycle of the product, from raw materials, through to production, distribution and the use of the product including the elimination or recycling of the product (Aoe 2007; Kobayashi 2006). In such cases, there is a previous phase where all the environmental aspects of the product design are taken into account. Enterprises consider that eco-design offers some advantage over competitors who have not implemented it in their business activities. Moreover, by analysing the life cycle of their product they can justify the impact of the product at public tenders. In the same way, when analysing environmental improvements in the design phase, they may identify ways to reduce the use of resources thereby achieving considerable savings. One of the problems, which one company highlights, is that eco-design can have a negative influence on the product launch time. Eco-design involves an exhaustive analysis of all the environmental impacts and as such delays the release of the product. Furthermore, when introducing new changes to improve environmental impacts, the cost of the product may increase and prevent it from being introduced in the marketplace, the time spent on these projects may also be considerable. Some enterprises are not able to implement eco-design as they are obliged to do what the clients require and cannot introduce changes.

19.3.3 Involvement of all Parts of the Company

The environmental activities in the company cannot be the sole responsibility of the environmental department. Each of the different departments of the business should participate. When there are changes in the product or in the process of the company, every member has to know why those changes have taken place. These explications have to be transmitted from the manager to the rest of the workers.

Higher Management Involvement All enterprises agree on the importance of management involvement. They consider that environmental improvements can only be carried out if the management of the organization is committed to the environment. Managers should participate in environmental meetings; they should take an interest in what is happening. If the management is not involved, environmental goals cannot be met. Some of the companies that were interviewed claimed that they could not go further in the environmental management of the company because higher management did not feel that it was important. In these cases, it is very difficult to implement environmental improvements, as insufficient resources are made available to achieve them.

19.3.4 Communication

It is very important to communicate the practices that are carried out in the company. All parties interested in the product or service should know the measures that are being taken in order to collaborate as much as possible. In many cases, communication is a prerequisite to achieve effective measures. If workers know why they are implementing different changes these changes will work better since workers will be more motivated.

Internal Environmental practices, which are carried out in the enterprise, should be transmitted among all the areas of the company, in such a way that everyone is conscious of the measures that are being taken and therefore feel that it is part of their work. Nonetheless, before changes are communicated to the parts of the company, the necessary means should be made available in order that the workers are properly prepared for the work they will be asked to do. Communication will be useless if necessary means to carry out environmental practices are not made available. Managers and technicians of some of the companies join together once a year to make new proposals to improve environmental impact. Moreover, workers may have their own ideas and in the case that they are accepted the worker is not only rewarded but also positively evaluated at annual meetings.

External communication is also very important. On the one hand, communication with suppliers is essential. If environmental measures are being taken in the enterprise, it is very important to inform suppliers so they can take the necessary measures to carry out what is planned. Often, when a life cycle analysis is done,

Table 19.1 Importance of the factors among the interviewed enterprises

		Business 1	Business 2	Business 3	Business 4	Business 5	Business 6	Business 7	Business 8	Business 9	Business 10
Incentives or constraints	Legislation	Y	Y	Y	Y	Y	Y	Y	Y	Y	Y
	Market	H	L	VL	L	H	H	M	L	VL	L
	Green brand	VL	L	VL	H	M	H	VL	VL	VL	L
	Horizontal pressure	M	L	VL	L	M	M	VL	VL	L	M
	Awarness	L	H	M	H	H	H	VL	M	L	H
Certifications and tools	Certifications	Y	Y	Y	Y	Y	Y	Y	Y	Y	Y
	Recycling	M	L	VH	VH	H	L	VH	M	L	H
	Indicators	H	H	M	H	H	M	L	L	M	H
	Eco-design	M	VL	VL	L	L	H	VL	VL	VL	VL
Involvement	All parts involvement	VL	H	VL	H	H	L	VL	L	VL	H
	Top management	M	H	L	H	H	M	L	M	L	H
Communication	Internal	M	VH	M	H	H	VH	L	L	L	H
	External	VL	VL	VL	H	H	L	VL	VL	VL	L

		Business 11	Business 12	Business 13	Business 14	Business 15	Business 16	Business 17	Business 18	Business 19
Incentives or constraints	Legislation	Y	Y	Y	Y	Y	N	Y	N	N
	Market	M	VL	M	M	VL	VL	H	VL	VL
	Green brand	H	L	H	H	VL	VL	H	VL	VL
	Horizontal pressure	L	L	L	L	VL	VL	L	VL	VL
	Awarness	H	M	H	H	L	L	H	VL	M

(continued)

Table 19.1 (continued)

		Business 11	Business 12	Business 13	Business 14	Business 15	Business 16	Business 17	Business 18	Business 19
Certifications and tools	Certifications	Y	Y	Y	Y	Y	N	Y	N	N
	Recycling	H	L	H	L	L	L	M	VL	L
	Indicators	H	H	H	H	H	VL	H	VL	M
	Eco-design	VL	VL	VL	L	VL	VL	VL	VL	VL
Involvement	All parts involvement	H	M	H	M	L	VL	M	VL	VL
	Top management	H	M	H	M	M	VL	M	VL	L
Communication	Internal	H	L	H	M	L	VL	H	VL	L
	External	M	L	M	VL	VL	VL	L	VL	VL

the introduction of different raw materials is suggested leading to a decrease in the environmental impact of the product. For this reason it is important to maintain communication between the enterprise and the suppliers. On the other hand, the communication with customers is very important as well. A lot of enterprises take advantage of their higher environmental credentials in comparison to other companies and make their improvements public. As there is ever greater environmental awareness in society and as many public organisms are promoting green procurement, enterprises may take advantage of the situation to gain market share.

19.3.5 Review of Factors Among Companies

In Table 19.1 a review of the factors mentioned above has been carried out.

Each factor has a score depending on the importance that it has for each enterprise. Two different scales are being used.

- Yes (Y) - No (N). Some of the factors cannot be rated, so this scale is being used to reflect if each enterprise has this variable implemented or not.
- Very low-Very high. The different scores are very low (VL), low (L), medium (M), high (H) and very high (VH), being very low when the company does not take into account the corresponding factor for its environmental management and very high when the company has that factor totally implemented in its activities.

19.4 Conclusions

Through the analysis of the environmental management in nineteen different companies, a study has been carried out to describe and examine the factors related to environmental management in those enterprises. Through this research, it has been demonstrated that there are factors that have an effect on every environmental management system. In regard to enterprises, knowledge of all the factors that may influence the environmental management in a company would be useful in order to evaluate all the relevant factors for their enterprise and try to improve them. Some of the factors may be influenced by others, therefore the following step in this research should be to find the interrelation among all these factors and analyse the evolution that each company is following in its environmental management to study if the evolution of every company towards environmental excellence has the same structure.

References

Abarca R, Sepúlveda S (2001) Eco-etiquetado: Un Instrumento para Diferenciar Productos e Incentivar la Competitividad. Comercio-Ambiente 17:1–65

Aoe T (2007) Eco-efficiency and ecodesign in electrical and electronic products. J Clean Prod 15(15):1406–1414

Epstein MJ (2008) Implementing corporate sustainability: Measuring and managing. Social and environmental impacts. Strategic Finance

Kobayashi H (2006) A systematic approach to eco-innovative product design based on life cycle planning. Adv Eng Inform 20(2):113–125

Ludevid M (2004) La gestión ambiental de la empresa. Ariel, España

Yin RK (2009) Case study research: design and methods, 4th edn. Sage, London

Chapter 20
Organizational Structure Shapes Performance in Dynamic Environments: Studying the Relationship Between Structure and Performance

Arsalan Nisar, Carlos Rodríguez-Monroy, Felipe Ruiz and Wu Yuxi

Abstract There has been a long-standing concern that strategy literature needs a better understanding of how organizational structure can affect performance and decision making within organizations. Previous research indicates that the relationship between organizational structure and performance is unexpectedly asymmetric. The balance between too much and too little structure is critical for an organization's high performance in dynamic environments. Organizations with too little structure lack enough guidance but can be flexible, while organizations with too much structure are too constrained and lack flexibility. In addition, less structured emergent strategies tend to be high performing in dynamic environments, whereas more structured prescriptive strategies tend to perform well in stable environments. Accordingly, the optimal amount of structure decreases with increasing environmental dynamism, a consistent finding in existing literature. In this paper a review of literature on how organizational structure influences performance within organizations and the role of the environment is presented. It also incorporates and signifies the role of the Strategy, Structure and Performance (SSP) paradigm, which has played a central role within strategy research. By means of this, we will be able to know how the core knowledge required for decision-making and the coordination challenges within firms drive their internal structures, and how the implication of the appropriate degree of strategy-structure fit on the performance of a firm. The aim here is to develop a more precise theory of the fundamental relationships among structure, performance and the environment.

A. Nisar (✉) · C. Rodríguez-Monroy · F. Ruiz · W. Yuxi
The Technical University of Madrid (UPM), Madrid, Spain
e-mail: arsalan.nisar@alumnos.upm.es

S. P. Sethi et al. (eds.), *Industrial Engineering: Innovative Networks*,
DOI: 10.1007/978-1-4471-2321-7_20, © Springer-Verlag London Limited 2012

20.1 Introduction

There is a long-standing concern that the strategy literature needs a better understanding of how organizational structure and decision-making affect organizational performance, especially in dynamic and turbulent environments. Csaszar (2008) indicates that this concern goes back at least to Cyert and March (1963), who used the following questions in motivating their theoretical enterprise: "What happens to information as it is processed through the organization? What predictable screening biases are there in an organization? [...] How do hierarchical groups make decisions?" But with a few exceptions, questions of this sort remain mostly unexplored in the strategy literature (Rumelt et al. 1994). This lack of knowledge concerning how structure affects organizational performance continually resurfaces in different areas of management.

Given its fundamental importance, this question has been frequently studied in a variety of research traditions, ranging from organizational studies (Burns and Stalker 1961; Hargadon and Sutton 1997) and competitive strategy (Rindova and Kotha 2001; Rothaermel et al. 2006) to network sociology (Uzzi 1997; Owen-Smith and Powell 2003) and, more broadly, the complexity sciences (Kauffman 1993; Anderson 1999).

However, strategy literature still needs an improved understanding of how organizational structure influences performance, establishes the relation between structure and performance and tries to assess the role of the environment. The literature that underlines the relationship between structure and performance is incomplete. Davis et al. (2009) suggest that the balance between too much and too little structure is challenging and crucial. Given the current situation, it has become imperative for organizations to find the right amount of tension between structure and performance. Moreover, the present turbulent environment requires organizations to be strategically agile and make the best use of existing resources in order to be competitive.

20.2 Organizational Structure and Performance

Organization structure may be considered as the anatomy of the organization, within which it functions. There are many definitions of structure, with varied attributes such as formalization (e.g., rules, routines), centralization (e.g., hierarchy, use of authority, verticality), control systems (e.g., span of control), coupling and structural embeddedness (e.g., tie strength, tie density), and specialization (e.g., role clarity) (Weber 1946; e.g., Burns and Stalker 1961; Pugh et al. 1963; Galbraith 1973; Scott 2003). While the definitions contain different attributes, they all share an importance in shaping the actions of organizational members. Entities are more structured when they shape more activities of their constituent elements and thus constrain more action. Conversely, entities are less structured when their constituent elements have more flexibility in their behavior. Accordingly, structure is broadly defined as constraint on action.

Davis et al. (2009) suggest that organizations with too much structure are too inflexible, while organizations with too little structure are too inefficient. However, this argument neglects key factors such as limited attention, time delays, and the fleeting and varied nature of opportunities that might influence this tradeoff. So, for example, the theory does not consider that, although less structure enables flexible improvisation, improvisation is an attention-consuming and mistake-prone process (Hatch 1998; Weick and Roberts 1993). As a result, the theory does not clarify precisely how structure influences efficiency and flexibility.

Davis et al. (2009) find that the performance—structure relationship is unexpectedly asymmetric and in that it is better to err on the side of too much structure, and that different environmental dynamic dimensions (i.e., velocity, complexity, ambiguity, and unpredictability) have unique effects on performance. Moreover, performance gradually declines with too much structure but drops catastrophically with too little. Relatively, efficiency and flexibility are distinct functions that change increasingly more slowly when structure is high. In contrast, efficiency and especially flexibility change more rapidly when structure is low, creating a more acute tradeoff between efficiency and flexibility. The implication is that it is safer to err on the side of too much structure (efficiency) than on the side of too little (flexibility).

Considerable research has been focused on how environmental dynamism influences the relationship between structure and performance. The arguments are that as the environment becomes more dynamic, it becomes more advantageous for the organization to be flexible; in contrast, as the environment becomes less dynamic, more efficiency and more structure are preferred. This argumentation finds extensive support in strategy literature that has found less structured emergent strategies to be higher performing in dynamic environments, whereas more structured deliberate strategies work better in stable ones (Mintzberg and McHugh 1985).

Fig 20.1 presents the four widely accepted dimensions of environmental dynamism and the available literature is not clear on how the different dimensions operate. The empirical literature also reflects this imprecision, as studies often mingle dimensions such as complexity, velocity, unpredictability, and ambiguity (Eisenhardt 1989; Pisano 1994) that may have distinct effects. Understanding the influence of different dimensions is important because they may have unexpected implications for theory and practice. For instance, it may be possible that only one or two dimensions shift the relation between structure and performance or that the structure-performance relation has distinctive shapes for different environments.

In general, there are some unresolved questions regarding the relation between structure, performance and environment, which can be further analyzed to improve our understanding and contribute to existing literature.

Velocity	Complexity
Ambiguity	Unpredictability

Fig. 20.1 Four environmental dynamism dimensions (Davis et al. 2009)

20.3 Strategy, Structure and Performance

With this chapter, we indicate that it is important to incorporate and signify the role of the Strategy, Structure and Performance (SSP) paradigm, which has developed a central role within strategy research. By this we understand that there is need to know how the core knowledge required for decision-making and the coordination challenges within firms drive their internal structures and the implication of the appropriate degree of strategy-structure fit on the performance of a firm (Wasserman 2008).

Rumelt (1974) laid the foundations for development of the strategy-structure-performance paradigm based on the early groundbreaking work of Chandler (1962). Focusing on the degree of diversification and the level of divisionalization, these and subsequent studies showed that the degree of 'fit' between firm strategy and internal structure had important implications for firm performance. Chandler (1962) studied the relationship between structure and strategy by examining linkages between organizational structure and diversification strategy. He argued that changes in strategy—i.e., in product-market diversification—required structural changes. Particularly, shifts from a strategy focused on a single product market to one that was vertically integrated and multi-business in scope were accompanied by shifts in structure from being functionally organized to being organized around divisions. Rumelt (1974) extended this argument to examine performance implications and found that the match between diversification strategy and divisional structure affects performance. Firms that follow strategies of controlled diversity while adopting divisional structures attained the highest level of economic performance, whereas firms that pursued unrelated diversification performed the worst (Wasserman 2008).

Further research in this area incorporated this groundbreaking work to frame the SSP paradigm that suggests that, rather than consider each strategy or structure alone as having an impact on performance, it is the linkage between them that is important (Lenz 1980; Miller and Friesen 1984; Miller 1996, p. 510). The paradigm posits that, when an organization's strategy and its structure are congruent, the organization's performance is likely to be higher than if they do not match.

20.4 Conclusions and Future Research

To conclude, the relationship between structure and performance still remains unpredictably asymmetric. Less structure can be associated with wider opportunities in the market-place as organizations with less structure are less rigid. Previous literature focuses on balancing efficiency and flexibility (Tushman and O'Reilly 1996; Brown and Eisenhardt 1997; Uzzi 1997; Rowley et al. 2000), whereas Davis et al. (2009) suggest that this balance is more accurately the flexible capture of widely varying opportunities versus efficient execution of specific opportunities. Less structure opens up the organization to the option of addressing a wider range of opportunities that occur by accident, but it also deters the rapid, mistake-free execution of those opportunities. Conversely, more structure allows the effective implementation of specific opportunities that can be foreseen. However, too much structure is more than just too rigid. It also restricts the range of likely opportunities, implying that structure is most valuable when many similar opportunities are accessible.

Although the literature and the models that were presented have provided many important insights on what is the impact of structure on performance, the field of organizations still lacks an empirically validated theory that, starting from structure at the level of individuals, is able to predict organization-level measures of performance relevant to firm strategy. The reviewed literatures do not provide such a theory, as they are not able to describe structure at the individual level of analysis and do not predict the measures of performance that can be useful to strategy. The work of Csaszar (2008) addresses these reasons empirically to some extent, but the results still remain limited to the findings based on data from mutual funds only.

Future research could address some of the ambiguities in existing literature that were highlighted, such as the need to understand and study the influence of different dimensions of the environment on organizational performance and the lack of knowledge on how decision-making structure affects organizational performance especially, in the context of ambidextrous organizations. Likewise, Raisch and Birkinshaw (2008) note that "far less research has traditionally been devoted to how organizations achieve organizational ambidexterity. Moreover, further research can be carried out to consider the influence of the quality of organizational structure on performance. There is very little literature compelling the idea of quality of structures in organizational setting.

References

Anderson P (1999) Complexity theory and organization science. Organ Sci 10:216–232

Brown SL, Eisenhardt KM (1997) The art of continuous change: linking complexity theory and time-paced evolution in relentlessly shifting organizations. Adm Sci Q 42:1–34

Burns T, Stalker GM (1961) The management of innovation. Tavistock, London

Chandler A (1962) Strategy and structure: chapters in the history of the American enterprise. MIT Press, Boston

Csaszar AF (2008) Organizational structure as a determinant of performance: evidence from mutual funds. The Wharton School, University of Pennsylvania
Cyert RM, March JG (1963) A behavioral theory of the firm. Prentice-Hall, Englewood Cliffs
Davis JP, Eisenhardt KM, Bingham CB (2009) Optimal structure, market dynamism, and the strategy of simple rules. Adm Sci Q 54:413–452
Eisenhardt KM (1989) Making fast strategic decisions in high-velocity environments. Acad Manag J 32:543–576
Galbraith J (1973) Designing complex organizations. Addison-Wesley, Reading
Hargadon A, Sutton RI (1997) Technology brokering and innovation in a product development firm. Adm Sci Q 42:716–749
Hatch MJ (1998) Jazz as a metaphor for organizing in the 21st century. Organ Sci 9:556–557
Kauffman S (1993) The origins of order. Oxford University Press, New York
Lenz RT (1980) Environment, strategy, organization structure, and performance: patterns in one industry. Strateg Manag J 1:209–226
Miller D (1996) Configurations revisited. Strateg Manag J 17:505–512
Miller D, Friesen PH (1984) Organizations: a quantum view. Prentice-Hall, Englewood Cliffs
Mintzberg H, McHugh A (1985) Strategy formation in an adhocracy. Adm Sci Q 30:160–197
Owen-Smith J, Powell WW (2003) Knowledge networks as channels and conduits: the effects of spillovers in the Boston biotechnology community. Organ Sci 15:5–21
Pisano GP (1994) Knowledge, integration, and the locus of learning: an empirical analysis of process development. Strateg Manag J 15:85–100
Pugh D, Hickson D, Hinings C, Macdonald K, Turner C, Lupton T (1963) A conceptual scheme for organizational analysis. Adm Sci Q 8:289–315
Raisch S, Birkinshaw J (2008) Organizational ambidexterity: antecedents, outcomes, and moderators. J Manag 34:375–409
Rindova V, Kotha S (2001) Continuous morphing: competing through dynamic capabilities, form, and function. Acad Manag J 44:1263–1280
Rothaermel FT, Hitt M, Jobe L (2006) Balancing vertical integration and strategic outsourcing: effects on product portfolios, new product success, and firm performance. Strateg Manag J 27:1033–1056
Rowley TJ, Behrens D, Krackhardt D (2000) Redundant governance structures: an analysis of structural and relational embeddedness in the steel and semiconductor industries. Strateg Manag J 21:369–386
Rumelt RP (1974) Strategy, structure, and economic performance. Harvard University Press, Cambridge
Rumelt RP, Schendel DE, Teece DJ (1994) Fundamental issues in strategy: a research agenda. Harvard Business School Press, Boston
Scott WR (2003) Organizations: rational, natural and open systems, 5th edn. Prentice-Hall, Upper Saddle River
Tushman M, O'Reilly CA III (1996) Ambidextrous organizations: managing evolutionary and revolutionary change. Calif Manag Rev 38(summer):8–30
Uzzi B (1997) Social structure and competition in interfirm networks: the paradox of embeddedness. Adm Sci Q 42:36–67
Wasserman N (2008) Revisiting the strategy, structure, and performance paradigm: the case of venture capital. Organ Sci 19(2):241–259
Weber M (1946) From Max Weber: essays in sociology (trans: Gerth HH, Mills CW, eds). Oxford University Press, New York
Weick KE, Roberts KH (1993) Collective minds in organizations: heedful interrelating on flight decks. Adm Sci Q 38:357–381

Chapter 21
Competing Through People: A Model for the Implementation of Employee Participation Systems

José Carlos Prado–Prado, Jesús García-Arca and Arturo Fernández-González

Abstract The main objective of this paper is to present a justified, well-supported model for the implementation of Employee Participation Systems (EPSs) in companies. The model is based on state-of-the-art literature, on the results of a research carried out among 40 Spanish leading companies regarding the implementation of Employee Participation Systems (EPSs) and, finally, on our experience in developing EPSs-related action research projects with companies for more than 20 years.

21.1 Introduction. Continuous Improvement and other Relevant Approaches

Companies under threat from a changing environment with increasing market demands (wider range of products, with good quality, a low profit margin and a high level of service) need to respond to the challenge, searching to increase their survival possibilities.

This search for improvements in the processes developed by companies can be based on the field of investment (in technologies, systems and installations), as well as on small improvements designed to gradually increase their efficiency. This latter improvement option, which requires almost no investment, is termed as "continuous improvement" or "Kaizen", and is the basis for different approaches, methodologies or philosophies, such as "Just-in-time", Lean production or Lean manufacturing, Total Quality (TQ), Total Quality Management (TQM), Total Quality Control (TQC) or Company Wide Quality Control (CWQC), Six Sigma, etc.

J. C. Prado–Prado (✉) · J. García-Arca · A. Fernández-González
Grupo de Ingeniería de Organización (GIO), E.T.S. I.I, 36310 Vigo, Spain
e-mail: jcprado@uvigo.es

S. P. Sethi et al. (eds.), *Industrial Engineering: Innovative Networks*,
DOI: 10.1007/978-1-4471-2321-7_21,

Many authors have dealt with this issue of continuous improvement. Among them we could cite Bessant et al. (1994); Lawler (1998); Berling (2000); Boer et al. (2000); Goh (2000); Lillrank (2003); Lycke (2003); Vora (2004). All have contributed their own definitions of continuous improvement but we would highlight that of Boer et al. (2000), according to whom continuous improvement is "the planned, organized and systematic process of ongoing, incremental and company-wide change of existing practices aimed at improving company performance". And this increase of their performance is the final reason for which companies introduce continuous improvement (Boer et al. 2000; Jorgensen et al. 2006; García and Prado 2008).

Among the most important internal resources that were indicated as a source of competitive advantage in the implementation of continuous improvement is the human factor (Teece and Pisano 1994), mainly due to its intangible characteristics, which are difficult to copy: knowledge, abilities and attitudes (Mueller 1996; Barney and Wright 1998); therefore, it is important to stress that the active participation of the whole staff of a company is a key element that constitutes the basis of continuous improvement, and this total and active participation is difficult to obtain (García and Prado 2002).

Therefore, we think that we can present a new definition of what is continuous improvement, based on the Boer et al. (2000) definition: Continuous improvement is the planned, organized and systematic process of ongoing, incremental and radical, company- and supply chain-wide, change of existing practices aimed at improving company competitiveness and social responsibility. In this regard, Employee Participation Systems (EPSs), which may materialize as improvement teams, quality circles, suggestion systems or workshops, among other forms, stand out for their effectiveness as a way to implement this continuous improvement.

21.2 Key Factors in a EPSs Implementation. Literature Review

From a review of all these contributions, it is clear that the authors have identified critical success factors (or variables for structuring an implementation methodology) in terms of a satisfactory implementation of EPSs based on a continuous improvement programme (see Table 21.1), drafted from bibliographical syntheses and reviews by García and Prado (2007, 2008); markedly, a greater or lesser satisfaction with the implementation of these participation systems should be related to the greater or lesser capacity of contributing to improve enterprise results (either directly or indirectly).

The lack of detailed treatment in the literature on the effect of structuring for the successful implementation of participation within a context of continuous improvement, as well as its potential impact on the economic and social results of companies (improvement of enterprise competitiveness), makes it a new issue for

Table 21.1 Key Factors in an EPS Implementation (García and Prado 2007, 2008)

N.	KEY FACTOR
1	Emphasis on the design and analysis of improvement prior to its implementation
2	Select the most appropriate time for implementing/promoting the EPS
3	Commitment of top management
4	Commitment of the middle management
5	Commitment/Support of trade unions/workers committee
6	Involvement of participants in the EPS
7	Participation of the proponent in the design, analysis and implementation of the improvements
8	Choice of an appropriate organizational structure to implement the EPS
9	Training participants in the EPS
10	In-house understanding of the "cultural change" associated with implementing continuous improvement
11	Adopting an adequate work dynamics based on indicators and objectives
12	Set priorities in implementing the improvements proposed
13	EPS based on multi-hierarchical and multi-departmental work teams
14	Adopt an appropriate systematic to manage and follow up the improvements proposed
15	Provide resources to swiftly respond to problems, queries or proposals
16	External assessment for implementing/maintaining/improving an EPS
17	Continuous communication of improvements and activities developed in the EPS to all organization levels
18	Systematize the improvements obtained by developing and implementing procedures/work instructions
19	Obtain results from the start in order to strengthen trust in the EPS
20	Adopt an acknowledgement and/or retribution system for those participating in the EPS

research, and all the more so when there is evidence that not all companies have had the expected success (or have simply failed) to implement it (Marín-García et al. 2008).

21.3 Key Factors from the Point of View of Spanish Companies

The authors have developed a field study in 40 Spanish companies, belonging to both the industrial and service sector, characterized by their commitment to implementing continuous improvement by means of employee participation systems (EPS). In this sample, 13 companies in the Committee for the Participation and Improvement of the Spanish Quality Association (AEC) are included, and a further 27 are companies from different sectors, especially acknowledged for their participation systems by the "Club of Management Excellence" and by AENOR (Spanish Association of Standardization and Certification).

The methodology followed in this study was based on carrying out personal in-depth interviews with those in charge of the company participation systems (mostly managers, heads of quality department or heads of human resources), based on a questionnaire with 128 (open- and closed-ended) questions, dealing with aspects of strategy, organization and methodologies to do with the implementation of the EPSs in companies.

In order to be able to evaluate and put in order of importance each of the 20 critical factors for successful implementation of EPS listed in Table 21.1, the companies studied were asked to value each of these factors in Table 21.1 by using the Lykert scale (1–5). It is concluded from an analysis of the opinions expressed by the companies that there are significant differences in the valuation of each factor. Significantly, the three factors with an average mark over 4.5 are those that the companies considered to be the most important: "Commitment of top management" (factor 3), "Commitment and motivation of the participants" (factor 6) and "Commitment of the middle management" (factor 4). Therefore, the companies ascribe more critical importance to successfully implementing their EPSs towards aspects of motivation and commitment throughout the organization (at all hierarchical levels); among these factors, the literature particularly mentions the importance of the involvement of management.

At a second level of importance, with no significant differences (statistically) is a series of 14 factors (1, 2, 7, 8, 10, 11, 12, 13, 14, 15, 17, 18, 19, 20) shown in Table 21.1. Some of these are given special mention in the literature, such as personnel participation in the design and implementation of improvements (factor 7), cultural change linked to continuous improvement (factor 8), team work (factor 13), the availability of resources (factor 15), communication (factor 17), standardization and documentation of processes (factor 18) or incentive systems (factor 20).

At another lesser level of importance (statistically) is the training of participants in EPS (factor 9); the relative little importance ascribed to this aspect by the companies is interesting, especially if we consider the abundance of references to this issue found in the literature. Finally, union backing (factor 5) and external assessment (factor 16) rank in the last group of factors, these being the least critical or important.

21.4 A Model for Implementation of EPSs

In this paper we can ultimately present a justified, well-reasoned model for the implementation of Employee Participation Systems (EPSs) in companies. The model is based on the previous state-of-the-art literature, on the results of a research carried out in Spanish companies and finally, on more than 20 years' experience working with companies on the development of action-oriented research projects in this field.

Our research group (GIO) has a long-standing experience in conducting collaboration projects with companies, particularly in the field of continuous improvement, applying the "action-research" approach.

As is known, the origins of action research lie in the works by socio-technical authors at the Tavistock Institute (among others, Emery and Thosrsud 1969; Trist et al. 1963) and is based on experiences in the English coal mines in the fifties. As Coughlan and Coghlan (2002) so rightly note in their revision of action research, "Action research is an approach to research that aims both at taking action and creating knowledge or theory about that action" (p.220).

Before presenting the model, a series of considerations on people in companies are put forward. We think, as many other authors, that people are the most important asset in a company. Nevertheless, we often hear, 'It's not the people...,' but it is: competent, capable, excellent people.

To help us structure a model, we have identified three kinds of people during our research within the companies we have collaborated with. We know the below may at times sound like a joke, but here is our invention: a virtual X-ray of personnel.

We know that it is difficult to classify or to categorize people, because everybody is different and not one company is equal to another. We called the first group of people the "fallen-in-love", or devoted workers. These are a small percentage of the total, between 5% and 20%. They exist in all companies and are people who always say yes to any positive proposal made by the company (they behave similarly in their own personal lives). If the company says, for example, 'Does anybody want to start an EPS with an improvement team?' they always say, 'Yes, of course, well done, fantastic idea, when do we start?' The same happens with a Lean or JIT initiative or with the implementation of ISO 14000.

The second group are called the apathetic (the 'I dońt care'/'I dońt mind'/ indifferent/dońt know/no answer). They make the largest percentage of the company, 60% to 80%, and they never know what to do –and usually do nothing.

The third group is "the irritable boys", or the disgruntled, or disaffected; speaking in Star Wars® jargon, *they are the forces of corruption*. They are usually the same percentage as the devoted. They always say no to everything. They are actually very dangerous and can pollute the whole organization.

Our model for continuous improvement needs the top management involvement because without it, the continuous improvement is impossible. Then, really, this is a previous condition for the implementation and not, strictly speaking, a factor of success.

In fact, even when making a selection of Spanish companies that traditionally develop their own EPSs, a small percentage of them has been identified (10%; 4) as having a decreasing level of EPSs that jeopardizes their very future survival, basically due to a lack of management's involvement.

On the other hand, we always start with devoted workers, due to the fact that, as we said before, it is very important to start with personnel committed to the project in order to ensure its success. Based on these two aspects, we divide the model for the implementation of EPSs shown in Fig. 21.1 into three kinds of factors: core, key and important. This is due to the fact that we consider that there are important differences among the factors, and some are more critical to the success of the process of implementing the continuous improvement (and EPSs) in the company.

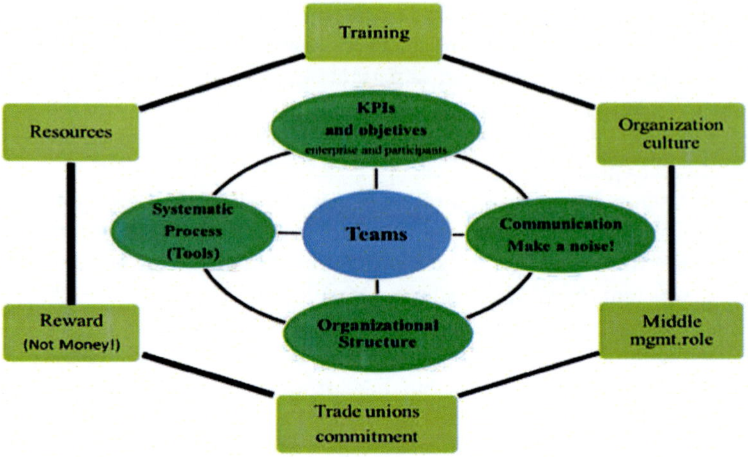

Fig. 21.1 A model for the implementation of EPSs

Our model sets out to prioritize the importance of the different factors present in the state-of-the-art literature and in the results of the field study in Spanish companies. The core of the process is the need to promote systems based on teams, better than individual systems. Suggestion systems are a timid way to implement continuous improvement. There is no doubt that they are the most extended systems of participation; they are easy to implement (Fairbank et al. 2003; Fairbank and Williams 2001). If well managed, and when proper incentives are provided, they can generate ideas which may be very profitable for the company as a whole (Marín-García et al. 2008; Rapp and Eklund 2007).

Nevertheless, the results are, in general, poor: the process to deal with a suggestion is too long, etc. As Marín-García and Bautista-Poveda (2010) put it, they have a certain tendency, in time, to lose momentum. They are advisable as an initial stage to develop a continuous improvement culture, before progressing to the implementation of team work improvement schemes. In fact, Spanish companies mostly choose the group systems option (87.5%).

Organizational structure is another key factor too (for defining strategy and objectives, set priorities, etc.). In our approach (Prado-Prado 2001), the organizational structure of the improvement process is based on two teams: the implementation team and the improvement team.

The implementation team is the spindle on which the process turns around. They are responsible for leading and taking good care of the process (their tasks including planning the process and monitoring the achievement of the goals being sought). The implementation team consists of a small number of people (2 or 3), clearly "devoted", and is kept all through the process (this guarantees future continuity of achievements and the continuous search for improvements). Since the key in any process is to engage top management, their participation in this team is highly recommended.

The improvement team, headed by the implementation team, is responsible for proposing and analysing problems and implementing improvements (but setting priorities) that contribute to achieving the desired goal. It can be seen that the participation of top management through the implementation team facilitates putting into practice the proposals made by the improvement teams.

Another key factor is that launching EPSs is a systematic process that covers aspects ranging from the selection of members, to structuring the meetings with a pre-established regularity, with a set day and time, as well as a pre-established duration, with a clear role for the members of the improvement team use of simple tools and obtaining results from the start in order to strengthen trust in the EPS, etc. (Prado–Prado 2001).

With regard to training, we think that a short session (about 4 h) is enough, because people learn by doing, and you can use the meetings of the improvement team (and other mechanisms) to correct inappropriate behaviour, clear up any queries, etc. As Edmondson (1999) says, effective team learning processes include experimentation and collaborative problem solving. When teams make changes to improve future performance, such as those of improvement teams, they learn. For this reason, training is only an important factor in our model.

The last key factor is communication, continuous communication ("make noise"!) with everybody in the organization (not only the participants in the EPSs) about the achievements, the results, the activities developed, etc. Communication is the key to be able to extend the continuous improvement process from the devoted to the apathetic workers, and even to the disgruntled.

This communication is not only made by word of mouth but also by making known how the project is progressing on a notice board or a project improvement board. Even better, present the results obtained by the improvement team in a meeting with all concerned personnel, including those of the section or area in which the improvement was made. And set up a board with the results so that the rest of the personnel are informed. Obviously, the good results obtained with the project are, by themselves, one of the most significant communication tools.

And finally, there are less important factors than the previously mentioned: the important factors. We identified six of them: training, reward, middle management role, trade union commitment, resources in the project and organizational culture.

21.5 Conclusions

This paper presents a justified model for implementing EPSs, which is based on a review of the literature -including all the key factors-, in the situation of Spanish companies with a considerable experience in EPSs and in our own experience in action-research projects with companies.

However, the implementation of the model is not easy. As highlighted in the study on Spanish companies with a considerable experience in EPSs, sometimes the key factors are not given the importance that they deserve. So, despite the fact that 85% of the companies have some kind of global indicator associated with EPSs, only 33% of them have a detailed system, and only 55% define improvement objectives associated with the working order of the EPSs. Furthermore, 25% state that they make no kind of communication about participation activities or their results during the life of the EPSs.

References

Barney JB, Wright PM (1998) On becoming a strategic partner. The role of human resources in gaining competitive advantage. Human Resource Manage 37(1):31–46

Berling C (2000) Continuous improvement as seen from groups and improvement agents. Total Quality Manage Bus Excellence 11 (4–5):484–489

Bessant J, Caffyn S, Gilbert J (1994) Mobilising continuous improvement for strategic advantage. In: Platts KW, Gregorj MJ, Neely A (eds) Operations Strategy and Performance, Manufacturing Engineering Group. University of Cambridge, Cambridge, pp 175–180

Boer H, Berger A, Chapman R, Gertsen F (eds) (2000) CI changes, From suggestion box to organisational learning, Continuous improvement in Europe and Australia. Aldershot

Coughlan P, Coghlan D (2002) Action research for operations management. Int J Operations and Production Manage 22(2):220–240

Edmondson AC (1999) Psychological safety and learning behaviour in work teams. Admin Sci Quart 44(2):947–958

Emery F, Thorsrud E (1969) Form and Content in Industrial Democracy, Ed. Tavistock

Fairbank JF, Williams SD (2001) Motivating creativity and enhancing innovation through employee suggestion system technology, creativity and innovation management, 10(2), p 68

Fairbank JF et al (2003) Motivating creativity through a computer-mediated employee suggestion management system. Behaviour and Information Technology 22(5), p 305

García A, Prado JC (2002) La participación del personal en la mejora continua de las empresas. Edited by AEC, Madrid

García J, Prado JC (2007) The implementation of new technologies through a participative approach. Creativity and Innovation Manage J 16(4):386–396

García J, Prado JC (2008) Logistics redesign through systematic personnel participation. A case study, Proceedings Book 2008 CINET Conference (Valencia)

Goh M (2000) Quality Circles: Journey of an Asian Public Enterprise. Int J Quality and Reliability Manag 17(7):784–799

Jorgensen F, Laugen B, Boer H (2006) Human resource management for continuous improvement, CINet 2006, pp 443–454

Lawler III EE, Mohrman SA, Ledford JRGE (1998) Creating high performance organizations, practices and results of employee involvement and total quality management in fortune 1000 companies. Jossey-Bass, San Francisco

Lillrank P, Shani AB, Lindberg P (2003) Continuous improvement: exploring alternative organizational designs. Total Quality Manage Bus Excellence 12(19):41–55

Lycke L (2003) Team development when implementing TPM. Total Quality Manage Bus Excellence 14(2):205–213

Marchington M (1992) Managing the team: a guide to successful employee involvement. Blackwell, Oxford

Marín-García JA, Pardo del Val M, Bonavía-Martín T (2008) Longitudinal study of the results of continuous improvement in a company. Team Performance Manage 14:55–69

Marin-Garcia JA, Bautista-Poveda Y (2010) The implementation of a continuous improvement project at a Spanish marketing company: a case study. Int J Manage 27(3):593–606

Mueller F (1996) Human resources as strategic assets: an evolutionary resource-based theory. J Manag Stud 33(6):757–785

Prado–Prado JC (2001) Beyond Quality Circles and Improvement Teams. Total Quality Manage Bus Excellence 12(6):789–798

Rapp C, Eklund J (2007) Sustainable development of a suggestion system: factors influencing improvement activities in a confectionary company. Human Factors in Ergonomics & Manufacturing 17(1):79–94

Teece D, Pisano G (1994) The dynamic capabilities of firms: an introduction. Industrial and Corporate Change 3(3):537–555

Trist E, Higgin GW, Murray H, Pollock AB (1963) Organization Choice. Tavistock Institute Publications, London

Vora KM (2004) Creating employee value in a global economy through participation, Motivation and development. Total Quality Manage 15(5–6):793–806

Chapter 22
On Dynamic Games in the European Electricity Market

Kateřina Staňková

Abstract In this paper we define the European electricity market liberalization problem as a game with electricity producers as players, while the consumers' electricity demand is exogenous. The model is based on real data. The producers maximize their profit by investing and choosing how much electricity they will produce by available means of electricity production. The aim of the research presented in this paper is to analyze different scenarios: a market with one electricity producer being a Stackelberg leader, a market with two electricity producers being Stackelberg leaders noncooperative among themselves, and a perfectly competitive market. As expected, in our case studies the perfectly competitive market yields the lowest electricity prices for the consumers.

22.1 Introduction

Since 1990s, the European electricity market has been undergoing a process of liberalization. The speed and state of liberalization vary among different European countries, from a near monopoly in central, eastern, and southern Europe (Koroneos and Nanaki 2007; Ganev 2009; Pollit 2009), to highly competitive markets in England, Wales (Green 2005; Newbery 2006), and Nordic countries (Amundsen and Gergman 1998; Lise et al. 2004; Bye and Hope 2005; Lévêque 2007; Van Eck 2007; Joskow 2008; Brakman et al. 2009).

K. Staňková (✉)
Delft Center for Systems & Control, Delft University of Technology,
2628CD Delft, The Netherlands
e-mail: k.stankova@tudelft.nl

S. P. Sethi et al. (eds.), *Industrial Engineering: Innovative Networks*,
DOI: 10.1007/978-1-4471-2321-7_22,

Extensive studies of electricity market liberalization have been carried out. In (Neuhoff 2005) the Belgian, Dutch, French, and German electricity markets are considered and the effects of market power among three static models are compared. One of these models is additionally studied in (Hobbs et al. 2004a; Hobbs et al. 2004b; Hobbs et al. 2005). In (Kromann 2001; Ekeberg et al. 2003) consequences of market power in the Nordic electricity market are discussed. In Lise et al. (2006) the market with eight European countries is considered, with decisions variables of electricity producers being market power mark-ups. In Staňková (2009, 2010) we have studied a game of electricity market with 1, 2, and 8 European countries and decision variables of individual producers being quantities of electricity produced. This paper proposes a dynamic extension this model.

The rest of the paper is organized as follows. In Sect. 22.2 we introduce a basic game-theoretic model of the European electricity market, based on Staňková (2010), and its dynamic extension. In Sect. 22.3 case studies are presented. Conclusions and future research are discussed in Sect. 22.4.

22.2 The Game-Theoretic Model

22.2.1 Basic Model

Let F and R be a set of firms and a set of regions, respectively. Let $F_r \in F$ be a set of firms located in region r. Let I be a set of technologies for electricity generation. Let $I_r \subset I$ be a set of technologies available in $r \in R$ and let $I_f \in I_r$ be a set of technologies available to firm f located in r. Let L be a set of possible load modes and let U_r be a set containing r and its neighbors. Let K be a set of emissions produced by all electricity production technologies considered. Firm $f \in F_r$ maximizes

$$J_f \triangleq \sum_{l\in L} h_l \sum_{r'\in R} p_{r',l} s_{f,r',l} - \sum_{l\in L} h_l \sum_{r'\in R} \sum_{i\in I_f} c_{i,r} q_{i,f,r',l}, \qquad (22.1)$$

where $s_{f,r,l}$ [GW] denotes the supply of electricity of load mode l from firm f into region r and $p_{r,l}$ [€/GW] denotes the unit electricity market price for region r and load mode l. Moreover, h_l $[h]$ denotes the number of hours belonging to load l per year, $c_{r,l}$ [€/GW] is the variable production costs of technology i in region r, while $q_{i,f,r',l}$ [GW] is the production of f with technology i for region r and load mode l. The electricity supply of load mode l per firm f to region r is $s_{f,r',l} \triangleq (1-\lambda_{r'}) \sum_{i\in I} q_{i,f,r',l}$, where $\lambda_{r'} \in [0,1]$ is the loss of electricity due to its transport. The total electricity supply $S_{r',l}$ [GW] for load mode l per region r is $S_{r',l} = \sum_{f\in F_{r'}} s_{f,r',l}$. The electricity production is limited to the electricity capacity owned by the firm. The demand function is a

Constant Elasticity of Distribution (Arrow et al. 1961; Armington 1969; Lise et al. 2006) and depends on the elasticity $\varepsilon_{r,l} > 0$, reference demand $d^0_{r,l}$[GWh], and reference price $p^0_{r,l}$ [€/GWh]: $d_{r,l}\left(p^0_{r,l}/p^0_{r,l} - p^0_{r,l}\right)^{-\varepsilon_{r,l}} = \sum_{f \in F} S_{f,r,l} = S_{r,l}$. The firms have opportunity to trade electricity with neighboring countries, the amount of electricity traded is $x_{r,r',l} = \sum_{f \in F_r} \sum_{i \in I} q_{i,f,r',l} - \sum_{f' \in F_{r'}} \sum_{i \in I} q_{i,f',r,l}$. This amount is complementary to the shadow price $\tau_{r,r',l}$, which applies if the trade reaches the trade capacity: $\tau_{r,r',l}\left(x_{r,r',l} - x^{\max}_{r,r'}\right) = 0,\ \tau_{r,r',l} \geq 0,\ x^{\max}_{r,r'} \geq x_{r,r',l}$, with the maximum amount of electricity $x^{\max}_{r,r'}$.We assume that the maximum production capacity is complementary to the shadow price $\mu_{i,f,l}$ [€/GWh], which applies if the production with technology i by firm f during load mode l reaches the production capacity:$\mu_{i,f,l}$ $\left(\sum_{r' \in R} q_{i,f,r',l} - q^{\max}_{i,f}\right) = 0,\ \mu_{i,f,l} \geq 0,\ q^{max}_{i,f} \geq \sum_{r' \in R} q_{i,f,r',l}$. Emissions can be also limited, an emission price K^k[€/GWh] applies if the current amount of emissions reaches permissible emission ceiling $E^k[g]$, $\kappa^k\left(\sum_{l \in L} h_l \sum_{r' \in R} \sum_{i \in I} \sum_{f \in F} \sigma^k_{i,r}\right.$ $\left. q_{i,f,r',l} - E^k\right) = 0,\quad \kappa^k \geq 0,\quad E^k \geq \sum_{l \in L} h_l \sum_{r' \in R} \sum_{i \in I} \sum_{f \in F} \sigma^k_{i,r} q_{i,f,r',l}$. Emission factors $\sigma^k_{i,r}$ [€/GWh] are associated with the region, in which f produces electricity.

Including all above mentioned constraints into the maximization problem (22.1) of firm f leads to

$$
\begin{aligned}
L_f \triangleq & \sum_{l \in L} h_l \sum_{r' \in R} \sum_{i \in I} \left(p_{r',l}(1 - \lambda_{r'}) q_{i,f,r',l} - c_{i,r} q_{i,f,r',l}\right) \\
& - \sum_{l \in L} h_l \sum_{i \in I_f} \sum_{i \in I} \left(\mu_{i,f,l}\left(\sum_{r' \in R} q_{i,f,r',l} - q^{max}_{i,f}\right)\right) \\
& - \sum_{l \in L} h_l \sum_{r' \neq r} \tau_{r,r',l} \left(\sum_{f' \in F_r} \sum_{i \in I_f} q_{i,f',r',l} - \sum_{f' \in F'_r} \sum_{i \in I'_{f'}} q_{i,f',r,l} - x^{max}_{r,r'}\right) \\
& - \sum_{k \in K} \kappa^k \left(\sum_{l \in L} h_l \sum_{r' \in R} \sum_{i \in I} \sum_{f'' \in F} \sigma^k_{i,r} q_{i,f'',r',l} - E^k\right).
\end{aligned} \tag{22.2}
$$

Three possible games among the electricity producers are considered (Staňková 2010): a perfect competition (P), a Stackelberg game with one leader per country (S), in which the other producers is perfectly competitive, and a Stackelberg game with two leaders per country, competitive among themselves (NS).

Perfect competition: The firms act as players on the same level. The problem (P) of an firm $f \in F_r$ for each $i \in I_r, l \in L, r' \in U_r$:

$$\min_{q_{i,f,r',l}} p_{r',l}(\cdot)$$
$$s.t.\ q_{i,f,r',l} \geq 0,\ q_{i,f,r',l} \leq q^{max}_{i,f,r',l},$$
$$q_{i,f,r',l}\left(c^{m}_{i,f,r',l} - (1-\lambda_{r'})p_{r',l}\left(1 - \frac{\pi_{f,r',l}}{\varepsilon_{r',l}}\right)\right) \geq 0. \tag{22.3}$$

Stackelberg game with one leader per region: There is a firm f_S in each region $r \in R$ acting as the leader, choosing $\left(q^{S}_{i,f,r',l}\right)_{i \in I_r, r' \in U_r, l \in L}$ so as to maximize L_f, whereas others are perfectly competitive followers. The problem (S) of the leader for each $i \in I_r, l \in L, r' \in U_r$:

$$\max_{q_{if_S,r',l}} L_{f_S}(\cdot)$$
$$s.t.\ q_{i,f,r',l} \geq 0,\ q_{i,f,r',l} \leq q^{max}_{i,f,r',l}$$
$$q_{i,f,r',l}\left(c^{m}_{i,f,r',l} - (1-\lambda_{r'})p_{r',l}\left(1 - \frac{\pi_{f,r',l}}{\varepsilon_{r',l}}\right)\right) \geq 0. \tag{22.4}$$

Stackelberg game with two leaders per region: There are two leading firms $f_S^{(1)}, f_S^{(2)} \in F_r$, acting first, being noncooperative among each other and choosing $\left(q^{NS}_{i,f_S^{(1)},r',l}\right)_{i \in I_r, r' \in U_r, l \in L}, \left(q^{NS}_{i,f_S^{(2)},r'',l}\right)_{i \in I_r, r'' \in U_r, l \in L}$ so as to maximize their profits $L_{f_S^{(1)}}$ and $L_{f_S^{(2)}}$. Other (perfectly competitive) firms choose their production amounts per load and technology after the leaders have made their choice. The problem (NS) of $f_S^{(1)}, f_S^{(2)}$ for each $i \in I_r, l \in L, r' \in U_r$:

$$\max_{q_{if_S,r',l}} L_{f_S}(\cdot)$$
$$s.t.\ q_{i,f,r',l} \geq 0,\ q_{i,f,r',l} \leq q^{max}_{i,f,r',l}$$
$$q_{i,f,r',l}\left(c^{m}_{i,f,r',l} - (1-\lambda_{r'})p_{r',l}\left(1 - \frac{\pi_{f,r',l}}{\varepsilon_{r',l}}\right)\right) \geq 0. \tag{22.5}$$

22.2.2 Dynamic Extension of the Model

If the time horizon of the model is extended one time period ahead, firms aim to maximize their discounted payoff by choosing the electricity quantity to produce with various technologies for each time period. Unlike in the static setting, firms have to accommodate the fixed costs by investments. The following equation expresses the payoffs in the next time period (with $\tilde{q}_{i,f}^{\max}$ defined below):

$$\tilde{J}_f = \frac{1}{1+\beta}\Big(\sum_{l\in L} h_l \sum_{r'\in R} \tilde{p}_{r',l}\tilde{s}_{f,r',l} - \sum_{l\in L} h_l \sum_{r'\in R}\sum_{i\in I} c_{i,r}\tilde{q}_{i,f,r',l} + \sum_{i\in I} V_i \tilde{q}_{i,f}^{max}\Big) - \sum_{i\in I} V_i q_{i,f}^{new}. \tag{22.6}$$

The variables in this equation are defined as in Sect. 22.2, with "$\sim$" identifying the variable for the next time period. The prices, supply, generation, and production capacity are assigned to the next period and discounted with interest rate β. Firms take into account the value of installed capacity in the next period, while new investments are needed to keep the production capacity at a desired level. Parameter V_i [€/kW] represents the value of technology i (equivalent to value of time in traffic applications (Staňková et al. 2009) while $q_{i,f}^{new}$ [GW] denotes the amount of the new production capacity of electricity (chosen by the producer). This means that the maximum production capacity is no longer fixed. The firms make their investment decisions in every time period based on the most recent information (cf. the feedback information structure, see (Basar and Olsder 1999). The capacity evolution depends on a technology-specific depreciation rate $\delta_i : \overline{q}_{i,f}^{max} = (1-\delta_i)q_{i,f}^{max} + q_{i,f}^{new}$. Because of environmental considerations and physical limitations on the capacity of the technologies used, we assume $\tilde{\varphi}_i\left(\sum_{f\in F} q_{i,f}^{new} + (1-\delta_i)\sum_{f\in F} q_{i,f}^{\max} - q_i^{\max abs}\right) = 0$, $\tilde{\varphi}_i \geq 0, \sum_{f\in F} q_{i,f}^{new}(1-\delta_i)\sum_{f\in F} q_{i,f}^{max} - q_i^{\max abs} \leq 0$. The shadow price $\tilde{\varphi}_i$ becomes nonzero once the planned expansion of capacity of a certain technology reaches the maximum allowable installed capacity $q_i^{\max abs}$ [GW]. Each firm maximizes the net profit L_f^{DYN}, by a joint choice of the investment decision and production of technologies for possible regions and both load types in the next period. Therefore, the net profit can be defined as:

$$\begin{aligned} L_f^{DYN} = \frac{1}{1+\beta}\Big(&\sum_{l\in L} h_l \sum_{r'\in R} \tilde{p}_{r',l}\tilde{s}_{f,r',l} - \sum_{l\in L} h_l \sum_{r'\in R}\sum_{i\in I} c_{i,r}\tilde{q}_{i,f,r',l} + \sum_{i\in I} V_i \tilde{q}_{i,f}^{max}\Big) \\ &- \sum_{l\in L} V_i q_{i,f}^{new} - \frac{1}{1+\beta}\sum_{l\in L} h_l \sum_{i\in I}\tilde{\mu}_{i,f,l}\Big(\sum_{r'\in R}\tilde{q}_{i,f,r',l} - \tilde{q}_{i,f}^{max}\Big) \\ &- \frac{1}{1+\beta}\Big(\sum_{l\in L} h_l \sum_{r'\in R}\sum_{r\in R}\tau_{r',r,l} \\ &\times \Big(\sum_{f'\in F_r}\sum_{i\in I}\tilde{q}_{i,f',r',l} - \sum_{g\in F_r}\sum_{i\in I}\tilde{q}_{i,f',r',l} - \tilde{x}_{r',r}^{max}\Big)\Big) \\ &- \frac{1}{1+\beta}\Big(\sum_{k\in K}\tilde{\kappa}_k \sum_{l\in L} h_l \sum_{r'\in R}\sum_{i\in I}\sum_{f''\in F}\sigma_{i,k}\tilde{q}_{i,f',r',l} - \tilde{E}_k\Big) \\ &- \frac{1}{1+\beta}\sum_{i\in I}\tilde{\varphi}_i\Big(\sum_{f\in F} q_{i,f}^{new} + \Big(1-\delta_i\Big)\sum_{f\in F} q_{i,f}^{max} - q_i^{maxabs}\Big). \end{aligned} \tag{22.7}$$

Derivative of L_f^{DYN} with respect to production yields the first-order optimality condition for the next time period, equivalent to those for the static case, derivative of L_f^{DYN} with respect to the investment in new capacity leads to the second first-order optimality condition. The shadow price of the capacity usage $\tilde{\mu}_{i,f,l}$ determines to what extent a firm would like to use a particular technology during a particular load period. The amount of production capacity is no longer constant and the following conditions written in terms of the decision variables holds:

$$\begin{aligned} &\tilde{\mu}_{i,f,l}\left(\sum\nolimits_{r' \in R} \tilde{q}_{i,f,r',l} - (1-\delta_i)q_{i,f}^{\max} - q_{i,f}^{new}\right) = 0, \\ &\tilde{\mu}_{i,f,l} \geq 0, \sum\nolimits_{r' \in R} \tilde{q}_{i,f,r',l} - (1-\delta_i)q_{i,f}^{\max} - q_{i,f}^{new} \geq 0. \end{aligned} \qquad (22.8)$$

22.3 Case Studies

To solve the problems (P), (S), (NS) in both static and dynamic setting we have used nonlinear programming (see Staňková et al. (2010) for more details). In the static game, we observed that the electricity prices became lower when cross-border electricity transmissions were allowed. As expected, the electricity prices were higher in Stackelberg games than in the situation with perfectly competitive market. In the perfect competition the producers tended to use cheaper and non-environmentally friendly means of electricity production. Emission restrictions implied higher electricity prices, especially in the countries with a low number of hydro and wind power plants. The resulting electricity prices and profits were remarkably lower than those reported for the current European electricity market.

The dynamic problem was solved the problem numerically, using dynamic programming (see Staňková (2010) for more details). The emission constraints were set to the same values as in Staňková et al. (2010). Only base load was considered. Figures 22.1, 22.2 and 22.3 show the prices computed for the dynamic (two-step) variants of the (P), (S), and (NS) problems, respectively, with $\beta = 0.045$. The left columns represent the electricity prices in the first year; the right columns represent the electricity prices in the second year. Clearly, the perfect competition leads to the lowest electricity prices, the highest electricity prices are attained when the Stackelberg game with one leader is played. As in the case of the static model we observe that emission restrictions lead to increase of electricity prices and that the cross-border transmissions of electricity lead to their decrease.

When applied iteratively, the dynamic model can be used for prediction of the market behavior. However, it is necessary to include the life expectancy of individual electricity production means if a very extended time horizon is considered.

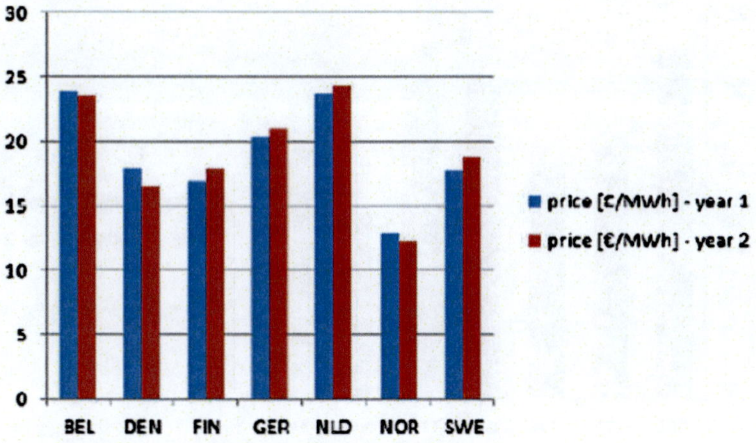

Fig. 22.1 Prices for two time steps of the dynamic variant of the (NS) problem

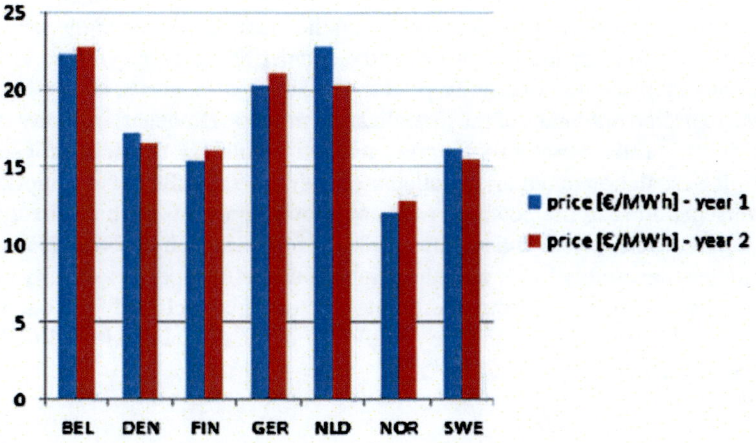

Fig. 22.2 Prices for two time steps of the dynamic variant of the (S) problem

22.4 Conclusions and Future Research

We have proposed a model of the liberalized European electricity market, consisting of 8 European countries. In this model emission restrictions can be set as well as maximal transmission capacities between the neighboring countries. The goal was to see how different the electricity prices will be in a situation with one leading producer per country, a situation with two leading producers per country, and a perfectly competitive situation. We have proposed the dynamic extension of the model in which investments of individual electricity producers are taken into account, which can be easily implemented for many years ahead.

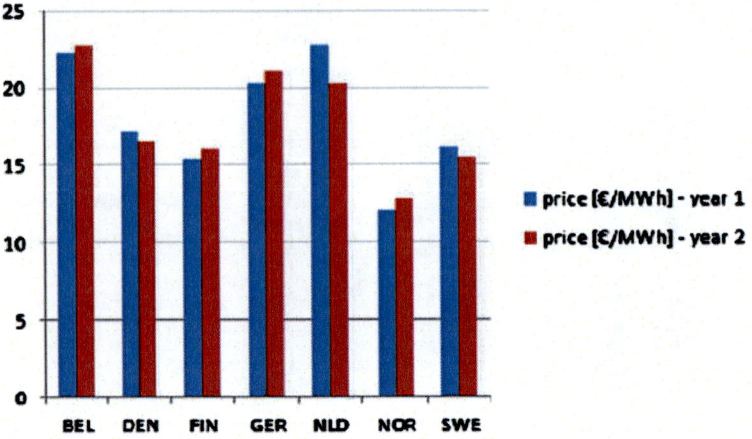

Fig. 22.3 Prices for two time steps of the dynamic variant of the (NS) problem

In the situations with Stackelberg leader(s) the model produced higher electricity prices than in the situation with perfectly competitive market(s). The emission restrictions increase the electricity prices. Outcomes of our case studies coincide with the expected outcomes of the liberalization process. However, in reality the risk exists that the fully competitive market without regulatory restrictions leads to a rapid price escalation and market collapse (Puller 2007), while our assumption that the overscheduling of the power lines is not allowed prevents such a situation. The resulting electricity prices and resulting profits of the electricity producers are in our case studies remarkably lower than those in the current European electricity market. We assume that there are some additional factors, not included into our model, influencing the price. Increasing the complexity of the model and its validation are subjects of the future research.

References

Amundsen ES, Bergman L (1998) The performance of the deregulated electricity markets in Norway and Sweden: a tentative assessment. Technical Report 0798, Department of Economics, University of Bergen, Norway

Armington PS (1969) A theory of demand for products distinguished by place of production. IMF Staff Papers 16:159–178

Arrow KJ, Chenery HB, Minhas BS, Solow RM (1961) Capital-labor substitution and economic efficiency. Rev Econ Stat 43(3):225–250

Basar T, Olsder GJ (1999) Dynamic noncooperative game theory. SIAM, Philadelphia

Brakman S, van Marrewijk C, van Witteloostuijn A (2009) Market liberalization in the European natural gas market—the importance of capacity constraints and efficiency differences. Working Papers 09–15, Utrecht School of Economics

Bye T, Hope E (2005) Deregulation of electricity markets—The Norwegian experience. Discussion Papers 433, Research Department of Statistics Norway

Ekeberg L, Sundahl L, Römpötti M, Halsos OS, Bryng PK (2003) A powerful competition policy: towards a more coherent competition policy in the Nordic market for electricity power. Technical Report 1/2003

Ganev P (2009) Bulgarian electricity market restructuring. Utilities Policy 17(1):65–75

Green R (2005) Electricity and markets. Oxford Rev Econ Policy 21(1):67–87

Hobbs BF, Rijkers FAM, Wals AF (2004a) Strategic generation with conjectured price responses in a mixed transmission pricing system–Part I: formulation. IEEE Trans Power Syst 19:707–717

Hobbs BF, Rijkers FAM, Wals AF (2004b) Strategic generation with conjectured price responses in a mixed transmission pricing system–Part II: application. IEEE Trans Power Syst 19: 872–879

Hobbs BF, Rijkers FAM, Boots MG (2005) The more cooperation, the more competition? a cournot analysis of the benefits of electric market coupling. Eng J 26(4):69–98

Joskow P (2008) Lessons learned from electricity market liberalization. Eng J 29(2):9–42

Koroneos CJ, Nanaki EA (2007) Electric energy sustainability in the Eastern Balkans. Eng Policy 35(7):3826–3842

Kromann MT (2001) Imperfect competition in the Nordic electricity markets. Working Paper 2001:03, Danish Economic Council, Copenhagen, Denmark

Lévêque F (2007) Competitive electricity markets and sustainability. Edward Elgar pub, Cheltenham

Lise W, Linderhof V (2004) Electricity market liberalisation in Europe—who's got the power? internally reviewed report R-04/03, Institute for Enviromental Studies, Free University, Amsterdam, The Netherlands

Lise W, Linderhof V, Kuik O, Kemfert C, Ostling R, Heinzow T (2006) The European electricity market–what are the effects of market power on prices and the environment? Eng Policy 34(5):2123–2136

Neuhoff K, Barquin J, Boots MG, Ehrenmann A, Hobbs BF, Rijkers FAM, Vázquez M (2005) Network-constrained cournot models of liberalized electricity markets: the devil is in the details. Eng Econom 27:495–525

Newbery D (2006) Electricity deregulation: choices and challenges. Econ J 116(512):333–335

Pollitt M (2009) Evaluating the evidence on electricity reform: lessons for the South East Europe (SEE) market. Utilities Policy 17(1):13–23

Puller SL (2007) Pricing and firm conduct in California's deregulated electricity market. Rev Econom Statist 89(1):75–87

Staňková K (2009) On Stackelberg and inverse stackelberg games & their applications in the optimal toll design problem, the energy market liberalization problem, and in the theory of incentives. Ph.D. thesis, Delft University of Technology, Delft, The Netherlands

Staňková K, Olsder GJ, Bliemer MCJ (2009) Comparison of different toll policies in the dynamic second-best optimal toll design problem: case study on a three-link network. Eur J Transp Infrastruct Res 9(4):331–346

Staňková K, Olsder GJ, Schutter BD (2010) On European electricity market liberalization: a game-theoretic approach. INFOR: Inf Syst Operational Res 48(4):267–280

Van Eck T (2007). A new balance for the energy sector: no longer a puppet in the hands of technology, public interests and market. Industrielinqs Pers en Platform, Amsterdam

Chapter 23
Contingency Between Elasticity of Demand and Bullwhip Effect in Logistics Chains

Francisco Campuzano-Bolarín, Marija Bogataj and Lorenzo Ros-McDonnell

Abstract This paper discusses how the elasticity of final demand influences the bullwhip amplification when sudden economic changes appear along the entire multi-level supply chain. Increased variability of prices does not affect added value substantially in a supply chain where the price elasticity of demand is small (around 0.1) but with an elasticity of 10 or more, high price variances may result in significant losses. A traditional model of dynamic supply chain structure is used for our study, based on the seminal work of Forrester. A simulation platform for supply chain management with stochastic demand has been developed to study such a phenomenon. Vensim® simulation software was used for developing the appropriate supply chain dynamic models. The aim of our study is to gain a deeper insight into the processes in a logistic chain, at different elasticities of price demand.

23.1 Introduction

The Bullwhip effect is a well-known phenomenon which has been studied intensively since the mid-20th century. It was previously known as "demand amplification" (see Forrester 1961), and was often taken into account because

F. Campuzano-Bolarín (✉) · L. Ros-McDonnell
Department of Business Economics, Universidad Politécnica de Cartagena, Campus Muralla del Mar., 30202 Cartagena (Murcia), Spain
e-mail: francisco.campuzano@upct.es

L. Ros-McDonnell
e-mail: lorenzo.ros@upct.es

M. Bogataj
Faculty of Economics, Ljubljana University, Kardeljeva ploščad 17, 1000 Ljubljana, Slovenia
e-mail: marija.bogataj@guest.arnes.si

S. P. Sethi et al. (eds.), *Industrial Engineering: Innovative Networks*,
DOI: 10.1007/978-1-4471-2321-7_23,

of its reduction of uncertainty in demand and in case of stochastic behaviour of lead times.

Most of the research work on demand amplification has been focused on demonstrating its existence, identifying its possible causes and developing methods for mitigating the negative effect of it. Lee et al. (1997) identified four main causes of amplification: wrong methods of demand forecasting, supply shortage anticipation, batch ordering and price variation. The purpose of our study is to get a deeper insight into the processes in a logistic chain at different elasticities of price demand in order to develop the analytical approach at a later stage. Price variations, the effect of promotions and other actions in a supply chain refer to the practice of offering products at reduced prices to stimulate demand. Higher price variations could suddenly appear, as in case of an economic crisis; but higher variations of prices create variations of demand on the final product, which could have a serious impact on the dynamics of a supply chain, especially on the added value.

23.2 Measuring the Bullwhip Effect

According to Fransoo and Woters (2000), the bullwhip effect at any particular level of a multi-level supply chain is measured as the quotient between the coefficient of demand variation at the level where the bullwhip effect is measured and the coefficient of demand variation received at this level.

$$Bullwhip = \frac{C_{orders}}{C_{demand}} \tag{23.1}$$

where

$$C_{orders} = \frac{\sigma_{orders}}{\mu_{orders}} \quad C_{demand} = \frac{\sigma_{demand}}{\mu_{demand}} \tag{23.2}$$

23.3 Model Construction

In this paper we will study the demand management process along a three-stage supply chain. The main characteristics of the system being considered are summarized in the following points:

- We consider a four stage supply chain system consisting of identical behaviour by agents; it means that each agent orders products only from its upper stage. It follows that there is a Customer, Retailer, Wholesaler and Manufacturer in our supply chain.

- Over a period of time, demand variation is affected differently by price variations in elastic and inelastic scenarios. Demand is assumed to follow an isoelastic demand function with small variations around the isoelastic regression curve.
- An agent ships goods immediately after receiving the order if there is a sufficient amount of inventory on-hand. The shipment is not influenced by prices at higher stages of the supply chain. The flow of goods is influenced by the changes of prices of the final product, which influence external demand.
- Orders may be partially fulfilled (every order to be delivered includes actual demand and backlogged orders—if they exist), and unfulfilled orders are backlogged. We assume that the ordering costs are constant per order, but transportation costs are constant per item.
- Shipped goods arrive with a transit lead time of goods. The sequence of lead time creates a certain lead time pattern.
- The last stage (manufacturer) receives the raw materials from an infinite source. The capacity is fixed in the way that it does not affect the size of the orders generated in our cases.

The ordering policy which was chosen for our analysis is a generalized order-up-to policy (Silver et al. 1988). The simulation approach is based on Systems Dynamics methodology. In developing our model, the first step was creating causal loops to integrate the key factors of the system and put the relations on the links between pairs of these factors. See Campuzano et al. (2008) for a detailed description of the model used.

23.4 The Hypotheses

The first hypothesis is that the ordering quantity dynamics and propagation upstream of a supply chain depend on the price elasticity of the final demand, and it behaves differently when sudden price perturbations occur. The extent of the amplification in an elastic case is higher than in an inelastic demand.

The second hypothesis is that the time delay pattern of a supply chain influences the amplification ratio of elastic and inelastic demand differently on the total time horizon, but especially when a sudden perturbation arrives.

23.5 The Example of Two Different Price Elasticities

The main purpose of this paper is to study the influence of price fluctuations on the behaviour of a supply chain considering the different elasticities of product demand. The question is how different elasticities of demand influence the characteristics of the supply chain when high perturbations in prices happen unexpectedly. Let us suppose a price pattern is like the one presented in Fig. 1.1.

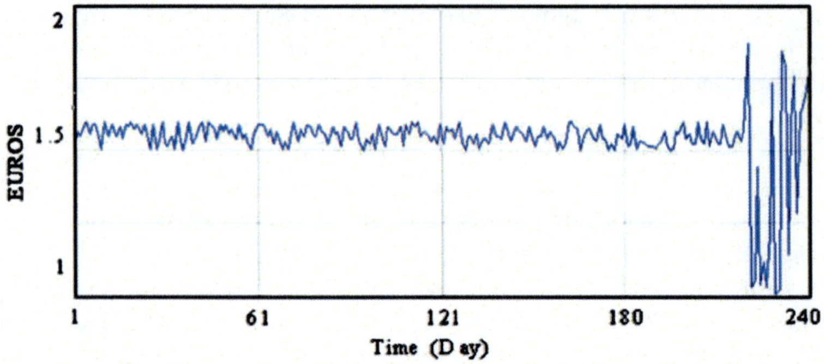

Fig. 23.1 Price pattern

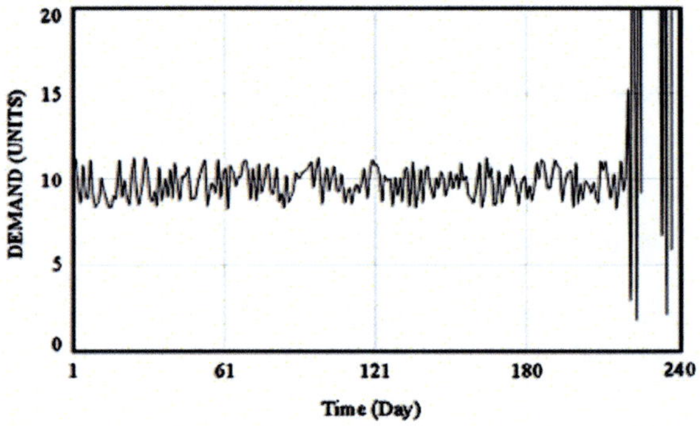

Fig. 23.2 The reaction of demand at the elastic demand pattern

The price pattern will firstly (in 220 periods) follow a normal distribution with a minimum value in the sample equal to 1.55 and a maximum value in the sample equal to 1.6; the average value is equal to 1.57 and standard deviation equals 0.1. The fixed seed in the simulation is equal here to 99. After 220 periods high perturbations appear, lasting for the following 20 periods with a minimum value in the sample equal to 1, maximum value in the sample equal to 1.9, average value equal to 1.57 and standard deviation equal to 1. The fixed seed for the simulation is also equal to 99 in this case (see Fig. 23.1).

At different elasticities the demand reacts differently on the price pattern, as can be seen in Figs. 23.2 and 23.3. The price variability amplification in the periods 221–240 increases the total cumulative demand for elastic cases (elasticity equal to 10), which becomes 1.6 times higher than the total demand, as opposed to the case of elasticity equal to 0.1. But it is especially important to know how different elasticities of demand influence variability of demand at sudden price increases, and how these changes influence the supply chain.

The behaviour of a supply chain not only depends on the demand pattern but also on its layout and other characteristics. For our study, the following values of supply chain parameters have been introduced, as presented in Table 23.1.

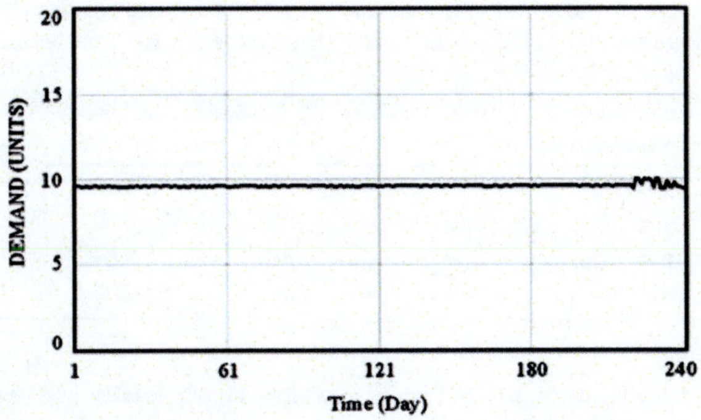

Fig. 23.3 The reaction of demand at the inelastic demand pattern

Table 23.1 Supply chain simulation parameters

Initial inventory on hand	50,100,200 units
Safety factor	$k = 2$, $Z \sim N(0, 1) \rightarrow$ $P(Z > k) = 0.0228$
Lead time (days) *(When there are no delays caused by stock-out periods or insufficient capacity constraints)*	First layout—lead time pattern $\vec{\tau} = (2, 3)$ Wholesaler to retailer 2 days Manufacturer to Wholesaler 3 days Second layout—lead time pattern $\vec{\tau} = (4, 1)$ Wholesaler to retailer 4 days Manufacturer to Wholesaler 1 day

Table 23.2 Bullwhip effect at retailer

Initial inventories	**50**		**100**		**200**	
Elasticity	**0.1**	**10**	**0.1**	**10**	**10**	**0.1**
Coefficient of variation	0.34	0.316	0.63	0.83	0.63	0.75

The bullwhip effect caused by the demands presented in Figs. 23.2 and 23.3 at the retailer stage are presented in Table 23.2.

The ratio between bullwhip measures calculated at wholesaler and retailer ($BWC_W/BCWC_R = W/R$) and the ratio between bullwhip measures at manufacturer and wholesaler (M/W) is presented in Table 23.3 for the lead time pattern $\tau = (2.3)$.

We can observe that the amplifications and central tendencies of the distribution created by using lead times 1 and 4 increase upstream at the elastic and the

Table 23.3 The ratio between bullwhip measures at wholesaler and retailer (W/R) and between manufacturer and wholesaler (M/W)—the contingency table for the first layout of supply chain lead times pattern $\tau = (2.3)$

	The ratio between Bullwhip measures	0.00–1.99	2.00–3.99	4.00–5.99	6.00 or more
W/R	Elasticity = 0.1	477	20	108	0
	Elasticity = 10	153	136	170	197
M/W	Elasticity = 0.1	238	54	184	161
	Elasticity = 10	44	206	206	182

Table 23.4 The ratio between the bullwhip measures at wholesaler and retailer (W/R) and between manufacturer and wholesaler (M/W)—the contingency table for the second layout of supply chain lead time $\tau = (4.1)$

	The ratio between Bullwhip measures	0.00–1.99	2.00–3.99	4.00–5.99	6.00 or more
W/R	Elasticity = 0.1	496	85	4	0
	Elasticity = 10	107	526	8	14
M/W	Elasticity = 0.1	240	352	47	0
	Elasticity = 10	13	144	186	253

inelastic demand. But the increase of amplification in inelastic cases is more significant ($\chi^2(3) = 275{,}65$) than in elastic demand $\chi^2(3) = 78{,}44$. Elastic demand has a higher amplification ratio on the first stage than in inelastic cases, but it looks as if a higher lead time on the second stage did not allow the amplification ratio to go over 6 in more cases than at the first stage.

We can also see that at lead time pattern $\tau = (4.1)$ the amplifications and central tendencies of distribution increase upstream at elastic demand $\chi^2 = 667.39$ and inelastic demand $\chi^2 = 286.61$. But at both stages the amplification ratio is higher in elastic demand, as the hypothesis supposed. Furthermore, we can see that a lower lead time on the second stage increased the propagation more than it is the case in Table 23.3, where lead time on the second stage was higher.

From Tables 23.3 and 23.4 we can also conclude that the modus of amplification ratio for inelastic demand is nearly 0 in case of lead time pattern $\tau = (2.3)$ at both stages and at the first stage (W/R) of lead time pattern $\tau = (4.1)$, but it is approximately 3 at the second stage (M/W) of the lead time pattern $\tau = (4.1)$, where the lead time is only 1.

Using a simple linear regression analysis of the data in Table 1.5 we can conclude that, on average, a higher lead time reduces the modus of amplification ratio in case of elastic demand by $b = -1.5$ per unit of lead time ($\rho = -0.94$, $N = 4$), but in case of inelastic demand this reduction is less significant ($b = -0.9$; $\rho = -0.77$, $N = 4$). When we are looking at the mean amplification in the total time horizon of inelastic cases, the conclusions are the following: higher lead time does not reduce mean amplification ratio significantly ($b = -0.2$; $\rho = -0.19$, $N = 4$), but in case of elastic demand this reduction is more significant ($b = -0.75$; $\rho = -0.84$, $N = 4$). The behaviour of the supply after

Table 23.5 The simple linear regression analysis of amplification ratio between bullwhip effects

Lead time	Elasticity 0.1:		Elasticity 10:	
	Modus	The average amplification rate estimation	Modus	The average amplification rate estimation
1	3	2	7	5.3
2	0	1	7	4
3	0	3.5	4	4.6
4	0	0.5	3	2.6
Regression coefficient	−0.9	−0.2	−1.5	−0.75
Correlation coefficient	−0.77	−0.19	−0.94	−0.84

Table 23.6 The amplification ratio between the bullwhip effects at both stages for the inelastic case from period 220 to 240

Lead time pattern	Stage	Initial inventories		
		50	100	200
$\vec{\tau} = (4, 1)$	M/W: 1	2.30	1.86	2.62
	W/R: 4	0.94	0.75	1.10
$\vec{\tau} = (2, 3)$	M/W: 3	0.97	3.64	3.55
	W/R: 2	0.98	2.02	1.89

perturbations is also included in Table 23.5. The analysis is made for the total time horizon.

When we are looking at the average amplification rate estimation only after high perturbations of inelastic cases, the conclusions can be derived from Table 23.6.

Even for the inelastic case, where amplifications and their growth upstream are lower, it can be concluded that after high perturbations of prices, from period 220 to 240 the amplification ratio between the bullwhip effects at wholesaler and retailer (W/R) was higher at the lead time 2 of the pattern $\tau = (2, 3)$ than at the lead time 4 of the pattern $\tau = (4.1)$, in all three different cases of initial inventories. Here the hypothesis that higher lead time influences lower amplification ratio also for the inelastic case at sudden unexpected perturbations in economy was accepted (P-value is 0.1), but could be refused for the amplification ratio between bullwhip effects at the manufacturer and the wholesaler.

23.6 Conclusions

In our research two different hypotheses have been investigated and put to the test, with the following results:

1. The increase of amplification in an elastic case is higher than in an inelastic demand. From Tables 23.3 and 23.4, where the ratio between bullwhip measures at wholesaler and retailer (W/R) and between manufacturer and wholesaler (M/W) is presented at different lead time patterns, we can conclude that in all cases amplification, and therefore propagation upstream, is higher for elastic demand.
2. The second hypothesis has been put to the test for the total time horizon but especially on the horizon after a sudden perturbation arrives. χ^2 values of contingency tables confirm this hypothesis. Also, the simple linear regression analysis of ratio between bullwhip effects in Table 23.5 presents this phenomenon on the total horizon. One can see that the average amplification rate estimation increases when decreasing lead times have a regression coefficient at elastic demand equal to -0.75 and a correlation coefficient equal to 0.84, while in inelastic demand the correlation of time delay with the average amplification rate estimation has not been significant. Therefore, the inelastic case was studied in detail for perturbed intervals. After high perturbations of prices, from period 220 to 240, the amplification ratio between bullwhip effects at wholesaler and retailer (W/R) was higher at the lead time 2 of the pattern $\tau = (2.3)$ than at the lead time 4 of the pattern $\tau = (4.1)$ in all three different cases of initial inventories. The paired t-test was used to work out whether there is a significant difference. Here the hypothesis that higher lead time influences a lower amplification ratio at sudden unexpected perturbations in economy was accepted (P-value is 0.1). From period 220 to 240 the average amplification ratio between bullwhip effects at the manufacturer and the wholesaler was not significantly dependent on time delay.

References

Campuzano F, McDonnell L, Lario FC (2008) Bullwhip effect consequences according to different supply chain management strategies: modelling and simulation. J Quant Methods Econ Bus Adm 5:49–66

Forrester J (1961) Industrial dynamics. MIT Press, Cambridge

Fransoo JC, Woters MJF (2000) Measuring the bullwhip effect in the supply chain. Supply Chain Manag Int J 5(2):78–89

Lee HL, Padmanabhan V, Whang S (1997) The bullwhip effect in supply chains. Sloan Manag Rev 38(3):93–102

Silver EA, Pyke DF, Peterson R (1988) Inventory management and production planning and scheduling. Wiley, New York

Chapter 24
Safety Management in Manufacturing and its Influence in Injury Rates: Evidences from Spanish National Safety Management Survey (2009)

Jesús A. Carrillo, Ventura Pérez and Luis Onieva

Abstract Occupational Safety is a complex issue and several factors are related to accident occurrence. Safety Management Systems have impact in safety performance. In this paper we used data from the National Survey of Safety Management (Instituto Nacional de Seguridad e Higiene en el Trabajo) to identify management and organizational predictive variables related to injury rates. As an improvement compared to some other previous studies, multivariate models are used. This study shows the importance of safety management and some characteristics of the successful implementation, at least in terms of injury rates, among manufacturing companies in Spain.

24.1 Introduction

Analysis of organization performance is a complex issue. One important aim of organization science is to identify which company management system characteristics should be considered. Safety management is one of the key aspects in every organization.

J. A. Carrillo (✉)
Dirección General de Seguridad y Salud Laboral,
Avda. Hytasa, 14 P-1, 41006 Sevilla, Spain
e-mail: jcarrillo@andaluciajunta.es

V. Pérez · L. Onieva
Escuela de Ingenieros de la Universidad de Sevilla,
CM de los Descubrimientos, s/n 41092 Sevilla, Spain
e-mail: ventura@gtiesi.us.es

L. Onieva
e-mail: onieva@esi.us.es

S. P. Sethi et al. (eds.), *Industrial Engineering: Innovative Networks*,
DOI: 10.1007/978-1-4471-2321-7_24,

In Spain, as part of the European Union, there are some mandatory elements of safety management, according to the Framework Directive 89/391/EEC, on the introduction of measures to encourage improvements in the safety and health of workers at work.

Safety Management Systems (SMS) is defined by International Labour Organization as "a set of interrelated or interacting elements to establish occupational safety and health (OSH) policy and objectives and to achieve them". According to OSHAS 18001 an occupational health and safety management system is "used to establish an OHS policy and to manage OHS risks".

An organization's SMS is one part of a larger management system. A management system, including an SMS, is a network of interrelated elements. These elements include responsibilities, authorities, relationships, functions, activities, processes, practices, procedures, and resources. A management system uses these elements to establish policies, plans, programs, and objectives and to develop ways of implementing these policies, plans, and programs, and achieving these objectives. In our paper we assume that the main purpose of SMS is to intervene through active actions in order to improve safety.

Previous surveys have been able to differentiate between adopters and non-adopters of SMS (Bottani et al. 2009), as an evaluation of factors identified as performance in main attitudes to do active preventive activities (such as defining and communicating goals, updating risk, assessing and evaluating risks and training of workers).

Safety Management relates to actual practices, roles and functions and it is an antecedent of safety climate as the system impacts employees' attitude and behaviour (Fernández-Muñiz et al. 2009). This safety climate is an intermediate outcome.

Safety behaviour is usually identify as a single construct (Vinodkumar and Bhasi 2011). However from a conceptual point of view we can differentiate organizational behaviour or safety management, interactional behaviour or safety activities and personal behaviour of each worker. Relationship between safety management and safety climate or culture is difficult to demonstrate as injury rates depend on other factors like risk exposure, technology and worker individual characteristics.

That is one of the reasons why studies performed looking for demonstrating effectiveness of safety management intervention have not founded enough evidences (Robson et al. 2007). Besides, most of them lack of an appropriate methodology.

Accident rate is not the only indicator for safety management evaluation (Sgorou et al. 2010), but it is a quite objective and relavant one. Other approaches proposed for evaluating safety performance have been safety climate (Zohar 1980) or safety audits (Øien et al. 2011).

24.2 Conceptual Model

Previous research has shown that certain aspects affect accident rate at company level (Cagno et al. 2011). Although the causes of each accident should be studied, a preliminary observation can be done through direct ecological studies which can identify explanatory variables.

Table 24.1 Practices expected in a safety management system

	Zohar (1980)	Shannon et al. (1997) Geldart et al. (2010)	Vredenburg (2002)	Fernández-Muñiz et al. (2009)	Bottani et al. (2009))	Micheli and Cagno (2010)	Vinodkumar (2010) Vinodkumar and Bhasi (2011)
Training	X	X	X	X	X	X	X
Safety planning							
Financial resources						X	
Accident analysis						X	
Risk assestment and safety audit		X		X	X	X	
Worker empowerment		X	X				X
Reward system			X	X			
Management commitment	X	X		X			X
Worker selection			X				
Safety communication				X	X		X

Table 24.2 Survey data

Companies (INE [a], Spain)	Workers (INE [a], Spain)	Rate (INSHT[b], Spain)	Companies surveyed	Workers in surveyed companies	Surveyed rate
133,062	2,123,441	89.4	1,283	138,404	95.7

[a] *INE* National Statistics Institute
[b] *INSHT* National Health and Safety Institute

National Safety Management Survey provides a cross-sectional analysis of Spanish Manufacturing and Industrial sector. This is not a study about injured workers and the circumstances of accidents, but about companies of which data is gathered at company level.

Although based on individual's opinions on some items, this survey provides important information as each case is a real company data and appropriate multivariate analysis can extract the relationships between injury rates and company management indicators. None of the variables can be considered a direct cause in terms of accident causation; but they can be interpreted as contributing factors in terms of environmental and organizational framework.

The set of possible factors is an open list so far. Not all of the factors can be identified as part of the accident factors, and we do not have the accident investigation to correlate factors and accidents among those companies surveyed.

Previous studies have identified some of the practices expected in a safety management system. We show some of those practices identified in Table 24.1.

24.3 Data

Data has been selected from the National Survey of Safety and Health Enterprises Management performed by the National Institute of Occupational Health and Safety (Instituto Nacional de Seguridad e Higiene en el Trabajo 2010). Some of the variables are quantitative data but some others are qualitative. We have categorized those that are quantitative when necessary (Table 24.2).

Most of the qualitative data gathered are an opinion of company representatives surveyed about management and it should be considered a latent variable of a construct that theoretically represents some of the company attitudes and safety culture.

Only cases with activity gathered from companies in manufacturing and industrial sector were considered. Therefore only companies with NACE-93 from 15 to 37, according to Council Regulation (EEC) No 3037/90 of 9 October 1990 on the statistical classification of economic activities in the European Community, are included in this study. Nine cases have been withdrawn because of inconsistency in answers and eighteen because they were not completely filled. Finally 1,265 cases are included in the analysis.

24.4 Methodology

This is not a longitudinal study as survey is only done once, so we do not know what happened before or after the survey and each conclusion would need further research as experimental check of relationships. Our analysis is oriented to predict what company organizational factors are related to high injury rates as an outcome value that can identify weak health and safety performance.

Dependent variable is calculated as company injury rate divided by mean rate of their NACE activity. Categories considered for these relative rates depend on whether the accident rate is higher than 150% of activity mean rate. This criteria is used to include a company in some promotion programs such as PAEMSA (Plan for assessment of high accident rate companies in Andalusia) so we name the categorical dependent variable as PAEMSA. Our aim is to identify which other variables predict whether a company would likely be included or not in PAEMSA.

Most of explanatory variables are categorical or can be easily categorized. In univariate analysis we compute odds rate using logistic regression with dependent variable PAEMSA and each possible predictive variable. In multivariate analysis, a model is adjusted with logistic regression, selecting those variables that are significant.

Most authors consider that 10 cases are needed for each included variable, as a rule of thumb. As we selected 12 explanatory variables, our sample is enough for fitting the model in regression analysis. In terms of statistical power, for small effects with power of 0.80 (Cohen 1992), assuming $\alpha = 0.05$, $n = 1{,}283$ ($N = 133{,}062$), and four degrees of freedom, we need a sample of 1,194 companies. In case of regression analysis with 8 variables, we need a sample of 757. So sample size is adequate for the analysis considered.

All statistical analysis has been done with SPSS v.18 and both ".sav" files and results are available upon request.

24.5 Univariate Analysis

In order to identify those explanatory variables that influence the injury rates declared by surveyed companies, we first checked univariate relationship using. Odds ratio calculated for being in PAEMSA are the following: to have less than 10 employees, to have less than 27% of women (below average), to have less than 3 quality tools, to not consider that there is any risk of accidents, to not warn workers, to not have been fined, not having worker representation, not having health and safety committee, not having worker participation, to have more than 4 prevention activities, to not train workers and to not have renew more than 10% of equipment. We show univariate odds ratio calculated in Table 24.3.

Table 24.3 Univariate analysis. Odds ratio for being in PAEMSA

Variable	OR 95% (sig.)	Values
Activity (NACE rev2)	(*)	From 15 to 37
Mean age (18–34; 35–44; >44)	(*)	Young/normal/mature
Company size (no of employees)	0.35–0.67 (0.0)	≥10/**<10**
% of immigrants in staff	(*)	<7%/≥7% (average is 7%)
% of women in staff	0.44–0.76 (0.0)	<27%/**≥27%** (average is 27%)
Strategy*	(*)	Cost, quality, prevention, productivity, environment, image, innovation
Number of quality tools used	0.36–0.62 (0.0)	Less than 3/**More than 2**
OSMS	(*)	Not/yes
Accident risk	0.20–0.38 (0.0)	Not/**yes**
Worker warnings	0.33–0.59 (0.0)	Not/**yes**
Fines	0.19–0.37 (0.0)	Not/**yes**
Indicted	(*)	Not/yes
Representation	0.34–0.60 (0.0)	Not/**yes**
Participation*	0.39–0.88 (0.1)	Not/**yes**
Preventive activities	0.30–0.59 (0.0)	Less than 5/**more than 4**
Training	0.27–0.58 (0.0)	Not/**yes**
% Renovation renew	0.42–0.70 (0.0)	<10%, **≥10%**

* Variables with no significant univariate relationship, $p > 0.05$. If company considers that there is an OSMS is an opinion (depends on what responders believe)

24.6 Multivariate Analysis. Logistic Regression Model

There is a possibility that a confounding variable could hide a true relationship that it is not significant in a univariate analysis. Thus multivariate approach is more realistic as accidents are multicausal.

Logistic regression deals properly with categorical explanatory variables. Also it is easy to explain logistic regression coefficients in terms of odds ratios. Hence, the model predicts what companies would have high accident rate compared to their activity classification. Although not presented in this paper, general linear regression model for injury rates as dependent variable identify the same significant explanation variables.

The adjusted multivariate model, using logistic unconditional regression and considering all available cases (1,265), have a Nagelke $R2 = 0.162$ and a global correct classification of 75%.

Those significant variables after adjusting with logistic regression model are included in Table 24.4. Companies with no fines, no accident risk perception, without worker representatives, with few quality activities and tools implemented (less than 3), with few equipment renewed (less than 10%) and with safety management system are likely not classified as high accident rate companies (PAEMSA).

Table 24.4 Logistic regression model adjusted

	Wald	Sig.	Exp(B)	I.C. 95% para EXP(B)	
				Inferior	Superior
Quality Activities (less than 3)	9.512	0.002	0.628	0.467	0.844
Accident risk (no)	34.042	0.000	0.363	0.258	0.510
Fines (not in last year)	27.523	0.000	0.383	0.268	0.548
Worker representation (no)	6.074	0.014	0.676	0.495	0.923
Equipment renovation (less 10%)	7.374	0.007	0.685	0.521	0.900
OSMS (yes)	7.897	0.005	0.675	0.513	0.888
Constant	4.163	0.041	1.473		

24.7 Discussion

Many of the explanatory variables have been identified in previous studies (Shannon et al. 1996; Micheli and Cagno 2010). Some personal characteristics of workers and, as a consequence, staff profiles influence injury rates. Other explanatory variables are related to company management, both general and safety, and to technical hazards.

Older workers seem to have more accidents (Salminen 2004) and companies with high number of immigrants have been identified with higher rates (Ahonen et al. 2007). However this factors lack of statistical significance in our multivariate model.

Female workers behavior and the increasing number of women in manufacturing is a phenomenon under study (Taiwo et al. 2008). According to our findings, companies with higher proportion of female workers have better relative rates. This is a hypothesis that needs further research.

Previous studies have considered the increased rates associated with small companies (Leigh 1989; Fabiano et al. 2004; Sorensen et al. 2007). Besides differences in technical equipment and worker profiles there are also differences in management systems (Cagno et al. 2011) as many relevant prevention activities are not likely done by micro and small manufacturing companies.

Several surveys have identified strong relationship between accident rate and certain preventive activities (Arocena et al. 2008; Geldart et al. 2010). According to the analysis, having a safety management system is one predictive variable of not having a high relative rate but not each of the preventive activities and elements. One explanation can be that, even in the same NACE activity, those companies with higher risks have more preventive activities.

Certification as a safety management performance indicator has not yet been proved (Vinodkumar and Bhasi 2011).

This study has some limitations. Data have been obtained from survey so some information may have errors and the perception of the person who was interviewed may affect answers (Huang et al. 2011). In further analysis it should be checked which people are interviewed in each company in order to reduce this bias.

The only outcome measure considered is injury rate. Other important safety indicator, as safety climate of independent risk assessment would provide a deeper view about safety performance of companies. Nevertheless, lack of safety management has been identified as a risk factor for companies having high relative injury rates.

Acknowledgments The authors thank INSHT. This work is based in Spanish National Safety Management Survey (2009), which microdata have been made public for researchers.

References

Ahonen EQ, Benavides FG, Benach J (2007) Inmigrant populations, work and health—a systematic literature review. Scand J Work Environ Health 33(2):96–104

Arocena P, Nuñez I, Villanueva M (2008) The impact of prevention measures and organizational factors on occupational injuries. Saf Sci 46:1369–1384

Bottani E, Monica L, Vignali G (2009) Safety management systems: performance differences between adopters and non-adopters. Saf Sci 47:155–162

Cagno E, Micheli GJ, Perotti S (2011) Identification of OHS-related factors and interactions among those and OHS performance in SMEs. Saf Sci 49:216–225

Cohen J (1992) A power primer. Psychol Bull 112(1):155–159

Fabiano B, Currò F, Pastorino R (2004) A study of relationship between occupational injuries and firm size and type in the Italian industry. Saf Sci 42:587–600

Fernández-Muñiz B, Montes-Peón JM, Vázquez-Ordás CJ (2009) Relation between occupational safety management and firm performance. Saf Sci 47:980–991

Geldart S, Smith CA, Shannon HS, Lohfeld L (2010) Organizational practices and workplace health and safety: a cross-sectional study. Saf Sci 48:562–569

Huang Y-H, Leamon TB, Courtney TK et al (2011) A comparison of workplace safety perceptions among financial decision-makers of medium- vs. large-size companies. Accid Anal Prev 43:1–10

Instituto Nacional de Estadística. Encuesta Industrial de Empresas. http://www.ine.es/inebmenu/mnu_industria.htm. Accessed 20 June 2011

Instituto Nacional de Seguridad e Higiene en el Trabajo (2010) Encuesta Nacional de Gestión de la Seguridad y Salud en las Empresas. Instituto Nacional de Seguridad e Higiene en el Trabajo. http://encuestasnacionales.oect.es/enge/EngeAnalisis.jsp. Accessed 20 June 2011

Leigh J (1989) Firm size and occupational injury and illness incidence rates in manufacturing industries. J Commun Health 14(1):44–52

Micheli GJ, Cagno E (2010) Dealing with SMEs as a whole in OHS issues: warnings from empirical evidence. Saf Sci 48:729–733

Øien K, Utne I, Herrera I (2011) Building safety indicators: part 1—theoretical foundation. Saf Sci 49:148–161

Robson LS, Clarke JA, Cullen K et al (2007) The effectiviness of occupational health and safety management systems interventions: a systematic review. Saf Sci 45:329–353

Salminen S (2004) Have young workers more injuries than older ones? an international literature review. J Saf Res 35:513–521

Sgorou E, Katsakiori P, Goutsos S et al (2010) Assessment of selected safety performance evaluation methods in regards. Saf Sci 48:1019–1025

Shannon HS, Waiters V, Lewchuk W et al (1996) Workplace organizational correlates of lost-time accident rates in manufacturing. Am J Ind Med 29:258–268

Shannon HS, Mayr J, Haines T (1997) Overview of the relationship between organizational and workplace factors and injury rates. Saf Sci 26(3):201–217

Sorensen OH, Hasle P, Bach E (2007) Working in smail enterprises—is there a special risk? Saf Sci 45:1044–1059

Taiwo OA, Cantley LF, Slade MD et al (2008) Sex differences in injury patterns among workers in heavy manufacturing. Am J Epidemiol 169(2):161–166

Vinodkumar M (2010) Safety management practices and safety behaviour: assessing the mediating role of safety knowledge and motivation. Accid Anal Prev 42:2082–2093

Vinodkumar M, Bhasi M (2011) A study on the impact of management system certification on safety management. Saf Sci 49:498–507

Vredenburg AG (2002) Organizational safety: which management practices are most effective in reducing employee injury rates? J Saf Res 33(2):259–276

Zohar D (1980) Safety climate in industrial organizations: theoretical and applied implications. J Appl Psychol 65(1):96–102

Chapter 25
Method for IT Governance Based on Enterprise Modelling

Eloy Hontoria, Antonio Fernández and M. Victoria de-la-Fuente

Abstract The continuous evolution of Information and Communication Technologies (ICT) implies that the governance paradigm lies in the processes of integration and collaboration which seek to integrate business processes in a more efficient way, aiming to increase the flexibility of the organization and to have information systems and technologies that interact easily with the resources of the organization (systems, personnel, etc.). In this evolving environment and as a response to the *concepts relating to IT governance*, this paper defines and designs a method that facilitates the change in an organization towards the new paradigms of IT design and management, and deals mainly in their interaction with inter-organization networks. The application of the defined method on the organization of the information and communications service of a University has facilitated the integration of different tools (for example GTI4U model, Maturity Model, Enterprise Integration Methodology) in the IT governance process, thanks to the modelization of the business processes and the flows related with this.

25.1 Introduction

Information technologies were questioned in the past, because of their lack of productivity (Roach 1987), or due to their limited importance in the organization (Carr 2003). Nowadays, they are considered to contribute value to organizations,

E. Hontoria (✉) · M. V. de-la-Fuente
ETSII-Universidad Politécnica de Cartagena, c/Dr. Fleming, s/n,
30202 Cartagena, Spain
e-mail: Eloy.hontoria@upct.es

A. Fernández
Escuela Superior de Ingeniería - Universidad de Almería,
Carretera de Sacramento s/n, 04120 Almería, Spain

S. P. Sethi et al. (eds.), *Industrial Engineering: Innovative Networks*,
DOI: 10.1007/978-1-4471-2321-7_25,

and they also provide a competitive advantage (Porter 2006). IT only increases performance in those organizations that are able to govern them in an appropriate way (Weill and Ross 2004).

As shown by Benemati et al. (1997), IT evolved quickly and, with its strategic impact on the operation of organizations becoming more and more important, it is crucial to deal with it appropriately. Olugbode et al. (2007) propose typifying the role of IT in an organization as the adjustment or alignment with the strategic objectives of the organization. IT can only prove truly useful if the infrastructure that has been used to implement the IT strategy is the right one. IT strategy must support both the organization's strategy and its business processes.

The administration and global management of an organization are based on corporate governance and IT governance (Hamaker 2003). Both governances are based on the same principles (Hamaker and Hutton 2004; De Haes and Van Grembergen 2005).

IT governance is the capacity of an organization, whose responsibility lies with the Board of Directors and the executive management, to drive IT design and implementation, thus assuring their alignment with the business strategy (Van Grembergen 2004).

In order to respond to the needs of its environment, the organization has to react to organization and administration models that will provide enough agility to implement immediate responses by means of the efficient and effective exploitation of their IT resources.

The organization obtains a return value in the form of costs saving, improved internal organization, users' satisfaction, corporate image and external projection. The main benefit obtained is an increment in the agility order to confront the changes that the organization must face (De Haes and Van Grembergen 2005).

25.2 Evolution of the IT Governance Paradigms

The concept of "IT Governance" was born in the mid-90 s and has grown ever since, embracing more topics and areas, until eventually becoming a discipline in itself. The term was initially defined by Loh and Venkatraman (1992), and later by Henderson and Venkatraman (1993), in order to describe the group of mechanisms which ensure the achievement of the necessary IT capacities for the optimal operation of business processes.

Good corporate government in an organization involves administration and management of its Information Technologies (IT). Directors of an organization should understand the opportunity to convert their IT into strategic elements associated with the rest of the organization.

According to Van Grembergen, IT Governance is the organisational capacity exercised by the Board, executive management and IT management to control the formulation and implementation of IT strategy and in this way ensure the fusion of business and IT (Van Grembergen 2004). The responsibility for IT governance is shared between these different groups.

Just like any another organization, Spanish Universities need to roll out IT governance systems, with the object of improving the organization's performance. Among the recent initiatives of the Conference of Principals (CRUE) is to promote the development of IT governance systems in Spanish Universities (Uceda 2008). In this specific case the steps to apply IT governance to an organization were explained by Van Grembergen and De Haes (2008):

- Educate - training the top management within the organization.
- Analyze and know the "maturity" of the initial IT governance situation.
- Design an action plan for implementing IT governance in the organization.

These steps above are very generic and do not address the process of analysis of the initial IT governance situation, let alone the design of an implementation plan.

In 2010, taking the referential framework of Van Grembergen and De Haes as a starting point, Fernández (2010) proposed an IT Governance Model for Universities (GTI4U), based on the ISO/IEC 38500: 2008 standard (ISO 38500 2008), which defines IT governance as a "system by means of which the use of current and future IT is managed and controlled". This entails control and measurability of the plans for IT use that support the organization, and monitoring this use in order to reach what is established in the plans. It sets out the principles for good IT governance in the view of Top Management.

The GTI4U model is centred on the second step specified by Van Grembergen and De Haes (2008): analyzing and ascertaining the "maturity" of the IT governance situation. A very detailed analysis process is incorporated in the model to evaluate the different IT aspects of the organization, and it also uses the concept of maturity model included in the ISO 38500 norm.

25.3 IT Governance

The fundamental objective of the research presented here is to improve IT governance, management and planning in Spanish Universities by means of the development of a method based on Enterprise Modelling that integrates the different models that have been developed, to highlight the following aspects:

- The absence of models for IT analysis, planning and governance in Spanish Universities, and its scarce number at an international level, which constitutes a narrow knowledge base.
- The current situation of Information Technologies, the culture of strategic planning and the initiatives in quality processes in Spanish Universities require new tools that help complex organizations, as in the case of Universities, to establish procedures that allow them to implement IT governance systems progressively.
- The use of the designed method aids the process of IT alignment and the organization's objectives. It also provides an analysis of the value added to the organizations by IT.

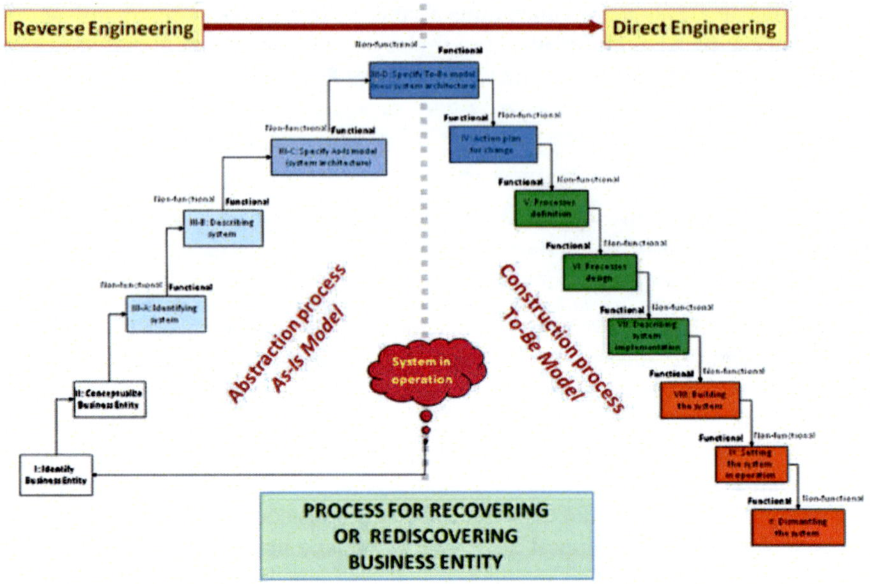

Fig. 25.1 Phases and stages of the ERE-GIO methodology

25.4 Methodology Used

For the development of the method, a methodology of Enterprise Modelling—ERE-GIO Methodology (de la Fuente et al. 2008, 2010) was used, which permits the re-discovery of any business entity by establishing two general work phases: reverse engineering and direct engineering. It is developed through a series of stages which permit both the study of the current situation of the organization and the subsequent plan of evolution towards a new model, or IT situation.

The Enterprise Modelling methodology is based on the application of the life cycles theory to business entities. It has been designed to be used as a tool that allows characterization and re-drawing of an organization, starting from the value of the elements that integrate it: resources, processes, technology, information, etc. It focuses on entreprise modelling in such a way that, by monitoring the life cycle for the development of new systems, it can be applied to a process of organizational evolution. (Fig. 25.1)

By using the IT governance maturity model, this methodology allows us to define the "As-Is" model. The construction of the "As-Is" model is developed by continuing the reverse engineering phase, which permits identification and measurement of the variables that characterize the current situation of the system, and their modelling.

In order to propose the "To-Be" model, that is, how the system is intended to work, we will start from the IT governance maturity model and will incorporate those aspects that the organization's IT managers want to be present in IT governance. Continuing with the direct engineering phase of the methodology, this "To-Be" model will permit to establish an implementation plan by means of the de-aggregation of the information carried out in the reverse phase.

25.4.1 Reverse Engineering Phase. GTI4U Model

In this phase the strategic framework of the business entity is defined, which in our particular case is IT governance. This facilitates the focus on processes, seeking customer satisfaction (external or internal of the entity) and eliminating those activities that do not contribute to this objective. This phase is centred on defining the sequence of activities that allow focusing on the improvement processes appropriately.

In the reverse engineering phase the processes carried out by the business entity selected are analyzed in order to develop its "As-Is" model, which is conceptualized based on the IT maturity model, with the aim of creating the proposed "To-Be" model.

To do so, we start from the GTI4U model designed and validated by Fernández (Fernández 2010). This model is based on ISO 38500 (ISO 38500, 2008), whose principles are Responsibility, Strategy, Acquisition, Performance, Conformance and Human Behaviour.

These principles are supplemented with a set of 17 objectives, which support the process of abstraction that permits the construction of the "As-Is" model of the organization.

The stages of the reverse engineering phase that must be followed with the support of the GTI4U model are:

1. *Identification of the business entity.* Select the portion of the company or business entity in which the analysis is to be developed.
2. *Conceptualisation of the business entity.* This stage provides the strategic framework within which the selected business entity is defined, and that allows to set the foundation for the further development of the "To-Be" model.
3. *Processes Analysis.* The objective of this stage is to describe business processes, focusing on those which are currently being used by the entity ("As-Is" model), and their most relevant features.

It is obvious that stage 3 is the fundamental stage of this reverse engineering phase, since it specifies the current processes ("As-Is" model), as well as the intended "To-Be" model of the business entity. This must not be confused with

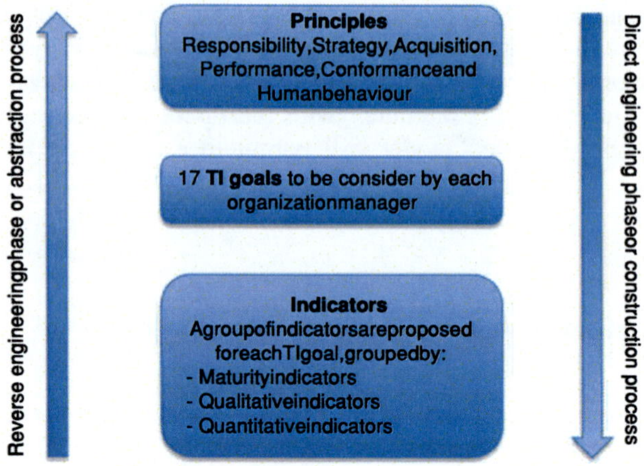

Fig. 25.2 Method for IT governance-based enterprise modelling

the definition and design of the new form of operating of the business entity, which is the main object of the direct engineering phase.

25.4.2 Direct Engineering Phase. Model PTI4U

The second phase of the methodology, denominated Direct Engineering phase, takes charge of the planned execution of the development, implementation and setting in progress of the processes (re-defined and/or new) of the new operation model ("To-Be" model) of the business entity. The GTI4U model uses the guidelines proposed by ISO 38500 (2008) as a basis in order to design their own catalogue of good practices (Fig. 25.2).

This Direct Engineering Phase consists of 6 stages:

1. *Action Plan for the Change.* An action plan is set up at this stage which can bridge the transition from one model to the other. The main purpose of this plan is to establish the environment and the conditions which will define the new behaviour of the new system and to point out how the change will be managed, through the following steps:
2. *Processes Definition.* Users and/or persons in charge of processes must define characteristics and activities to be developed in each of those processes.
3. *Processes Design.* This stage is about designing all the activities which will allow us to meet the requirements defined in the previous stage, through the following steps:
4. *System Implementation Description.* A selection is made of the most suitable techniques, tools and technologies to provide the components to be used in an integrated, effective realization of the system.

5. *System Construction*. After selecting the appropriate elements to carry out the required processes, the development and installation of these elements is performed in the selected entity.
6. *System Start-Up*. This is a comprehensive phase which includes execution of all the entity's activities specified in previous stages with the resources settled in the previous stage in order to achieve the desired objectives (Fig. 25.1).

25.5 Conclusions

IT organizations are constantly evolving and adapting to market requirements. In order to control the process of change, new methods are necessary which permit the analysis of current systems, and help the design and construction of new IT activities.

There are several methods for business processes modelization available, depending on which stages or needs of the organization we want to represent. In this paper a specific methodology has been used: ERE-GIO methodology, because of its application to the modelization process of IT governance (Fig. 25.2).

Due to its definition of two engineering phases, the ERE-GIO methodology easily adapts to the analysis of both the current governance process and the process that we intend to implement.

The GTI4U model offers an abstraction procedure that gives support to the construction of the "As-Is" governance model in the organization considered.

The GTI4U model, together with the guidelines of ISO 38500, has allowed us to build a source of good practices, which in turn will permit the direct model and the construction of the new IT governance system.

References

Benemati J, Lederer A, Singh M (1997) Changing information technology. Inf Techno Manag 31(5):276–288

Carr NG (2003) It doesn't matter. Harvard Business Review 81(5):41–49, 128

De Haes S, Van Grembergen W (2005) IT governance structures, processes and relational mechanisms—achieving IT—Business alignment in a major Belgian financial group. Proceedings of the 38ª Hawaii International Conference on System Science. IEEE. C.S

De la Fuente MV, Ros L, Cardós M (2008) Integrating forward and reverse supply chains: Application to a metal-mechanic company. Int J Prod Econ 111:782–792

De la Fuente MV, Ros L, Ortiz A (2010) Enterprise modelling methodology for forward and reverse supply chain flows integration. Comput Ind 61:702–710

Fernández A (2010) Gobierno de las TI para Universidades. CRUE

Hamaker S (2003) Spotlight on governance. Information System Control Journal, Volume 1

Hamaker S, Hutton A (2004) Principles of IT governance. Inf Syst Control J , vol. 2. from: www.isaca.org

Henderson JC, Venkatraman N (1993) Strategic alignment: leveraging information technology for transforming organisations. IBM Syst J 32(1):4–16
ISO 38500 (2008). ISO/IEC 38500:2008 Corporate Governance of Information Technology. ISO/IEC. 2008, from: www.iso.org
Loh L, Venkatraman N (1992) Diffusion of information technology outsourcing: influence sources and the Kodak effect. Inf Syst Res 3(4):334–359
Olugbode M, Richards R, Biss T (2007) The role of information technology in achieving the organisations strategic development's goals: a case study. Inf Syst 32(5):641–648
Porter M (2006) Tecnología y ventaja competitiva. Estrategia y ventaja competitiva Ed. Deusto, Madrid.
Roach S (1987) American's technology dilemma: a profile of the information economy. Morgan Stanley Special Economics Study, April 1987. New York
Uceda J (2008) Estrategia de la CRUE en relación a la implantación de modelos de gobierno de las TI en las Universidades Españolas. Seminario gobierno de las TI en las Universidades Españolas. Sectorial TIC de la CRUE. Universidad Politécnica de Madrid
Van Grembergen W (2004) Strategies for information technology governance. Idea group
Van Grembergen W, De Haes S (2008) Implementing Information Technology Governance. Models, Practices and Cases. IGI Publishing, New York
Weill P, Ross JW (2004) IT Governance: How Top Performers Manage IT Decision Rights for Superior Results. Harvard Business School Press, Boston

Part IV
Manufacturing

Chapter 26
Making Product-Service Systems in Collaborative Networks: Implications in Business Processes

Faustino Alarcón, María del Mar Eva Alemany and Ángel Ortiz

Abstract An interesting possibility for the differentiation of the industrial activity of enterprises is to evolve from the production of goods to the provision of systemic solutions consisting of product-service systems (PSS). This change can mean the creation of a collaborative network and reveals significant implications at strategic and operational levels. In this context, and from an operational perspective, a conceptual framework has been developed to facilitate the identification, classification and study of the implications in business processes of using the PSS concept. The conceptual framework proposed allows identifying, classifying and determining implications in business processes when a PPS is going to be marketed commercially in a collaborative environment.

26.1 Introduction

In the last few years, firms have attempted to be different by devising new offers for customers that combine products and services. This movement, which was termed by Vandermerwe and Rada (1988) as "servitization of business", is clearly a powerful new feature of the total market strategy. In this way, the total product received by the customer also includes services that aim to support not only the product, but also the customer in using the product.

Some authors call this form of offers integrated "bundles" (Vandermerwe and Rada 1988) or Product-Service System (PSS) (Manzini and Vezzoli 2003; Morelli 2006; Pawar et al. 2009). A PSS may be considered as a marketing concept that

F. Alarcón (✉) · M. M. E. Alemany · Á. Ortiz
Centre d'Investigació de Gestió i Enginyería de Producció (CIGIP),
Universitat Politècnica de València, Camí de Vera s/n, 46022
València, Espanya
e-mail: faualva@omp.upv.es

S. P. Sethi et al. (eds.), *Industrial Engineering: Innovative Networks*,
DOI: 10.1007/978-1-4471-2321-7_26, © Springer-Verlag London Limited 2012

extends the traditional functionality of a product by incorporating additional services. According to Mont (2002), a PSS is "a system of products, services, supporting networks and infrastructures that is designed to be competitive, to satisfy customer needs and to have a lower environmental impact than traditional business models". Other works describe it as "total care products" or "functional products" (Alonso-Rasgado et al. 2004), defining it as integrated systems that comprise hardware and support services.

Firms that wish to progress toward the PSS concept, either from a "servitization of products" or a "productization of services", could apply various solutions. One of these solutions is to search for collaborative models with other firms by forming supply chains (SCs) or networks in which the various components of PPS design, preparation/manufacturing and commercialization are de-centralized and shared as agreed. Recently, several research works have shown the pros and cons of using collaboration-based management models (Barratt 2004; Manthou et al. 2004) as opposed to traditional models with a single industrial entity. In such collaborative models, the complementors' association emerges (Lejeune and Yakova 2005), defined as a group of entities involved in different stages of the supply chain (e.g., a supplier and a manufacturer) that add value to the other entities' products and make them more attractive to customers.

Other conceptual terms similar to PSS have been used to design the products and services created and offered by firms in these collaborative business contexts (e-commerce). For example, Eschenbächer and Zwegers (2003) used the term "extended products" to designate products that improve basic products with additional intangible services, which has been identified as a new phenomenon in industrial engineering. Alemany et al. (2008) used the "product pack" concept, defined as a group of products with complementary functions which can be jointly commercialized, and which constitutes a tool to achieve product differentiation based on integral design and customer orientation. Although these authors initially defined "product pack" as a group of products, later work indicated that the product pack components might be both products and services. To reflect this fact, these authors considered it appropriate to use the expression "product-services package" (PS-P) in a later work (Alarcón et al. 2009).

In short, the use and study of the PSS concept is broad and multidisciplinary (Pawar et al. 2009), and it connects the literature of several disciplines, including services marketing, service operations management, design and engineering.

However, commercializing a PSS in collaborative contexts has significant implications for the collaborating entities, especially when we consider that the shift from selling products to providing a PSS is a strategic decision (Mont 2002). In this sense, Vandermerwe and Rada (1988) state that "servitization" has enormous strategic implications for the firm and for others (i.e., suppliers or distributors). Despite the significance of the implications that using this concept involves, so far little attention has been paid to it at the policy and operational levels (Mont 2002). Along these lines, Pawar et al. (2009) suggest that both the financial and environmental benefits of a PSS have been identified, but a gap remains in the operational implications: how to make such transitions remains unclear.

Even though the study of the design or structural (strategic) implications will be very important for collaborative networks to operate properly, this work focuses on the study of another type of implications: management or functional-organizational implications. As far as the functional-organizational implications of providing a PSS to the customer at an operational level are concerned, one of the most important matters is to study the implications in the business processes carried out in collaborative entities. This work proposes a conceptual framework which helps identify, classify and study the implications involved from using the PSS concept in business processes carried out in collaborative networks. The proposed framework contemplates two states: single product (or service) and PSS. A comparison of both states for each element of the framework identifies implications when using a PSS in any process.

From a business process perspective, the conceptual framework proposed helps identify and classify the operational implications of using a PSS, thereby helping bridge the gap identified in the literature reviewed (Mont 2002; Pawar et al. 2009). As a simple tool for each firm, it also helps practitioners know the implications involved in commercializing a PSS in their processes and to finally decide whether to participate in a collaborative business form.

We now go on to include the literature review which was considered relevant for this work. After this review the proposed conceptual framework follows, and the work ends with the conclusions section.

26.2 PSS and its Implications

For the commercializing of some PSS versions, companies will often need to evolve toward enterprise networks, thus supporting the provision of new products, functions or services. In addition, companies may also face novel challenges in reshaping relationships with the other relevant stakeholders operating outside supply chains. In this sense, Mont (2002) considers that new networks may be needed to develop PSSs and indicates that companies will need to adapt their traditional structures and overcome some psychological barriers. Johnson and Mena (2008) also stress that a servitized supply chain requires the careful synchronization of product and service supply chains in order to deliver a complete product-service proposition to the customer.

Along these lines, Alonso-Rasgado et al. (2004) indicate that a functional product provides the opportunity to develop increased intellectual knowledge, generates a "smoothed" cash flow in the company and provides the supplier with long-term business stability. These authors consider that some of the most important implications are those created in the design area, whose methodology and objectives must not only undergo considerable modification, but also adapt to the new characteristics of functional products.

Pawar et al. (2009) demonstrate the need to consider the "organization" or network of firms involved in defining, designing and delivering value through the PSS. This is conceptualized as a product-service-organization (PSO). They argue that the organizational implications of PSS extend beyond the boundaries of a single firm. Developing a PSS often needs competencies, resources and capabilities which may be new to the company and require collaboration with other partners. They propose that the organization should be designed simultaneously with the product-service combination.

Despite most of the implications identified in the above works being organizational-functional in nature, none refers directly and specifically to business processes. In fact, there are very few papers mentioning the business processes issue in the PSS context. For example, the work of Alarcón et al. (2009) indicates that the definition of a PS-P implies collaboration and coordination in different activities or functions (processes), and at different levels. Collaboration will be more necessary in the phase that determines the requirements, design and planning of both the business and PS-P, and their manufacturing. In contrast, coordination will be more important in the more operative or activity phases (production, commercialization/sales, storage, distribution, etc.). Finally, Alemany et al. (2008) suggest that commercializing a product pack has important implications, not only in product design but also in a key process, the Order Promising Process (OPP).

26.3 The Conceptual Framework Proposal

From the previously cited works, we understand that for a conceptual framework to identify, classify and study the implications of using a PSS in its processes, it should contain the main concepts relating to these implications in order to permit and facilitate their identification, description, organization, understanding and analysis in the supply networks context. Thus, the main concepts considered to be related to the *implications* of a *PSS* in *processes* are precisely the descriptions of the terms "implications", "PSS" and "processes" themselves. In our work, "implications" is interpreted as *changes* or *modifications*. Implications are said to exist when a PSS alters the process; that is, when a change takes place in the process while using the PSS.

Having reviewed the literature, a "PSS" can be defined as a system formed by a series of complementary products and/or services which involves uniting or merging the individual product/services of each selling chain making up a Collaborative Selling Chain (CSC).

Moreover, regarding the definition of "process", Melao and Pidd (2000) indicate that most of the literature simply uses (or adapts) the approximate definitions that re-engineering pioneers have used: a business process is a collection of related activities which offer the customer value. However, other authors argue that a business process may be better viewed as the transformation of inputs, derived from suppliers, into outputs which address customers, and that this transformation may be hierarchically broken down into sub-processes and activities. One definition that links both ideas is that of Hammer and Champy (1993): a process is the collection of activities that uses one type of inputs, or more, and which creates an output which offers the customer value. According to these definitions and the works consulted above, *process* is defined as a set of activities which uses inputs and obtains outputs to achieve specific objectives.

By taking this definition of *process* as a starting point, we may state that *process* will be generally defined when its activities, inputs and outputs are known. But detailed knowledge of its activities implies not only the description of the activities to be done (what?), but also the identification of the resources that have to either participate or be developed while performing these tasks (who?), and also the description of when and how they are to be done (when and how?). For a full description of a process, it is also important to consider its objectives and the performance indicators which are used to control the process functioning and to meet its objectives (Alarcón et al. 2009). In summary, these seven aspects or elements will help define and organize the implications, or changes, that using a PSS in a process brings about. So, for the purpose of providing a complete description of the process, it must be identified at each process activity level; that is, what must be done, who must do it, and when and how it must be done.

From all this, we gather that using a PSS will have implications in a process if any of its elements are modified: a) the activities to be done (what?), b) the resources to be developed or those who participate in performing tasks (who?), c) when and how they have to be done (when/how?), d) process inputs, e) process outputs, f) process objectives and g) performance indicators. Next, we indicate the kinds of modifications which may take place in all seven elements of the process:

What?–Any modification brought about by using a PSS in what has to be done or in any of the activities of the process, is considered an implication. This type of implication will change the description of the affected activity. In addition, those implications which involve adding or removing one or some activities to or from the process will fall into this category.

Who?–Using a PSS is considered to have an implication in this category whenever it modifies something, by either substituting, including or putting aside one or more resources which are responsible for or participate in the performance of any activity in the process.

When/how?–This category includes those aspects relating to the time at which activities are done. Such aspects help to determine the start time, the length and the end time of each activity. The duration, start times and end times of the activities may be temporary (e.g., begin or finish to fill the order at 10:00am), or may be conditioned or determined by events (e.g., begin or finish to fill the order when the lorry arrives). In addition, the sequence in which the activities are performed will also be relevant for the process to be completed. However, this sequence is not modified if the start or the end time of at least one of the activities in the process is not altered. So this aspect is put to one side on a second level.

Process inputs–two main types are notable: information and/or materials/ products. Another important aspect of inputs is the time at which they are available to be used by any part of the process, and the resource(s) which they have to generate or contribute, known as process suppliers. So a PSS has implications in the process inputs when it modifies: the amount of inputs, their characteristics or content, the time at which inputs are available, or the resource(s) to be generated and used by the process.

Process outputs–as with inputs, there are two main types: information and materials/products. Other aspects are needed to differentiate outputs, such as the amount and characteristics of the outputs, the time at which they are available for the customer, for whom the process is performed, and the identification of the customer who is to use these outputs.

Process objectives–an objective describes a business aspect and the value or status to be achieved in that business aspect at any given time. These objectives are interpreted when a change takes place in the objectives relating to a process, when their number increases or decreases, or there are changes in: the business aspect to which they refer, the value or status to be achieved and the time at which it is to be reached (available time for its achievement). Should the same process seek to achieve several objectives simultaneously, they may be prioritized to avoid blocking the process in case of a given conflict between them.

Performance indicators–obtaining values for performance indicators (measurements) will enable decisions to be made in relation to meeting objectives (control) and operation processes. Changing the responsible resource or liability for measurement, and the frequency or time with/at which measurement is taken, are considered changes made in relation to the performance indicators.

All the aforementioned aspects for each basic process element will form a list of aspects which will facilitate identifying the implications of using a PSS in processes; once they have been arranged, they will form the conceptual framework, as seen in Table 26.1. Table 26.1 includes three columns; one for each aspect: one describing the aspect in a process executed in a traditional products/services scenario (AS-IS column); one describing the aspect of a process executed in a PSS scenario (TO-BE column); and one reflecting where a modification takes place; that is, if there are implications.

In the present-day industrial environment, some firms attempt to be different by devising new offers for the customer which combine products and services. Such new offers have received various names in the literature we consulted. Some works are currently applying these very concepts to collaborative environments, which they term “extended products”, “product pack” or “product-service pack”.

However, putting these concepts to use in collaborative contexts reveals significant implications for collaborating entities at both the strategic and operational levels. Nevertheless, this last issue has not been covered sufficiently in the literature we consulted. Thus, identifying the implications of commercializing a PSS in business processes is considered vital from an operational perspective, and tools must be developed for this purpose.

26.4 Conclusions and Future Lines of Research

This work has concentrated on the study of the implications involved in providing the customer with a PSS at an operational level, these being management or

Table 26.1 Conceptual framework proposed

					Traditional unique product (AS-IS)	PSS (TO-BE)	Any implications?
INPUTS			1	Characteristics/content			
			2	Quantity			
			3	Time availability			
			4	Supplier			
PROCESS ACTIVITIES	WHAT?	Activities	5	Description			
			6	Add			
			7	Remove			
	WHO?	Responsible entity	8	Replace			
			9	Include			
			10	Put aside			
		Participants	11	Replace			
			12	Include			
			13	Put aside			
	WHEN/ HOW?	Fixed-conditioned	14	Start time			
			15	Duration			
			16	End time			
		Time-event	17	Start time			
			18	End time			
		Value	19	Start time			
			20	End time			
			21	Duration			
OUTPUTS			22	Characteristics			
			23	Quantity			
			24	Time availability			
			25	Client			
OBJECTIVES			26	Increase			
			27	Decrease			
			28	Change the aspect referenced			
			29	Value / state			
			30	Time for achieving			
			31	Priority			
PERFORMANCE			32	Resources responsible			
			33	Measuring frequency			
			34	Measuring time			

functional-organizational implications. From the literature reviewed, we propose a conceptual framework that helps identify, classify and study the implications derived from using the PSS concept in business processes carried out in collaborative networks. The conceptual framework proposed may be useful not only for research into the identification, organization and analysis of the implications in a process, but for practitioners as a simple tool to study the implications derived from various general business actions that may affect any given process.

Future lines of research basically relate to the application of the proposed conceptual framework to business processes carried out in collaborative environments. In addition, developing a methodology that defines the steps taken to apply the conceptual framework, and using examples, could complement this work by making the conceptual framework more useful for practitioners and researchers alike.

References

Alarcón F, Alemany MME, Ortiz A (2009) Conceptual framework for the characterization of the order promising process in a collaborative selling network context. Int J Prod Econ 120(1): 100–114

Alemany MME, Alarcón F, Ortiz A, Lario FC (2008) Order promising process for extended collaborative selling chain. Prod Plan Control 19(2):105–131

Alonso-Rasgado T, Thompson G, Elfstrom BO (2004) The design of functional (total care) products. J Eng Des 15(6):515–540

Barratt M (2004) Understanding the meaning of collaboration in the supply chain. Supply Chain Manag Int J 9(1):30–42

Eschenbächer J, Zwegers A (2003) Collaboration in value creating networks: the concepts of collaborative commerce. In Jagdev HS, Wortmann JC, Pels HJ (eds.) Collaborative systems for production management. Kluwer Academic Publishers, Eindhoven

Hammer M, Champy J (1993) Reengineering the Corporation: a Manifesto for Business Revolution, New York

Johnson M, Mena C (2008) Supply chain management for servitised products: a multi-industry case study. Int J Prod Econ 114(1):27–39

Lejeune MA, Yakova N (2005) On characterizing the 4 C's in supply chain management. J Oper Manag 23(1):81–100

Manthou V, Vlachopoulou M, Folinas D (2004) Virtual e-Chain (VeC) model for supply chain collaboration. Int J Prod Econ 87(3):241–250

Manzini E, Vezzoli C (2003) A strategic design approach to develop sustainable product service systems: examples taken from the 'environmentally friendly innovation' Italian prize. J Clean Prod 11(8):851–857

Melao N, Pidd M (2000) A conceptual framework for understanding business processes and business process modelling. Inf Syst J 10(2):105–129

Mont OK (2002) Clarifying the concept of product-service system. J Clean Prod 10(3):237–245

Morelli N (2006) Developing new product service systems (PSS): methodologies and operational tools. J Clean Prod 14(17):1495–1501

Pawar KS, Beltagui A, Riedel JCHK (2009) The PSO triangle: designing product, service and organisation to create value. Int J Oper Prod Manag 29(5):468–493

Vandermerwe S, Rada J (1988) Servitization of business: adding value by adding services. Eur Manag J 6(4):314–324

Chapter 27
Use of Value Mapping Tools for Manufacturing Systems Redesign

Luis F. Romero Dessens and Zuhara I. Chávez López

Abstract Researchers and practitioners have found that focusing on mapping value is a key issue. Several mapping tools have been used for improving and redesigning manufacturing systems making them to become more competitive, flexible and efficient in order to face market economic challenges in their manufacturing environment. This chapter analyzes mapping techniques their evolution, strengths, weaknesses, key aspects and to consider how they have been adapted to real environments with different characteristics.

27.1 Introduction

Nowadays, the success of the manufacturing industry is largely determined by its ability to rapidly respond to market changes and to immediately adjust to customer needs. This has resulted in an increasing requirement, need for deployment of systems that can cope with agility and efficiency to meet these demands (Lau and Mak 2004). Companies must respond with product designs or completely new products, modifying production processes seeking lower times, more flexibility in their production systems. It has been studied by several authors that, in order to respond to customer's requirements, flexible systems imply studies about the quickness with which these systems can be modified, the required new configuration which will turn the flow to become more agile, in such a way that the production systems will technically be viable and suitable for different purposes.

L. F. Romero Dessens (✉) · Z. I. Chávez López
Universidad de Sonora, Blvd. Luis Encinas y Rosales S/N, Col. Centro, Hermosillo, Son, 83000 Mexico
e-mail: lromero@industrial.uson.mx

S. P. Sethi et al. (eds.), *Industrial Engineering: Innovative Networks*,
DOI: 10.1007/978-1-4471-2321-7_27, © Springer-Verlag London Limited 2012

Lau and Mak (2004) underline the need for an effective tool in order to develop a manufacturing system or to modify an existing system for a specific requirement at a reasonable time frame and cost. There have been developed several tools that presumed to support actions such as: process mapping based on flow maps, IDEFO, GRAI, material and flows' modeling and simulation software, mapping techniques based on Toyota production system defects and Value stream mapping (VSM) has been defined by Rother and Shook (1998), as a powerful tool that not only highlights process inefficiencies, transactional and communication mismatches but also provides guides about the improvement and redesign of manufacturing environments. Such useful tool has been evolving through its different applications, under many cases of study, under different environments and contexts.

This chapter is an ongoing research to obtain a master degree and focuses in the analysis of mapping techniques and shows how these have been used for redesigning manufacturing systems. An existing gap in the literature review has been identified, where a framework is required which may be used as stream mapping support tool to decide which technique is more suitable to the manufacturing environment in which the study is going to be deployed. This may increase or provide more resources that can be used in the same or other tasks.

27.2 Manufacturing Systems Redesigning Tools

There are many redesign and improvement tools and methods that have been used to evaluate productive systems helping production people to adjust to changes, speeding up and making materials and information flows more agile, among them it can be found the following ones (Serrano et al. 2008):

- Process mapping based on flow maps.
- IDEFO.
- GRAI.
- Material and information flows modeling and simulation software.

Their characteristics have been evaluated and analyzed, but they have been founded to be too generic, they do not work so well to the modeling of manufacturing systems, do not give the value and importance to the numeric data as a manufacturing system requires since most of them are based on business processes and in the case of the last one mentioned the resources that it implies: training, time and effort that at the end all of them translate to money, make it not so practical and not that used for companies of any field and size, even the quantitative character of it, risk reduction and potential benefits that it could generate.

There are also seven mapping tools that Hines and Rich (1997), studies and presents in terms of the seven wastes that are commonly accepted for the Toyota production system: Process activity mapping, Supply chain response matrix,

Production variety funnel, Quality filter mapping, Demand amplification mapping, Decision point analysis and Physical structure (a) volume (b) value. According to Hines and Rich (1997), the first part of the process on using these tools is to identify the specific value stream to be reviewed in order to choose the proper one for the case we are going to analyze. Second, through a series of preliminary interviews with managers in the value stream, it is necessary to identify the various wastes that exist in the value stream, that managers believe can and should be removed having as reference the seven wastes mentioned earlier. In addition, it is relevant to gain the views of managers on the importance of understanding the complete industry structure, irrespective of which wastes are to be removed, considering what is the most valuable issue for the company at a determined time. It is also recommended that more than one of these tools is used, when there are several wastes to attack, they should be addressed by tools with which they are highly.

27.3 Value Stream Mapping and its Evolution

Value stream mapping (VSM) is a technique relatively recent that gives answer to needs expressed by manufacturers in order to develop value chains more competitive, efficient and flexible; with the ones they can solve their economic difficulties that market changes may bring (Serrano 2007). Both creators and researchers have founded that VSM fills the characteristics and properties needed in a tool to be used to redesign a productive system (Rother and Shook 1998). Unlike traditional process mapping tools, VSM is a mapping tool that maps not only material flows but also information flows that signal and control the material flows. This visual representation facilitates the process of lean implementation by helping to identify the value adding steps in a value stream and eliminating the non value adding steps, or wastes (Khaswala and Irani 2001). In applying VSM, waste is identified at a high level along the value stream in the form of all elements that prohibit or hamper flow and in the form of inventory (raw materials, work-in-process (WIP) and finished goods). In future state design, major issues that create waste in the process are addressed. The future state map forms the basis for the implementation plan, for focused improvement initiatives (such as set-up reduction) (Goubergen and Landeghem 2005). VSM developers recognize that many production systems have multiple streams coming together that can complicate and increase the time needed for mapping (Khaswala and Irani 2001). Despite success stories in the application of VSM, there have been generated variants of the original technique of VSM as a result of applications in different environments and contexts, Value stream macro mapping (VSMM), and Value network mapping (VNM).

Techniques that at the end are looking for to trace the value stream in a productive environment allowing to eliminate what are called non value adding activities. In the case of the first mentioned above Fontanini and Picchi (2004), present a practical case where VSM is applied and used for a building construction project. VSMM is an extension of VSM that allows not only to see waste and flow inside a company or what is known "dock to dock" but it allows understanding material and information flows in a full supply chain involving several companies. One can expect the possibility of lean concepts and techniques application inside of VSMM, aiming the entire flow improvement and not isolated tools application or isolated initiatives with limited results.

VNM was developed to eliminate the limitations imposed on the traditional methodology when "many value streams have multiple flows that merge" (Khaswala and Irani 2001). It is able to map the complete network of the flows in a value chain that belongs to a complex product, with complex bill of material and several levels of assembly. Also, it utilizes algorithms for clustering of similar manufacturing routings and design of facility layouts to identify families of similar routings for which a single composite Current State Map could be developed. In addition, these algorithms utilize special data structures that capture the complete assembly structure of the product instead of extracting the key components only as suggested to be done with VSM. The development and benefits of this approach have been demonstrated using results from a pilot study done in a job shop where the production is made by order. Braglia et al. (2006) also tested this approach in a real job shop environment showing satisfactory results.

In the cases where VSM application is analyzed, it is highlighted the following shortcomings of the original tool, which also match with several statements made by Braglia et al. (2006); Khaswala and Irani (2001):

- Fails to map multiple products that do not have identical material flow maps.
- Fails to relate Transportation and Queuing delays, and changes in transfer batch sizes due to poor plant layout and/or material handling, to operating parameters such as machine cycle times and measures of performance such as takt time of the manufacturing system.
- Lacks any worthwhile economic measure for "value" (profit, throughput, operating costs, inventory expenses).
- Lacks the spatial structure of the facility layout, and how that impacts inter-operation material handling delays, the sequence in which batches enter the queue formed at each processing step in a stream, container sizes, trip frequencies between operations, etc.
- It is based on manufacturing systems with low variety and high volume.

There is also a need to show the impact that the inefficiencies: product travelling long distances, not integrated flows, lack of independency in a process, lack or fail of communication between parts that integrate the complete manufacturing system (because of a none existing protocol or not following an existing one) have on creating bigger amounts of WIP, operation expenses, downtime and leisure (to mention the ones seem to be the most commonly seen in manufacturing systems). The other one not less important need, is detect situations where the ergonomic aspect is poor or exist potential improvements to be done in this aspect, since it also may create an impact on employee's productivity reducing the efficiency on creating value during a formal job shift. It has not been founded any paper or study that links ergonomic aspects to creating value, is normally seen as a separated thing when if we analyze the situation with a different perspective we could find that being work force still a must in a manufacturing system, therefore it's performance have an impact on the system performance, even we have available the most modern equipment this cannot be inherent to the system, therefore should not be excluded from the re design or development of a manufacturing system. Table 27.1 below, shows a comparison between the different mapping tools mentioned above, analyzing certain aspects founded as key on the literature review, such comparison pretend to enhance strengths and weaknesses on each one of the tools.

27.4 Context Facing Issues

Many organization in their drive to become "lean", face a major obstacle when they see the need to manually map and analyze the flow paths of anywhere between 100–5,000 + routings being produced in the typical custom parts manufacturing facility (Shahrukh 2000). This will be the case of manufacturing facilities that work under the scheme Make-To-Order and Engineer-To-Order that process a large variety of products and must design their facilities for flow. Being the context more related to a project environment that to a serial production type, there is a need to choose the proper tool, in order to be able to see the complete flow and trace the exact routing for each value stream. According to the literature analyzed it is noted that before mapping there should be defined characteristics as: complexity of Bill of materials, levels of assembly, number of processes, production strategy or scheme, business type and demand, before choosing an specific tool to be applied in order to gain the most at the journey and do not truncate the efforts.

27.5 After Mapping

Based on the analysis of what is called Current state map, one then develops a Future state map by improving the value adding steps and eliminating the nonvalue adding steps known as waste or Muda. Serrano et al. (2008) summarized the guidelines needed for the definition of the future state map bases on lean thinking principles which match the statements from (Rother and Shook 1998):

- The production rate must be imposed by the product demand. Take time is the concept that reflects such a rate.
- Establishment of continuous flow where possible (unique product transfer batches).

Table 27.1 Comparison between mapping techniques

Mapping technique/ Key aspects to evaluate	Use of quantitative data (processing times, takt time, lead time, inventory, travel distances, processing quantities, etc.)	Flow through plant layout	Influence of ergonomics in value creation	Job sequencing	Material handling	WIP buildup	Capacity constraints	Economic investment	Context of practical application
Process mapping based on flow maps	H	NC	NC	NC	L	NC	NC	L	SP, JS
IDEFO	L	NC	NC	NC	NC	NC	NC	L	SP,BP
GRAI	M	NC	NC	NC	NC	NC	NC	L	SP, BP
Material and information flows' modeling and simulation software	H	M	NC	H	L	H	H	H	SP,JS,BP
Process activity mapping	H	NC	NC	NC	L	NC	NC	L	SP,JS
Supply chain response matrix	H	NC	NC	NC	NC	NC	NC	L	SP,P
Production variety funnel	H	NC	NC	NC	NC	NC	NC	L	SP
Quality filter mapping	H	NC	NC	NC	NC	L	NC	L	SP,JS
Demand amplification mapping	H	NC	NC	NC	NC	NC	NC	L	SP
Decision point analysis.	M	NC	NC	NC	NC	NC	L	L	JS
Physical structure (a) volume (b) value.	M	NC	NC	NC	NC	NC	M	L	SP

(continued)

Table 27.1 (continued)

Mapping technique/ Key aspects to evaluate	Use of quantitative data (processing times, takt time, lead time, inventory, travel distances, processing quantities, etc.)	Flow through plant layout	Influence of ergonomics in value creation	Job sequencing	Material handling	WIP buildup	Capacity constraints	Economic investment	Context of practical application
VSM	H	NC	NC	NC	NC	NC	NC	L	SP,P,B
VSMM	H	NC	NC	NC	NC	NC	NC	L	SP,P
VNM	H	H	NC	NC	NC	NC	NC	L	SP,JS,P,B

Low (L), Medium (M), High (H), Not considered (NC), Serial production (SP), Job shop (JS), Project (P), Business processes (BP)

- Employment of pull systems between different work centres when continuous flow is not possible.
- Only one process, called the pacemaker process, should command the production of the different parts. This process will set the pace for the entire value stream.

In lean thinking is usual to be value mapping based that support strategies that precede any specific mapping linked to tools such one-piece flow, visual control, kaizen, cellular manufacturing, inventory management, Poke yoke, standardized work, workplace organization, among others that share the same purpose: to deploy a production system close as possible to an agile system which makes an appropriate use of resources available.

27.6 Conclusion and Further Research

Between all the mapping techniques analyzed, VNM has shown to have an approach that seems to be more suitable for the context we see now is predominant in companies structures and in their production strategies, which are contexts more like a project environment than a serial production scheme and budged for tool application is limited; despite the improvements that VNM brings to the original VSM, there have been defined several deficiencies in it:

- Lacks including information on lot sizing, cycle time, job sequencing at each process and work in process buildup at each process due to queuing delays, required to better design of the future state map.
- Do not consider capacity limitations of the system when multiple assemblies require to use capacity of a shared process.
- Lacks detailed analysis of the material handling systems and processes connecting different pairs of process boxes.

The mentioned above suggest that there is still much research work to be done. There is research in process focused on a case study application under a job shop environment; it is intended to propose a redesign on the system that be able to improve the manufacturing system's performance and also make an academic contribution to value mapping tools by working on attack the deficiencies of the analysis of the material handling and plant layout.

Acknowledgments Chávez Z., Thanks her thesis tutor for the support and guide given so far and for what will be needed in future through the whole research process.

References

Bhim S, Sharma SK (2009) Value stream mapping as a versatile tool for lean implementation: an Indian case study of a manufacturing firm. Meas Bus Excell 13(3):58–68

Braglia M, Carmignani G, Zammori F (2006) A new value stream mapping approach for complex production systems. Int J Prod Res 44(18–19):3929–3952

Fontanini PS, Picchi FA (2004) Value stream macro mapping: a case study of aluminum windows for construction supply chain. In: Proceedings of the 12th annual conference of the international group for lean construction (IGLC-12). Elsinore, Denmark

Goubergen DV, Landeghem HV (2005) Using value stream mapping to redesign engineering project work. Virginia polytechnic institute and state university.Van Goubergen PandM gcv

Gunasekaran A (1999) Agile manufacturing: a framework for research and development. Int J Prod Econ 62:87–105

Hines P, Rich N (1997) The seven value stream mapping tools. Int J Oper Prod Manag 17:146–164

Khaswala ZN, Irani SA (2001) Value network mapping (VNM): visualization and analysis of multiple flows in value stream maps. In: Proceedings of the lean management solutions conference. St. Louis, MO

Lau H, Mak K (2004) The design of flexible manufacturing using an extended unified framework. J Manufact Technol Manag 15(3):222–238

Marchwinski C, Shook J (2003) Lean lexicon: a graphical glossary for lean thinkers. Lean Enterprise Institute, Massachusetts

Rother M, Shook J (1998) Learning to see: Value stream mapping to add value and eliminate Muda. Lean Enterprise Institute, Massachusets

Serrano I (2007) Tesis doctoral: Análisis de la aplicabilidad de la técnica Value Stream Mapping en el rediseño de sistemas productivos. Universitat de Girona, España

Serrano I, Ochoa C, De castro R (2008) An evaluation of the value stream mapping tool. Bus Proc Manag J Vol 14(1):39–52

Shahrukh AI (2000) Value stream mapping in custom manufacturing and assembly. The Ohio Manufacturer, Vol 9, pp 12–13

Chapter 28
Developing a Plug&Lean-CiMo a Model for Improving Manufacturing Operations

Martha-Patricia García, Javier Santos, Mikel Arcelus and Elisabeth Viles

Abstract Manufacturers continue facing the never ending challenge of finding the best way to increase productivity levels in order to remain competitive. Significant level of productivity are given by the efficient utilization of production equipment, thus manufacturing companies use the overall equipment effectiveness (OEE) rate as an indicator to control equipment utilization related to the maintenance activity; however they do not take into account other valuable OEE-information to improve their entire manufacturing operations. This paper presents a continuous improvement model Plug&Lean-CiMo, whose aims are the accurate calculation of the OEE indicators, the appropriate classification of losses, and the systematic integration of lean manufacturing philosophy tools in an improvement methodology. The model has the advantage of using a portable wireless system to support the automated collection, of data to make the continuous improvement process easier. A case study is presented to illustrate the validation of the model.

28.1 Introduction

In this competitive age manufacturers have been facing the never ending challenge of finding the best way to increase productivity levels in order to remain competitive. Most manufacturing companies use the overall equipment effectiveness (OEE) rate

M.-P. García (✉)
Industrial Engineering Department, Instituto Tecnológico de Chihuahua II, México,
Av. De las Industrias 11101, 31136 Chihuahua, México
e-mail: pgarcia@tecnun.es

J. Santos · M. Arcelus · E. Viles
Industrial Management Department, TECNUN School of Engineering,
University of Navarra, San Sebastian, Spain

S. P. Sethi et al. (eds.), *Industrial Engineering: Innovative Networks*,
DOI: 10.1007/978-1-4471-2321-7_28, © Springer-Verlag London Limited 2012

as an indicator to monitor and control performance, equipment efficiency and productivity.

Relevant research (Ericsson and Dahlén 1997) reports two important findings: one, that accurate equipment performance data is essential to long-term effectiveness of operational activities; two, if the extent of equipment failures and reasons for production losses are not entirely understood, improvements actions cannot be positive to solve major problems or arrest deteriorating performance.

Given today's complex manufacturing environment, many electronic systems, intelligent systems and software tools supported by information technologies (IT) are available for equipment monitoring and control, which includes automatic data collection, OEE calculations and a variety of key performance indicators (KPI); However, many manufacturing companies are using hand-collect data because of both; they fear the costly and the drastic management changes brought by the adoption of these systems. Thus, to collect accurate data, to understand and build appropriate classification of losses and to adopt a proper continuous improvement methodology are significant factors to be considered for this research study. After analyzing these factors, two questions motivate this study:

How can we make OEE a valuable metric in order to identify opportunities to make improvements in the right place at the right time?

How can we develop a continuous improvement (CI) strategy while utilizing the OEE metric in such way to increase the effectiveness of entire manufacturing operations?

This paper present a research study for developing and constructing a continuous improvement model (Plug&Lean-CiMo), whose aim is the accurate calculation of the OEE indicators and the systematic integration of lean manufacturing philosophy tools in an improvement methodology for manufacturing operations. The model has the advantage of using a portable wireless system named Plug&Lean (composed of a wireless tool and management system) (Santos et al. 2011) to support the automated collection, calculation, and graphic presentation of data. The research methodology utilized is a case study.

28.2 Literature Review and Background

28.2.1 TPM and OEE

The implementation of total productive maintenance (TPM) methodology has been expanding significantly within manufacturing companies specially companies aspiring to become competitive because it really provides tangible results. Total Productive Maintenance (TPM) is an improvement methodology developed by Nakajima during the 1970 s based on two major concepts: (1) maximizing equipment effectiveness and (2) people involving. Additionally, Nakajima (1988) provides the overall equipment effectiveness (OEE) as a measure of manufacturing improvement because attempts to reveal hidden losses. The OEE calculation is split into three components: availability, performance and quality, these components identify the six big losses; a brief description is given as follow:

The main words in all headings (even run-in headings) begin with a capital letter. Articles, conjunctions and prepositions are the only words which should begin with a lower case letter.

- Availability: the effectiveness of equipment available for performing production activity. This rate indicates the operating time after time losses.
- Performance: the effectiveness of the production activity during the period of time that equipment is available and able for performing production activities. This rate indicates when a machine operates at less than the optimum conditions.
- Quality: the effectiveness of the production activity to produce units that meet quality specifications during the time that equipment is performing production activity.
- OEE: is measured in terms of these six big losses, which is essentially the result of the product of the multiplication of availability, performance and quality rate.

There are several researches about OEE concept resulting in different opinions about its potential. For example, Dal et al. (2000) explored the use of OEE not only as an operational measure but also as an indicator of process activities concluding that OEE should be balanced by other traditional measure for being more popular in into the performance research community. Jeong and Phillips (2001) proposed a new interpretation of OEE referring to time efficiency, speed efficiency and quality efficiency and consequently to discuss a new categorization of losses. Ljungberg (1998) discussed the importance for determining the understanding by personnel of the magnitude and reason for machinery losses in order to provide an appropriate base for planning improvement activities; Dilorio and Pomorsky (2003) proposed a TPM loss analysis model for generating accurate and non theoretical OEE metrics.

28.2.2 Lean Manufacturing

Lean manufacturing is an operational management philosophy focusing on reduces waste in a manufacturing system. The lean literature defines waste as everything that increases cost without adding value for customer (Womack 1996). The origin of this philosophy comes from the Toyota production system in 1950 s in Japan, and today is widely recognized and adopted by world class manufacturing companies. It success is because the elimination of waste impact on several issues: (1) the reduction of cost, time and defective products, (2) equipment effectiveness, (3) increase productivity; this philosophy is composed by principles, methodologies and tools, and its functioning is on the basis of continuous improvement and workers involved. Grishnik and Winkler (2008) examined the flaws and limitations of Lean: and they said that manufacturers are suffering "manufacturing myopia" because companies avoid understanding all the components of an entire Lean implementation and they have been seen manufacturing in a very narrow way: cost reduction; consequently they are missed gold opportunities to grow a better competitive position.

28.2.3 Continuous Improvement

Continuous improvement can be literally defined as continuous incremental improvement of the standard way of work (Imai 1986). Additionally, in the field of environmental manufacturing it can be defined as the constant measure of the effectiveness of its processes to eliminate difficulties to achieve standard objectives with less effort.

The continuous improvement (CI) methodology has evolved over the years and now, modern CI is considered to be a hybrid methodology rested under several initiatives, such as Lean Manufacturing, Total Productive Maintenance, Total Quality Management and Six Sigma, (Bhuiyan and Baghel 2005). Imai (1986) uses a Kaizen (Japanese) concept as synonym of CI; he states that CI can take place at three different levels within the organization: at the top management, group, and individual levels. At the top management level, CI is associate with the organization's strategy, group level CI comprise decisions and problem-solving activities, while individual level CI involve improvement on a day a day tasks. Hence, the practice of continuous improvement represents a new dimension in labor because it involves people at all hierarchical levels and encourages them to act and to creatively contribute to make their work better and to reduce unnecessary human efforts.

28.3 Research Methodology

The methodology utilized in this study is a case study, which has been recognized as being particularly suitable for theory refinement (Voss et al. 2002). This study focuses on the development of continuous improvement activities as a systematic way for improving equipment efficiency, hence our key factors are equipment efficiency and continuous improvement, and the link between them conceptualize our model for continuous improving.

28.4 The Model

The model is a functional model based on lean manufacturing philosophy and supported by the wireless Plug&Lean-system (Fig. 28.1). There are four specific objectives of the model which are described as follow:

- To automate collecting performance equipment data in real time. Its aim is the capture of accurate manufacturing data, to turn the data into information through the execution of the OEE calculation and to provide the availability, the performance and the quality rates and other customized key performance indicators (KPI). This is based on wireless technology.
- To display graphical information and charts regarded to indicators about performance. Its aim is to turn the information into knowledge through both, (1) Graphical charts in a parallel way displaying levels of actual and targeted OEE indicators, and (2) Graphical charts showing in detail the weak points that mean causes of variation in machinery or process performance. Additionally this stage allows accumulate the information for displaying graphically historical indicators and its tendency (trend) by the time. This is based on the Nakajima's theory of losses.
- To determine the root cause of losses. Its aim is to turn knowledge into action through the localization, classification and categorization of the sources of losses for making intervention prioritization activities in order to improve. This is based on the root cause analysis (RCA) and Lean tools.
- To implement a structured methodology to improve. Its aim is to confirm improvements through planning and developing CI activities. This is based on Lean tools.

The operation of the overall model depends on its processes, procedures and activities and the interrelation among they. The emphasis of this model is: to do with less effort in an easy structured way, bottom-line improvements.

The Plug&Lean is a portable electronic system for monitoring and measuring, manufacturing equipment performance in real time. It is composed by two lements (1) the Plug&Lean-tool, as a core element, built by small wireless devices, including a PDA and a bar-code system, and (2) a management platform. Its objective is facilitating the data collection, the OEE calculation and the reporting information from a selected production process. There are two attributes: (1) the design allows the tool to be easily plugged into a designated area or single piece of equipment in order to capture performance behavior, during a selected production period of time; (2) the management system allows data processing focusing on main equipment constraints for reporting graphical information, in such way that just a glance operators can identify losses (Santos et al. 2011).

28.4.1 Model Operation

The model operation section describes the procedures, activities and resources required for each process. Figure 28.2 shows an outline of the structured model, it allows see the articulation of the four processes which description is given in the subsequent paragraphs.

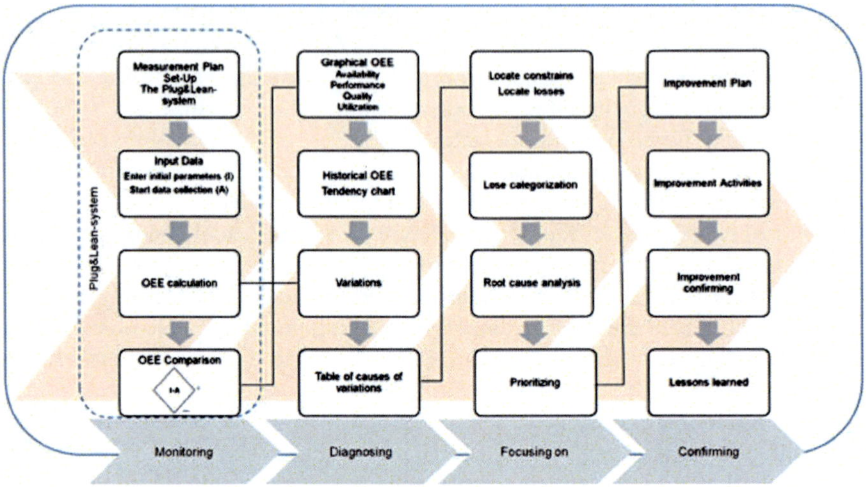

Fig. 28.1 The Plug&Lean-CiMo model design

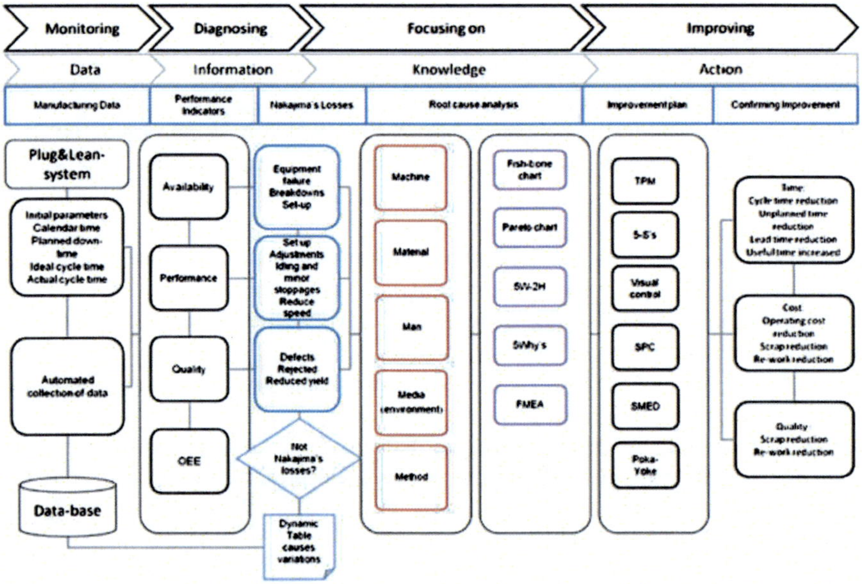

Fig. 28.2 Plug&Lean CiMo model structure

28.4.1.1 Monitoring, Diagnosis, Focusing on, and Improving

It is the procedure for monitoring the performance of equipment or production process during a period of time. This stage is composed of 4 processes: (1) data input, (2) OEE-KPI calculation, (3) comparison, and (4) classification. This monitoring process includes the use of the Plug&Lean system for supporting the operation of the model. It is important to mention again that the core element of the Plug&Lean system is the Plug&Lean-WiTool.

The Diagnosing stage is a procedure for reporting through graphics and charts actual levels of OEE to allow the examination of the availability (A), performance (P), quality (Q), utilization (U) and OEE indicators. This stage comprises four processes: (1) presentation of a current graphical report related to the OEE indicators and other customized KPIs; (2) presentation of cumulative and historical OEE indicators with trend; (3) graphical presentation of OEE discrepancies between ideal and actual rates; and (4) table of causes of variation.

The Focusing On stage is a procedure for localizing and analyzing causes of performance variation in equipment or process, this procedure takes the information from the diagnosing stage. It is composed of four processes: (1) constraint location; (2) lose categorization; (3) root-cause-analysis (RCA); (4) intervention and prioritization. It is supported by the seven quality tools, RCA methodology and lean tools. This stage is based on the total quality management TQM.

The improving stage is the fourth procedure of the model regarding to develop a structured continuous improvement plan in order to implement actions to eliminate or mitigate losses. This stage takes source of losses information from the focusing on stage for working along the four procedures.

28.5 Case Study

A case study is presented in order to validate the study. We selected a manufacturing company located in the Basque region of Spain which is comprised of three semi-automated and one automated continuous production lines. The company is interested in increasing efficiency and productivity, and it is committed to continuous improvement practices. The Plug&Lean-methodology was tested from May to December 2009, in the automated line. During the study, and because the new methodology proposed, the company faced a dilemma relating to be precise in the identification of the planned and non-planned downtime, as well as the identification of re-work as a quality loss. They held meetings and were involving personnel from the quality, the maintenance and the production areas for a brainstorm in order to make decisions to solve the dilemma; this meeting and action was capitalized and a first multidisciplinary team in the company was integrated. Because the confidentially of the company's information, we present only partial results. Figure 28.3 shows one of the reporting graphs, which presents two graphics for comparing indicators levels before and after implement our model; each graphic shows detailed charts with the OEE indicators,

The company could see how increase or decrease the levels of efficiency, by the time. This case study showed the company's acceptance of the OEE metrics, consequently they adopted the methodology of the model in the other three

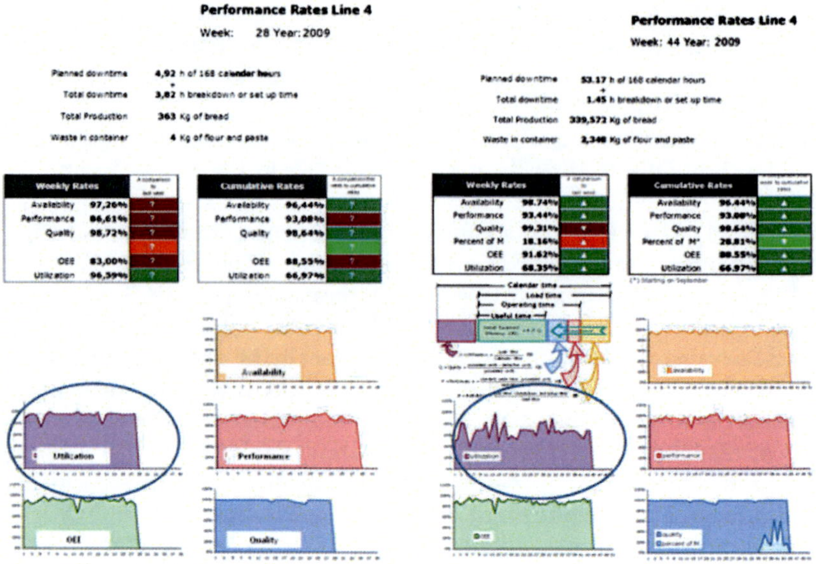

Fig. 28.3 Case study reporting graph

semi-automated production lines, resulting in a motivation for working because of the improvements they saw.

28.6 Conclusions

In this paper we proposed a continuous improvement model supported by wireless technology to provide a new way to improve the manufacturing operation process. In the manufacturing scenario, measuring and controlling are significant factors that are considered for successful improvements; measuring is based on having accurate data, and controlling is based on having appropriate classification of losses. The proposed model is driven by two key factors: (1) the OEE indicators and (2) the continuous improvement methodology. The advantage of the model is that it operates with a structured CI methodology and an accurate information processing platform thanks to the integration of the Lean manufacturing tools and the Plug&Lean system. The study's contributions are: (1) the integration of a cutting edge wireless device in the field of operations management, not only for maintenance but also for continuous improvements activities; (2) a valuable affordable procedure for businesses, especially small ones, that increases efficiency in the entire manufacturing operation; and (3) a valuable training methodology for academia.

References

Bhuiyan N, Baghel A (2005) An overview of continuous improvement: from the past to the present. Manage Decis 43(5):761–771

Dal B, Tugwell P, Greatbanks R (2000) Overall equipment effectiveness as a measure of operational improvement. A practical analysis. Intl J Oper Prod Manage 20(12):1488–1502

Dilorio S, Pomorski T (2003) Semi equipment performance and productivity measurement methods, algorithms and standards. Brooks-PRI automation, Asia user conference, Shanghai

Ericsson J, Dahlén P (1997) A conceptual model for disruption causes: a personnel and organization perspective. Int J Prod Econ 52:47–53

Grishnik K, Winkler C (2008) Make or break. How manufacturers can leap from decline to revitalization. McGraw-Hill, U S A

Jeong KY, Philips DT (2001) Operational efficiency and effectiveness measurement. Int J Oper Prod Manage 21(11):1404–1416

Imai M (1986) Kaizen. The key to Japan's competitive success. Random House Inc., Canada

Ljungberg O (1998) Measurement of overall equipment effectiveness as a basis for TPM activities. Intl J Oper Prod Manage 18(5):495–507

Nakajima S (1988) Introduction to TPM. Productivity Press, Cambridge

Santos J, Garcia MP, Arcelus M, Viles E, Uranga J (2011) Development of a wireless Plug&Lean system for improving manufacturing equipment diagnosis. Int J Comput Integr Manuf 24(4):338–351

Voss C, Tsikriktsis N, Frohlich M (2002) Case research in operations management. IJOM 22(2):195–219

Womack J, Jones D (1996) Lean thinking. Simon and Schuster, New York

Chapter 29
Order Fulfilment Strategies in the Capital Goods Sector. An Empirical Research

Raquel Sanchis, Eduardo Saiz, Eduardo Castellano and Raul Poler

Abstract The capital goods sector is characterized by complex and heavily engineered products. Moreover, customers demand customized features of such products. For this reason, the choice of the most appropriate Order Fulfilment Strategy and the position of the Customer Order Decoupling Point are key aspects to manufacture products in such a sector. This study analyzes the different strategies used by capital goods sector companies depending on the customization degree of the products manufactured. The research is based on an empirical study among capital goods sector companies and explores relationship between four types of products, depending on their customization degree (standard, configurable, customizable and unique) and four different Order Fulfilment Strategies: Make-to-stock, Assemble-to-order, Make-to-order and Engineer-to-order.

R. Sanchis (✉) · R. Poler
Centro de Investigación en Gestión e Ingeniería de Producción (CIGIP),
Universidad Politécnica de Valencia, Plaza Ferrándiz y Carbonell 2,
03801 Alcoy – Alicante, Spain
e-mail: rsanchis@cigip.upv.es

R. Poler
e-mail: rpoler@cigip.upv.es

E. Saiz · E. Castellano
Ikerlan-IK4. Centro de Investigaciones Tecnológicas,
Paseo JM Arizmendiarrieta 2, 20500 Arrasate-Mondragón, Spain
e-mail: esaiz@ikerlan.es

E. Castellano
e-mail: ecastellano@ikerlan.es

S. P. Sethi et al. (eds.), *Industrial Engineering: Innovative Networks*,
DOI: 10.1007/978-1-4471-2321-7_29,

29.1 Introduction

Markets have become global, increasing the range of choice. Customers place two major pressures on many companies: they want products that fit their specific needs but they are not willing to pay high premiums for these customized products, even when up against competing standard products in the market. These two forces can be related to the traditional trade-off: flexibility and productivity. The positioning of the Customer Order Decoupling Point (CODP), sometimes referred to as the Order Penetration Point (OPP), is the stage in the manufacturing value chain where a particular product is linked to a specific customer order (Olhager 2003) and involves identifying the optimal balance between the productivity and flexibility forces (Rudberg and Wikner 2004).

CODPs are used to classify value-adding activities in terms of Order Fulfilment Strategies (OFS) that encompass the material processing activities concerned with complying with customer instructions and the control of these activities. Brabazon and MacCarthy (2005) reviewed the literature and identified four structural forms which may be useful to choose the appropriate OFS:

Fulfilment from Stock

Make-to-stock (*MTS*): this is a build-ahead production approach in which production plans are based on information of historical demand, along with sales forecast information. MTS is considered good for high volume products where the demand is either seasonal or easily predicted, or both. Therefore, manufacturing and assembly occur before receiving orders. Orders are supplied from stocks of finished products.

Fulfilment from a Single Fixed CODP (Amaro et al. 1999)

Engineer-to-order (ETO): Products are manufactured to meet a specific customer's needs and so require unique engineering design or significant customisation. Thus, each customer order results in a unique set of part numbers, bill of materials, and routing.

Make-to-order (MTO): Most or all the operations necessary to manufacture each specific product only take place after the receipt of a customer order. In some situations even materials and component parts may have to be procured upon receipt of a particular order.

Assemble-to-order (ATO): The final products offered to customers, although presenting some degree of customisation, are produced with (common) standardised parts, which can be assembled in a number of different options. The receipt of an order initiates the assembly of the particular finished product.

They also analyzed the fulfilment from one of several fixed CODPs and fulfilment from several locations with floating CODPs. However, these two types of OFS fall outside the scope of this study.

Many authors have studied the factors that affect the positioning of the CODP. Olhager (2003) divides the most important factors in three categories: (i) market-related, (ii) product-related and (iii) production-related factors. Yang and Burns

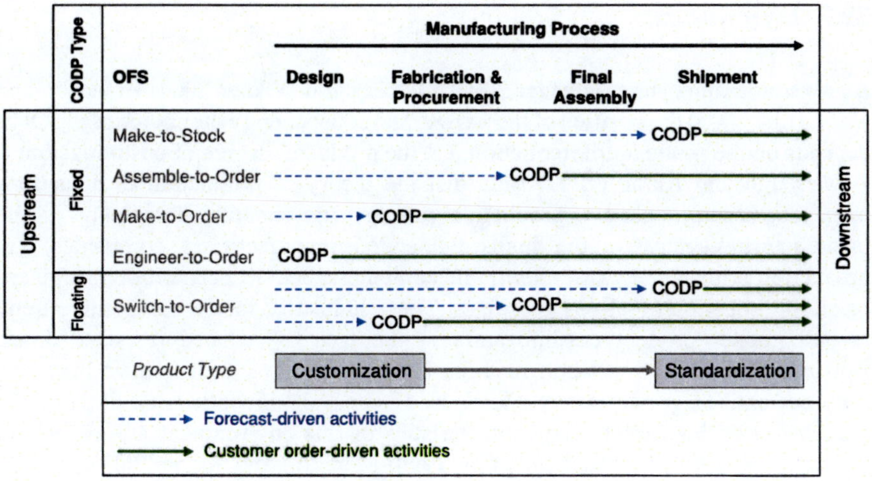

Fig. 29.1 OFS and its related CODPs (adapted from Olhager 2003)

(2003) explain that the position of the CODP depends on "a balance between the product type, market, process and stock characteristics". The product type is a key variable to choose the most appropriate OFS and position the CODP, depending on the degree of customization.

Rudberg and Wikner (2004) provided an analysis of the role of the CODP within a mass customisation context. The further downstream, or more specifically, the closer to the consumer the CODP is positioned, themore delayed or "postponed" the differentiation of the product builds will be (standard or configurable product types). Similarly, the further upstream the CODP is positioned, the earlier the differentiation of the product (customized and/or unique product types). Figure 29.1 illustrates this situation.

For a company, the decision to choose the right OFS is a pressing need to meet their customer requirements. The main objective of this paper is to analyze the different positions of the CODPs and the diverse OFS implemented in companies belonging to the capital goods sector, depending on four different types of products. These product types have been classified as follows, depending on their degree of customization:

- Non Customizable Standard product (NCSP).
- Configurable Product from a catalogue with Standard Options (CPSO).
- Customisable Product from a catalogue with standard options including partial design and production Engineering Options (CPEO).
- Unique Product with Complete Design and production engineering (UPCD).

29.2 Hypotheses

As several authors state (Olhager 2003; Rudberg and Wikner 2004; Brabazon and MacCarthy 2005) the position of the CODP, and consequently the choice of the OFS, depends on the product differentiation, i.e. the products' degree of customization.

Meredith and Akinc (2007) state that the theory of production in operations management has lacked a production strategy for one major segment of the manufacturing industry: the high-value-added manufacturing companies, i.e. production of machine tools, mainframe computers, heavy construction machinery or plastic injection moulding machines. These industries face a competitive trend toward greater product customization in the face of reduced response times. But in their case, competition requires manufacturers to deliver their high-cost semi-customized products with total lead times significantly shorter than the manufacturing lead time. The characteristics of this environment are as follows (Raturi et al. 1990; McCutcheon et al. 1994; Meredith and Akinc 2007):

- The product is large, heavily engineered, and very expensive.
- Volumes of products with any particular combination of features are small, and their demand is difficult to predict.
- The time to purchase components and manufacture the product (the build time or lead time) is much greater than the delivery time that customers expect.
- Customers desire customized features of the product that must be established early in its build cycle.
- Holding some critical stock inventories is virtually impossible, both financially and physically.

As aforementioned, the position of the COPD in the capital goods sector is crucial to manufacture products in mass customization scenarios. Based on the characteristics of the capital goods sector, the following joint analysis has been developed:

- For NCSP, the OFS considered in this study have been MTS, ATO and MTO.
- For CPSO, ATO and MTO have been analyzed.
- In the case of CPEO, the study has been focused on MTO and ETO.
- Finally, for UPCD, the main OFS analyzed has been ETO.

Taking into account these considerations and based on the previous literature review, the following hypotheses are proposed:

H1. There is a direct relationship between the product type (considering this as the different customization degree) and the different OFS chosen by the capital goods sector producers.
H2. Capital goods sector producers mainly manufacture CPEO and/or UPCD, and in order to turn out their products, they mainly use MTO and/or ETO strategies.

29.3 Results and Discussion

A quantitative study was performed by means of the development of an online survey[1]. The main objective of the survey was to gather information from a representative sample of European capital goods manufacturing companies in order to focus on the real OFS of the sector and in the different positions of the CODP.

To facilitate the distribution of the questionnaire to companies, as well as its accessibility, and to simplify its completion, it was drawn up in five different languages (English, German, French, Italian and Spanish). Additionally, a glossary of terms was included to ensure the understanding of unusual concepts.

The dissemination of the online survey was done through contacts with the main European associations of capital goods manufacturers, database queries, e.g. CNAE and DUNS databases in Spain, and direct email contacts. At the end of the analysis, each company received on demand a report comparing its company with the rest of participants. Comparative results were added taking into account factors such as size, sector and country of the company, to provide different levels of stratified information.

The profile of the companies surveyed includes small and medium-sized enterprises, as well as large companies. The sample of companies surveyed can be used to represent the whole range of companies within the capital goods sector (Fig. 29.2)

In order to contrast hypothesis H1, linear scales for product type (NCSP = 1, CPSO = 2, CPEO = 3, UPCD = 4) and OFS (MTS = 1, ATO = 2, MTO = 3, ETO = 4) have been generated, transforming the percentages of both variables into values between 1 and 4. Then the Pearson's correlation coefficient analysis was carried out over the product type and OFS scales, the result of which is shown in Table 29.1:

The confidence level was set at 95% to coincide with the 5% convention of statistical significance in hypothesis testing. In this case, the statistical significance (Sig.) parameter is lower than 5%.

The Pearson's correlation coefficient shows that there is a 0.345 positive correlation between the variables used: product type and OFS.

This result confirms hypothesis H1, since there is a correlation between the product types and the different OFS.

Due to the intrinsic characteristics of the sector (described in Sect. 29.2), most of the capital goods products are CPEO and UPCD (70% in absolute terms), as it is shown in Fig. 29.3. These product types present tailored complex characteristics and for this reason it is necessary to design and engineer to order, to satisfy the order's precise and/or unique product specification.

Moreover, ETO is the most widely used approach, and represents, in absolute terms, 55%. These results are graphically represented in Fig. 29.4. The relationship between the product types and the OFS is represented in Fig. 29.5.

In the case of NCSP, the MTO strategy is the most used, followed by ATO. It was foreseeable that MTS was one of the most chosen strategies. However, the

[1] Survey URL: http://www.remplanet.eu/ResilienceSurvey

Table 29.1 Pearson's correlation coefficient

Pearson's correlation coefficient (r)	0.345
Statistical significance parameter (Sig.)	0.042

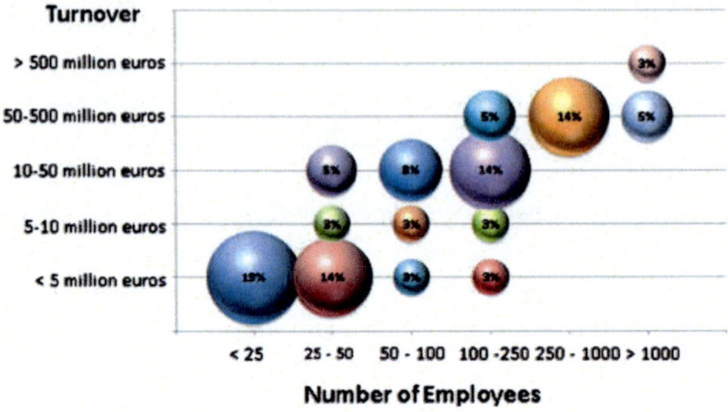

Fig. 29.2 Profile of capital goods companies surveyed

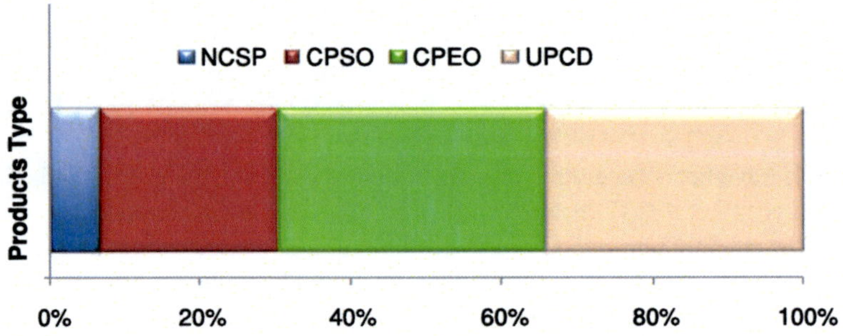

Fig. 29.3 Percentages of product types in the capital goods sector

percentage of the MTO strategy for NCSP is higher than in the case of MTS. This may be due to the fact that capital goods producers of NCSP do not base their manufacturing process on historical data forecasts, or even due to the high inventories costs, since the capital goods sector products are usually large and for this reason, they do not initiate their production until the customer order is received. It is also important to highlight that some of them use ATO in order to reduce response times.

On the other hand, the capital goods producers of CPSO use as much ATO approaches as MTO. The adoption of ATO may be due to the fact that this type of products is composed by standard elements through a set of predefined rules and customer interaction. Therefore, CPSO producers chose this strategy to give a quick response to the customer by reducing the manufacturing lead time. However, other producers prefer using MTO in order to reduce the risks of having a large quantity of fixed assets.

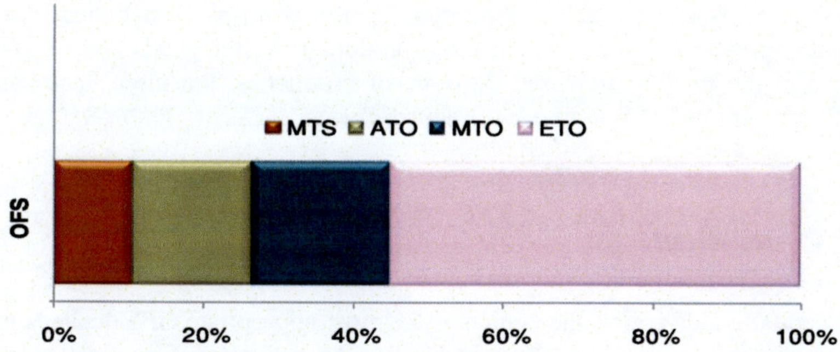

Fig. 29.4 Percentages of OFS in the capital goods sector

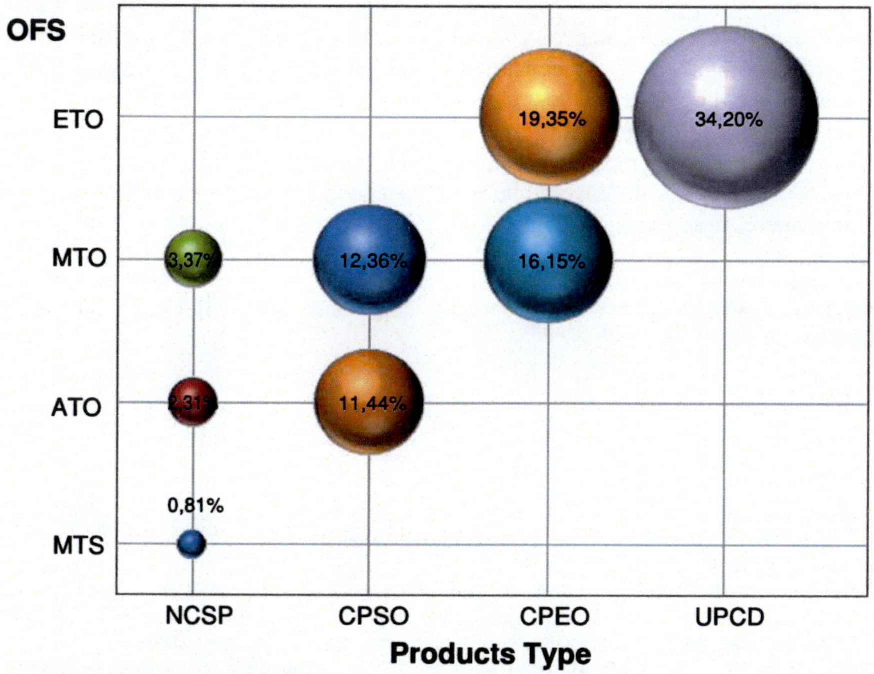

Fig. 29.5 OFS versus Products Types in the capital goods sector

The CPEO are products from a catalogue with standard options including partial design and production engineering options. In the case that the design and production engineering options are few, most of the CPEO producers use the MTO approach. However, if this type of production system has to be partially engineered, the CPEO capital goods sector producers adopt an ETO approach. Finally, the UPCD are heavily engineered products and for this reason, ETO is the main strategy used.

Therefore, hypothesis H2 is supported by the descriptive statistical analysis performed previously because, the most produced capital goods are CPEO and UPCD and the more customized the type of product is, the more upstream the CODP is positioned (MTO and ETO approaches).

29.4 Conclusions

In the capital goods sector, the choice of the most appropriate OFS depends on the type of product, with as much consideration to the type of product as to the different customization degree. The empirical research has corroborated that a correlation exists between the degree of product customization and the order fulfilment strategy selected by the companies. Moreover, the literature states that the products fabricated in companies such as producers of machine tools, mainframe computers, heavy construction machinery or plastic injection moulding machines are large, heavily engineered, very expensive and with various customized features. The results of the empirical study shows that the capital goods sector products are mainly CPEO and UPCD with design and production engineering and the producers chose MTO and ETO as the main strategies to manufacture these types of products.

Acknowledgments The research leading to these results has received funding from the European Community's Seventh Framework Programme (FP7/2007-2013) under grant agreement n° NMP2-SL-2009- 229333.

References

Amaro G, Hendry L, Kingsdam B (1999) Competitive advantage, customisation and a new taxonomy for non make-to-stock companies. Int J Oper Prod Man 19:349–371. doi:10.1108/01443579910254213

Brabazon PG, MacCarthy BL (2005) Review of order fulfilment models for catalogue mass customization. In: Mass Customization: concepts, tools, realization, International Mass Customization Meeting 2005 (IMCM'05). Hamburg, Austria 2–3 June 2005

McCutcheon DM, Raturi AS, Meredith JR (1994) The customization-responsiveness squeeze. Sloan Manage Rev 35:89–99

Meredith JR, Akinc U (2007) Characterizing and structuring a new make-to-forecast production strategy. J Oper Manag 25:623–642. doi:10.1016/j.jom.2006.04.006

Olhager J (2003) Strategic positioning of the order penetration point. Int J Prod Econ 85:319–329. doi:10.1016/S0925-5273(03)00119-1

Raturi AS, Meredith JR, McCutcheon DM, Camm JD (1990) Coping with the Build-to-Forecast Environment. J Oper Manag 9:230–249. doi:10.1016/0272-6963(90)90097-W

Rudberg M, Wikner J (2004) Mass customization in terms of the customer order decoupling point. Prod Plan Control 15:445–458. doi:10.1080/0953728042000238764

Yang B, Burns ND (2003) The implications of postponement for the supply chain. Int J Prod Res 41:2075–2090. doi:10.1080/00207544031000077284

Chapter 30
On the Exact Calculation of the Fill Rate for Repairable Parts: Application to an Airline Company

Marta Palmer, Eugenia Babiloni, Manuel Cardós and Sofía Estellés

Abstract This paper focuses on the improvement of inventory policies of repairable parts of the airline company Air Nostrum. The company uses the sales replacement policy to manage their repairable parts and the base stock is determined by means of an approximated expression of the fill rate assuming Poisson distributed demands. However, this paper shows that real data may not always be modelled with the Poisson distribution and the only exact method available in the literature fall into significant deviations when other discrete distribution is used. According to that, this paper derives an exact method to compute the fill rate in the presence of any stationary, discrete and *i.i.d* demand pattern for the policy the company uses. Illustrative examples show that deviations which arise from using approximate methods can lead to overestimate the fill rate so the inventory policy does not reach the target fill rate.

30.1 Introduction and Literature Review

Repairable parts are a type of spare part that eventually can be repaired and used again. Nowadays the optimization of the inventory of repairable parts has become a strategic issue in industries such as airlines or power stations since: (i) they serve

M. Palmer · E. Babiloni (✉) · M. Cardós · S. Estellés
Dpto. Organización de Empresas, Universidad Politécnica de Valencia,
Camino de Vera, s/n, 46022 Valencia, Spain
e-mail: mabagri@doe.upv.es

M. Palmer
e-mail: marpalga@doe.upv.es

M. Cardós
e-mail: mcardos@doe.upv.es

S. Estellés
e-mail: soesmi@doe.upv.es

S. P. Sethi et al. (eds.), *Industrial Engineering: Innovative Networks*,
DOI: 10.1007/978-1-4471-2321-7_30, © Springer-Verlag London Limited 2012

an installation with a high turn off cost; (ii) a high inventory level is needed in order to assure a high availability of the installation served; (iii) repairable parts are usually expensive and therefore to repair them becomes into a good option to save costs; and (iv) the resulting inventory value is usually very large. That is why during the last years researchers dedicate to propose models to improve the management of these inventories. A complete literature review about it can be found in Guide and Srivastava (1997) and Kennedy et al. (2002). However, there is only a handful of works that suggest exact methods to compute the fill rate, as the fraction of demand that is met immediately from self, even when knowing the availability of a repairable part is crucial is most cases.

In practice, the most common policy used to manage reparable parts is the base stock sales replacement policy ($S-1$, S). In this context, Feeney and Sherbrooke (1966) develop an exact method to compute the fill rate as the expected amount of demand that can be filled immediately from the on hand stock, assuming compound Poisson distributed demands. Some years later, Muckstadt and Thomas (1980) follow the same approach for Poisson distributed demands using the following easy to compute expression:

$$\beta_{M\&T} = F_L(S-1) \tag{30.1}$$

where S is the base stock level and $F_L(\cdot)$ is the cumulated distribution function of demand during the reparation lead time. Expression (30.1) is later used by De Haas and Verrijdt (1997) to compute the fill rate for repairable parts on the aircraft industry. Sherbrooke (1968) develops the multi-echelon model known as Metric which objective is the minimization of total costs of the system using the number of expected backorders as constrain. In this case demand is modelled with a compound Poisson distribution. This work shows that for a single-echelon system, the fill rate leads to the same results that the expected number of backorders in terms of stock level. Muckstad (1973) and Sherbrooke (1986) adapt the Metric model in order to assume Poisson distributed demands while Kutanoglu and Mahajan (2009) and Olsson (2010) introduce in the model the possibility of allowing transhipment between bases.

However, to the best of our knowledge, no research has been focused on deriving a general model to compute the exact value of the fill rate when demand is modelled with any discrete distribution function. This paper focuses on derive it when repairable parts are managed by the sales replacement policy and demand is uncorrelated, discrete and stationary.

The paper is organized as follows. Next Section is devoted to describe the problem. After that the ($S-1$, S) policy and the notation used in the paper is detailed. The general model for the exact estimation of the fill rate given any discrete demand is derived in the following Section. Then, some illustrative examples are used to compare the different fill rate estimations. Finally, conclusions and further research are briefly commented.

Table 30.1 Performance of the fill rate estimation method used by the company

Category	Average estimated fill rate	Average real fill rate
A	0.970	0.928
B	0.975	0.846
C	0.978	0.491
Total	0.976	0.637

30.2 Description of the Problem

The Spanish airline company Air Nostrum is involved in a research project to optimize its inventory system, specifically repairable part inventories. Air Nostrum is one of the most important European regional airlines, and is ranked between the 15 greatest regional airlines of the world. This company operates 450 flights and more than 150 routes in over 60 destinations located in 9 European and African countries. Repairable parts are so critical for the company that a stockout occasion can make the airplane to stay on ground, leading to very high costs and poor customer service. Moreover, these parts cannot be easily replaced by a similar part due to international regulations. A rule of thumb in this industry states that a reasonable inventory level of the spare parts is about 5% of the value of the fleet in service, being the repairable parts an important part of these inventories.

The company uses the base stock sales replacement policy to manage their repairable parts that consists on examining the inventory status dairy and launching a replenishment order of sufficient magnitude to raise the inventory position to the base stock, *S*. To determine *S* the company establishes periodically a target fill rate as a trade-off between the quality of the service offered to their customers and the inventory costs. Then, they compute the base stock with the objective of reaching a target fill rate while assuming that the part can always be repairable and the number of failures per period follows a Poisson distribution. The expression the company uses to estimate the fill rate is as follows:

$$\beta_C = F_L(S) \qquad (30.2)$$

However with the aim of improving the inventory system of the company, we have identified two problems in the estimation of the fill rate as in expression (30.2). First, we have observed that computing the base stock level using β_C provides a base stock level that can be one unit below the one from $\beta_{M\&T}$, so the part does not reach the target fill rate. In fact, the analysis of the historical data of failures that took place during 911 consecutive days shows that the estimated and the real fill rate (that should be at least equal that the target) differ greatly as Table 30.1 shows. The repairable parts have been classified using an ABC analysis based on the number of failures.

The second problem is related to the assumption that the failures follow a Poisson distribution function. A statistical analysis of historical data shows that this assumption is acceptable for 80% of the parts but it cannot be accepted for the

Table 30.2 Distribution functions of historical data failures

	Poisson	Negative binomial		
Category	# Parts	Failures	# Parts	Failures
A	36	3,244	56	15,997
B	182	2.322	93	1,263
C	531	1.060	43	154
Total	749 (80%)	6,626 (28%)	192 (20%)	17,414 (72%)

rest 20% of the parts which represent the 72% of the total number of failures. Moreover, these parts fit better using a negative binomial distribution as Table 30.2 summarizes.

Therefore, to improve the management of the repairable parts the company requires a method to estimate the fill rate that: (1) guarantees the achievement of the target fill rate, i.e. an exact method; and (2) is applicable to any discrete distribution function. The following Section dedicates to derive such method.

30.3 An Exact Method to Compute the Fill Rate for the (S−1, S) System

30.3.1 Notation and Assumptions

The sales replacement policy $(S-1,\ S)$ consists on examining the inventory position of the part every day. Then, if it is below the base stock level, S, a replenishment order of sufficient magnitude to raise the inventory position to S is launched [see for example (Fogarty et al. 1991; Silver et al. 1998)]. Hence, the inventory position is always equal to S at the beginning of each day. Obviously the policy follows the complete backordering model, since any demand, when out of stock, is backordered and filled as soon as possible.

We assume that: (i) the time required to repair a part is constant; (ii) the replenishment order is added to the inventory at the end of the period in which it is received; (iii) demand during a period is fulfilled with the inventory at the beginning of that period; and (iv) the demand process is considered discrete, stationary and independent and identically distributed.

The notation used in the rest is as follows:

S	=	order up to level (units),
L	=	lead time due to the reparation process (periods),
Z_t	=	net stock at t (units),
D_m	=	demand during m consecutive periods (units),
$f_m(\cdot)$	=	probability mass function of demand during m consecutive periods,
$F_m(\cdot)$	=	cumulative distribution function of demand during m consecutive periods.

30.3.2 Derivation

For the shake of simplicity we compute the fill rate based on the net stock (on hand stock minus backorders) when positive at the beginning of the replenishment cycle. Then, the value of the expected fill rate in that cycle can be expressed as:

$$\beta^* = \sum_{i=1}^{S} \beta^*(Z_0 = i) \cdot P(Z_0 = i) \tag{30.3}$$

By definition, the net stock is equivalent to the inventory position minus the on-order stock. Then the net stock balance at the beginning of the cycle is:

$$Z_0 = S - D_L \tag{30.4}$$

Based on (30.4), we can compute the probability of every feasible value of the net stock at the beginning of a period as follows:

$$P(Z_0) = P(D_L = S - Z_0) = f_L(S - Z_0) \tag{30.5}$$

Note that the fill rate is defined as the fraction of demand that is immediately fulfilled from shelf and therefore, cycles that do not show any demand should not be taken into account. Therefore to calculate the fill rate it is necessary to include the condition of having positive demand during the cycle. Given a positive net stock level at the beginning of the period, demand during that period can be: (i) lower or equal than the net stock and therefore the fill rate will be equal to 1; or (ii) greater than the net stock and therefore the fill rate will be the fraction of that demand which is satisfied by the on hand stock at the beginning of this cycle. Then:

$$\begin{aligned}\beta^*(Z_0 = i) &= 1 \cdot P(D_1 \le i | D_1 > 0) + \sum_{j=i+1}^{\infty} \frac{i}{j} \cdot P(D_1 = j > i | D_1 > 0) \\ &= \frac{P(0 < D_1 \le i)}{P(D_1 > 0)} + \sum_{j=i+1}^{\infty} \frac{i}{j} \cdot P(D_1 = j > i | D_1 > 0), \quad i > 0\end{aligned} \tag{30.6}$$

that expressed through the probability mass and cumulative distribution functions, $f_t(\cdot)$ and $F_t(\cdot)$ respectively results on

$$\beta^*(Z_0 = i) = \frac{F_1(i) - F_1(0)}{1 - F_1(0)} + \sum_{j=i+1}^{\infty} \frac{i}{j} \cdot \frac{f_1(j)}{1 - F_1(0)} \quad i > 0 \tag{30.7}$$

Therefore, applying expression (30.7) to every positive net stock level at the beginning of the period, the exact fill rate is estimated as follow:

$$\beta^* = \sum_{i=1}^{S} f_L(S - i) \cdot \left(\frac{F_1(i) - F_1(0)}{1 - F_1(0)} + \sum_{j=i+1}^{\infty} \frac{i}{j} \cdot \frac{f_1(j)}{1 - F_1(0)} \right) \tag{30.8}$$

that can be applied when demand follows any discrete distribution function.

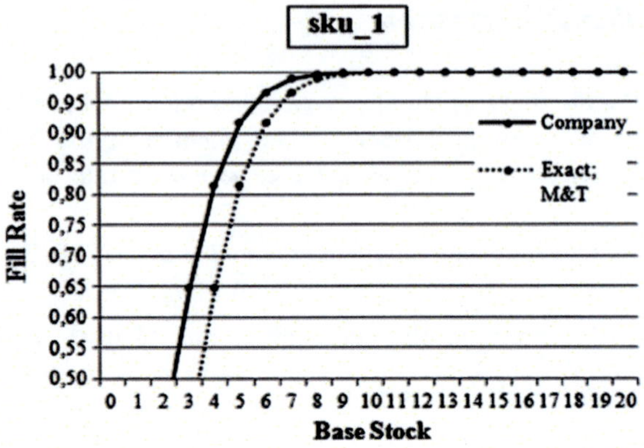

Fig. 30.1 Comparison between the company, M&T and the exact fill rate for sku_1

30.4 Illustrative Examples and Discussion

This section compares the value of the fill rate when using β_C and $\beta_{M\&T}$ against the exact β^* by means of two illustrative examples taken from the reparable parts that the company manages. The first is named sku_1 in the rest and its demand follows a Poisson distribution with demand rate $\lambda = 0.1$. The second one (sku_2 in the rest) has to be modelled with a negative binomial distribution with parameters $r = 0.002$ and $\theta = 0.092$.

Figure 30.1 compares for sku_1. In this case, both β^* and $\beta_{M\&T}$ are equal and exact since its demand is Poisson distributed. As expected, the estimation procedure the company uses is always overestimating the fill rate, and therefore, the base stock that results from the company estimation can be a unit less than required (e.g. for a target fill rate equal to 0.95, the company procedure provides $S = 6$ whereas the exact procedure gives $S = 7$).

Figure 30.2 shows the same comparison for sku_2. In this case, β_C and $\beta_{M\&T}$ show the same pattern as above but both not only overestimate the exact β^* but also show significant deviations. Note that $\beta_{M\&T}$ is only exact when demand is Poisson distributed. For example, if the base stock of the policy is set to 3, the $\beta_C \approx 1$, the $\beta_{M\&T} \approx 0.97$ when in fact the exact fill rate is about 0.75. When the company set a target fill rate equal to 0.95, the base stock from the company approach is $S = 2$, using the M&T is $S = 3$ whereas using the proposed exact method the base stock should be equal to 10 to guarantee the target service level.

Both illustrative examples show why the company does not always reach the target fill rate and why managers think that the system is more protect against stockouts than it really is. Moreover, this effect is even more evident when the repairable part can not be modelled by the Poisson distribution. Note that according to the analysis of the historical data showed above, the Poisson is not acceptable for 28% of the total of parts that represent 72% of the total of failures.

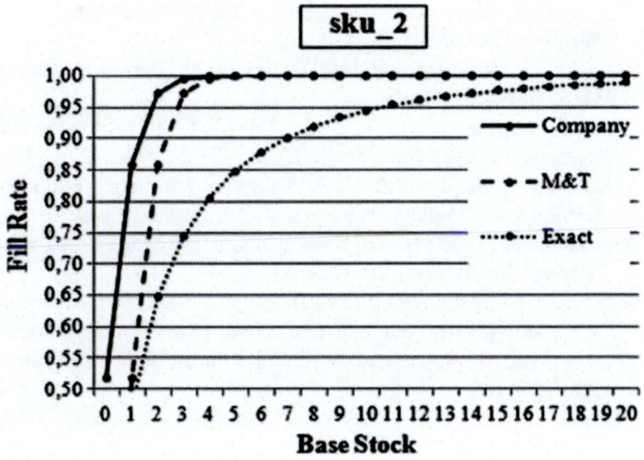

Fig. 30.2 Comparison between the company, M&T and the exact fill rate for sku_2

30.5 Conclusions

This paper focuses on the improvement of the inventory policies used to manage the repairable parts of the airline company Air Nostrum. We find that the heuristic method used by the company to compute the fill rate does not always lead to the base stock that guarantees the target service level and identify two main reasons: (i) the method itself is approximated when compare with the only exact method available on the literature to manage repairable parts (Muckstadt and Thomas 1980), although it assume Poisson demands; (ii) the demand of some of the repairable parts can not be modelled by the Poisson distribution and a negative binomial distribution has to be used instead. In fact the sum of the demand that do not follow a Poisson distribution is most of the total demand.

As a consequence of this, we provide an exact method to compute the fill rate able to cope with any discrete demand distributions, and not only with the Poisson one. The illustrative example for sku_1 shows that deviations between the company approach and the exact methods may have important implications for the system to be ignored. However, implications of modelling the demand of repairable parts with a Poisson distribution instead of using an appropriate distribution may have dramatic implications in terms of stockout costs when the customer service is not reached as shown Fig. 30.2.

Further research will be focused on exploring new approaches in order to reduce the total average repairable inventories while maintaining a target fill rate.

References

De Haas HFM, Verrijdt JHCM (1997) Target setting for the departments in an aircraft repairable item system. Eur J Oper Res 99:596–602

Feeney GJ, Sherbrooke CC (1966) The (S−1, S) inventory policy under compound Poisson demand. Manage Sci 12:391–411

Fogarty DW, Blackstone JH, Hoffman TR (1991) Production and Inventory Management. South-Western Publishing Co, Cincinnati

Guide VDR, Srivastava R (1997) Repairable inventory theory: models and applications. Eur J Oper Res 102:1–20

Kennedy WJ, Patterson JW, Fredendall LD (2002) An overview of recent literature on spare parts inventories. Int J Prod Econ 76:201–215

Kutanoglu E, Mahajan M (2009) An inventory sharing and allocation method for a multi-location service parts logistics network with time-based service levels. Eur J Oper Res 194:728–742

Muckstad JA (1973) Model for a Multi-Item, Multi-Echelon, Multi-Indenture inventory system. Manage Sci B-Appl 20:472–481

Muckstadt JA, Thomas LJ (1980) Are Multi-Echelon inventory methods worth implementing in systems with low-demand-rate items. Manage Sci 26:483–494

Olsson F (2010) An inventory model with unidirectional lateral transshipments. Eur J Oper Res 200:725–732

Sherbrooke C C (1968) Metric—a Multi-Echelon technique for recoverable item control. Oper Res 16:122–144

Sherbrooke CC (1986) Vari-metric—improved approximations for Multi-Indenture, Multi-Echelon availability models. Oper Res 34:311–319

Silver EA, Pyke DF, Peterson R (1998) Inventory Management and Production Planning and Scheduling. Wiley, New Jersey

Part V
Supply Chain

Chapter 31
Supply Chain Demand Forecasting: Towards an Integrated Approach

Juan R. Trapero and Diego J. Pedregal

Abstract Demand forecasts are crucial to drive supply chains and enterprise resource planning systems. Improved accuracy in forecasts directly affects all levels of the supply chain, reducing stock costs and increasing customer satisfaction. Usually, this problem is faced by testing various time series methods with a different level of complexity to find out which one is the most accurate. From our point of view, the problem should be re-addressed. In this sense, the effort should be focused on incorporating more efficient sources of information that are frequently overlooked. This paper explores different sources of information (apart from past observations) that might enhance the capability of a company to produce accurate forecasts. Such sources are: (i) Judgmental forecasting at SKU level and (ii) Information sharing. Additionally, new models are proposed to integrate such information well. Data collected from a manufacturer of household cleaning products and a major UK grocery retailer are used to illustrate the procedure.

31.1 Introduction

Companies working within supply chains use demand forecasts to direct purchasing and supply chain management. Accurate forecasts can affect positively the operational management of companies, leading to significant monetary savings, greater

J. R. Trapero (✉)
Facultad de Ciencias Químicas, Universidad de Castilla-La Mancha,
13071 Ciudad Real, Spain
e-mail: JuanRamon.Trapero@uclm.es

D. J. Pedregal
ETSI Industriales, Universidad de Castilla-La Mancha,
13071 Ciudad Real, Spain
e-mail: Diego.Pedregal@uclm.es

S. P. Sethi et al. (eds.), *Industrial Engineering: Innovative Networks*,
DOI: 10.1007/978-1-4471-2321-7_31, © Springer-Verlag London Limited 2012

competitiveness, enhanced channel relationships and customer satisfaction, lower inventory investment, reduced product obsolescence, improved distribution operations, more efficiently-scheduled production and distribution, and better profitable financial decisions. For most of these companies, a particular type of Decision Support System, known as Forecasting Support System (FSS) is employed to prepare the forecasts (Fildes et al. 2009). These FSS integrate a statistical forecasting approach with managerial judgment from forecasters in the organization. Managers may have access to information that is difficult to include in a statistical model, for example the effects of a promotion campaign. Thus, judgmental adjustments can incorporate that information into the model in order to improve the forecast accuracy.

The objective of this paper is twofold: firstly, we will investigate potential models that properly combine the adjustments made by managers and the statistical forecasts. In this first objective, our analysis is focused on a single company. Usually, companies interact with other firms that belong to the same supply chain. So should we not consider that fact when we propose our forecasting models? With this in mind, our second objective is to analyze the advantages of information sharing between companies improving the forecasting process.

This article is organized as follows: Sect. 2 studies the influence of judgmental adjustments on forecasting accuracy; Sect. 3 explores the impact of information sharing on supplier demand planning; and finally, main conclusions are shown in Sect. 4.

31.2 Judgmental Forecasting

In Franses and Legerstee (2009) a case study was presented where the experts adjusted the statistical forecast in 89.5% of the cases. Even though managers make frequent adjustments, the literature devoted to studying the effect at the stock keeping unit (SKU) level is scarce (see Fildes et al. 2009).

Recent literature suggests the existence of a bias towards making overly positive adjustments (see Fildes et al. 2009). In Mello (2009) the biases introduced are analyzed by means of forecast game playing, defined as the intentional manipulation of forecasting processes to gain personal, group or corporate advantage.

In order to correct the presence of the bias several works have modelled the appropriate weight that statistical forecasting and judgmental forecasting should have. For instance, Blattberg and Hoch (1990) used a 50% Model + 50% Manager heuristic approach that improved forecast quality. An optimal adjust model based on linear regression classifying the data depending on the sign of the adjustment was proposed in Fildes et al. (2009). In contrast to Blattberg and Hoch (1990), it was found that negative adjustments were more precise than positive ones. This discontinuity between positive and negative adjustments may indicate the desirability of adopting non-linear models to describe the judgmental process. In fact, such non-linearities can be considered in the design of the FSS to mitigate the worst effects of such biases.

The first objective of this paper is to analyze the non-linear effect of adjustments on the final forecast accuracy on a manufacturing company database containing one-step-ahead forecasts and actual sales. A State Dependent Estimation (SDP) approach is used to study the non-linearities involved in the manager's adjustment. SDP non-linear estimations belong to a family of methods within the Data-Based Mechanistic modelling (DBM) developed by Young and co-workers (see Young et al. 2001). The SDP technique uses recursive methods such as Fixed Interval Smoothing (FIS) combined with special data re-ordering and "backfitting" procedures which show —in a non-parametric way, i.e. through a graph, the state dependency between the parameter under study and an associated state variable.

31.2.1 A State-Dependent Parameter Estimation Approach

The Optimal Adjust model, proposed by Fildes et al. (2009) aimed at optimally combining two of the sources of information available to the forecaster, the system forecast and the forecaster's subjective adjustment, in order to deliver a more accurate forecast. It is given by:

$$y_{i,t} = \alpha_1 SF_{i,t} + \alpha_2 Adj_{i,t} + v_{i,t} \tag{31.1}$$

where $y_{i,t}$ stands for the actual sales for product i at time t. Regressors are $SF_{i,t}$ and $Adj_{i,t}$, which represent the System Forecast and Adjustments. Adjustments are computed as follows:

$$Adj_{i,t} = FF_{i,t} - SF_{i,t} \tag{31.2}$$

where $FF_{i,t}$ is the Final Forecast employed by the FSS. The error term is given by $v_{i,t}$. In order to evaluate the impact of judgmental adjustments on the forecasting accuracy, a non-linear version of the optimal model is proposed.

The SDP[1] optimal adjust model is defined as:

$$y_k = \alpha_1 SF_k + \alpha_2(Adj_k)Adj_k + v_k \tag{31.3}$$

Since the parameters are expected to vary depending on the adjustment size and sign, the data are sorted with respect to the adjustments. Accordingly, the data can be re-indexed by $k = 1,\ldots, N$, where N is the sample size. The main difference with model (31.1) relies on the parameter α_2, which depends on the state (Adj_k). Such a dependency can be stochastically modeled by using a particularization of the *Generalized Random Walk* proposed by Jakeman and Young (1984).

---

[1] SDP algorithms are avaible in the MATLAB toolbox: CAPTAIN http://www.es.lancs.ac.uk/cres/captain/

Table 31.1 MAE for the normalized validation dataset

Adjustments	System forecast	Final forecast	SDP	Optimal adjust	Blattberg-Hoch
Positive	0.93	0.72	0.69	0.73	0.71
Negative	0.75	0.46	0.46	0.45	0.56
None	0.58	0.58	0.56	0.56	0.58
Total	0.76	0.61	0.59	0.61	0.63

31.2.2 Case Study

A company specialized in household and healthcare products has been chosen to carry out the experiments. A dataset with 413 SKU was collected between 2004 and 2007. The dataset was broken down in the following variables: (i) system forecasts; (ii) adjustments; and (iii) actual values. 7,544 triplets monthly sampled formed the set of data under study.

31.2.3 Model Validation

In order to validate the proposed model, 20% of the data constituted by the last months of each SKU, which were not used for the parameter estimation of the models, were employed as the hold-out sample. Mean Absolute Error (MAE) was chosen as the error measurement system. Note that variables were normalized with respect to its SKU standard deviation. Table 31.1 shows the MAE for the hold-out experiment. In this same table we can observe that final forecasts are more accurate than System Forecasts. This indicates that experts improve the forecasts. In fact, experts are more precise for negative adjustments. On the other hand, the SDP model shows a lower forecasting error than the rest of linear models, particularly in relation to positive adjustments.

31.3 Information Sharing

Since the beginning of the 20th century, one of the main problems that Supply Chain Management has had to face is the phenomenon known as "the bullwhip effect" (Geary et al. 2006), which consists of demand variability amplification when moving upwards in the supply chain (Lee et al. 1997). Among the consequences of this amplification, for instance, we might find excess inventory, poor customer service and poor product forecasts. Demand signal processing, rationing game, order batching, and price variations were the four bullwhip effect sources

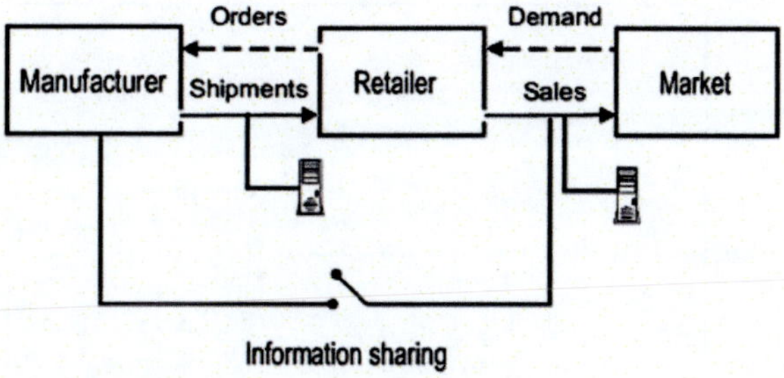

Fig. 31.1 Materials flow (——) and information (- -) for a two-level serially linked supply chain

analyzed in Lee et al. (1997). Therefore, if we pursue to reduce the forecasting error, the target can be achieved by reducing the bullwhip effect.

In order to avoid the bullwhip effect, two main methodologies can be found in Operations Management literature: (i) analysis of the ordering decisions and (ii) supply chain collaboration (see Cannella and Ciancimino 2009). This work is focused on the latter approach. The idea behind supply chain collaboration is to find a global optimal solution for all its components instead of different sub-optimal solutions for each one. Information sharing is a way to accomplish such collaboration.

The second objective of this paper is to use automatic system identification procedures to select the adequate structure for the supplier's sales by using the retailer sales information. The results show that ARX can improve the supplier's forecast performance by using the market sales information shared by the retailer. Other univariate techniques such as ARIMA, exponential smoothing and Moving Averages where used as benchmarks.

31.3.1 Case Study

The supply chain system consists of a serially linked two-level supply chain (see Fig. 31.1). There is a flow of information from the market towards the manufacturer and a reverse material flow. Market sales and shipments from the manufacturer are the measured variables, indicated by the sensors in Fig. 31.1. There is also a switch that represents the option of sharing information. When the switch is off it means that we are considering the traditional supply chain case. When it is on, market sales information is available for the manufacturer.

Data from a manufacturing company specialized in household products have been collected. These data were sampled weekly between October 2008 and October 2009. This manufacturing company provides products to one of the largest retailers in UK. The data consist of two time series per SKU: (i) shipments received by the retailer from the manufacturer; and (ii) customer demand measured by the retailer sales.

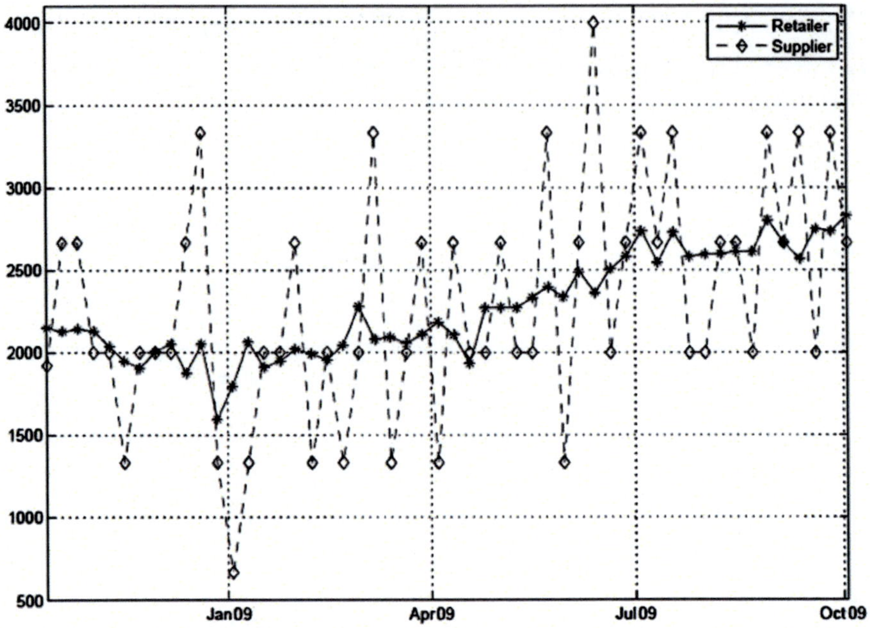

Fig. 31.2 Example of a typical SKU. Retailer sales are represented by a *solid line* (——) and Volume received by a *dashed line* (- -)

In summary, the dataset under study comprises 43 Stock Keeping Units (SKU) with 52 observations per SKU. An example is depicted in Fig. 31.2. In this figure we can clearly observe the Bullwhip effect, i.e., the amplification of the demand variability when comparing retailer sales with supplier sales with regards to the same SKU.

31.3.2 Models

Two kinds of models have been analyzed to find out whether retailer sales information is useful for suppliers seeking to improve the accuracy of their forecasting. On the one hand, we have used univariate models, such as Single Exponential Smoothing (SES), Autoregressive (AR), Moving Average (MA) and Autoregressive Integrated Moving Average (ARIMA) models, as well as a Naïve method. These methods rely exclusively on past supplier sales information to forecast. Therefore no information sharing is accomplished. On the other hand, a multivariate ARX model has also been applied, where supplier sales and retailer sales are used as dependent and explanatory variables, respectively.

Table 31.2 Mean of the MAPE for the considered forecasting methods

Method	Naïve	AR	MA	SES	ARIMA	ARX
MAPE %	47.05	38.20	36.28	34.43	34.50	26.62

31.3.3 Model Validation

In this section predictive validation is used to compare models. For this purpose, 20% of the data constituted by the last 10 weeks of each SKU are kept for comparing the performance of the proposed methods, as a hold-out sample. These last 10 weeks are not used for the parameter estimation of the models. All forecasts considered are of the one-step-ahead type.

The percentage errors of these forecasts are used to calculate the Mean Absolute percentage Error (MAPE) of each individual SKU across the different time origins, which are afterwards aggregated in dataset average figures, obtaining the Mean (MAPE) as an overall error measure through all SKUs. In fact, Table 31.2 shows the Mean (MAPE) obtained for each forecasting method considered. The main result is that the ARX model, which was based on information sharing, is the one that provides a lower forecasting error. Across the univariate techniques analysed, Single Exponential Smoothing obtained the lowest error. However, SES was just slightly better than ARIMA.

31.4 Conclusions

This paper addresses the need for some companies to forecast demand by using an integrated approach. Traditionally, numerous articles on forecasting have been devoted to show the advantages/disadvantages of different models which use past observations. In this case, we believe that the focus of the problem should be changed and new sources of information should be found and integrated in the models properly. The present work reports two sources of potential improvement in forecasting performance: (i) Judgmental adjustments and; (ii) Information sharing.

Here, the use of SDP estimation was exploited in a new application in order to combine the statistical forecast with the judgmental adjustments. This technique allowed us to obtain an error reduction. Additionally, collaboration between companies was analyzed, coming to the conclusion that information sharing improves forecasting performance. That result was based on the benchmarking of multivariate against univariate models using a real dataset, based on a serially linked supply chain.

We expect that further works will appear integrating the aforementioned as well as new aspects in models implemented in more advanced FSS.

References

Blattberg RC, Hoch SJ (1990) Database models and managerial intuition: 50% model + 50% manager. Manage Sci 36:887–899

Cannella S, Ciancimino E (2009) On the bullwhip avoidance phase: supply chain collaboration and order smoothing. Int J Prod Res 1:1–38

Fildes R, Goodwin P, Lawrence M, Nikolopoulos K (2009) Effective forecasting and judgmental adjustments: an empirical evaluation and strategies for improvement in supply-chain planning. Int J Forecast 25:3–23

Franses PH, Legerstee R (2009) Properties of expert adjustments on model-based SKU-level forecasts. Int J Forecast 25:35–47

Geary S, Disney SM, Towill DR (2006) On bullwhip in supply chains-historical review, present practice and expected future impact. Int J Prod Econ 101:2–18

Jakeman AJ, Young PC (1984) Recursive filtering and the inversion of ill-posed causal problems. Utilitas Mathematica 35:351–376

Lee HL, Padmanabhan V, Whang S (1997) The bullwhip effect in supply chains. Sloan Manage Rev 38:93–102

Mello J (2009) The impact of sales forecast game playing on supply chains. Foresight Int J Appl Forecasting 13:13–22

Young PC, McKenna P, Bruun J (2001) Identification of non-linear stochastic systems by state dependent parameter estimation. Int J Control 74:1837–1857

Chapter 32
Logistic Management Optimization for a Container Port Terminal

Elvira Maeso, Manuel Ballesteros and Javier Caballero

Abstract This article extracts a set of simple tools from the performance modelling of the elements which make up the logistic system of a container terminal in a hub port. These tools will allow us to give a heuristic solution to sea operations by optimizing the assigned resources and gauging their suitable quantity in order to minimize idle times and maximize productivity. Our work is focused on dealing with the problem of each independent element, by achieving a suitable performance level for each of them without causing waiting times in the rest. The process of resources assignation has been denominated Logistic Optimization Model (simplified). The application of the model could increase the global productivity of the terminal by approximately 5%.

32.1 Introduction

Sea traffic has undergone an important growth in the last few decades, but the increase in containers trade has been even bigger, and has shown annual rates above two digits. These data reflect the increasing connection between the different world economies and the economic advantages derived from the current completely mature technology (Drewry 2006).

The increase in the demand for sea transport, together with the increase of oil prices, has promoted the use of larger-sized ships. In view of the fact that most of the large container ships lack in on-board cranes, the whole load and unload

E. Maeso (✉) · M. Ballesteros · J. Caballero
Cátedra de Gestión del Transporte. Escuela Técnica Superior de Ingenieros Industriales, Universidad de Málaga, C/Ortiz Ramos s/n, 29071 Málaga, España
e-mail: emaeso@uma.es

S. P. Sethi et al. (eds.), *Industrial Engineering: Innovative Networks*,
DOI: 10.1007/978-1-4471-2321-7_32, © Springer-Verlag London Limited 2012

operation is assumed by the port infrastructure. Investment in ships is genuinely important and services are needed to be quick and efficient in order to minimize the return times.

On the other hand, the merger procedures of the shipping companies have entailed the concentration of 60% of the transport capacity in the 10 largest companies (UNCTAD 2008). These shipping companies join their services in several sea routes, in such way that the port operators have to compete for a few potential customers in each area. This situation forces the port operators to take part in an important concurrency of price and service offered.

Therefore, it is essential for port terminal operations to maximize their productivity and cost-effectiveness rates.

The development of a comprehensive system for the optimization of resources required in the operation, i.e. trucks, RTG cranes and STS cranes, allows for the optimization of the productivity rates and eliminates idle times.

32.2 Logistic System Features of a Container Terminal

On the whole, large container terminals consist of the pier, the yard and the gate. Its configuration must be prepared to adapt to the constant growth of ships and be able to keep its competitiveness (Gómez and Camarero 2010).

For the design and development of the system, we start from a research of the different existing configurations of container terminals and their logistic systems. The logistics is based on a set of interrelated cycles which are carried out by each operating element until the worklist is fulfilled (Palacio 2001).

The pier is the sea interface of the terminal on which a set of STS (Ship-to-Shore) cranes operate. The yard is divided into several areas, or boundaries, for the temporary storage of containers with the same destination and/or technical features. The gate is the inland interface of the terminal. The logistic system of the terminal reflects the established configuration to handle the containers in the yard of the terminal.

The present paper is focused on a container terminal with the yard configuration operated with RTG (Rubber-Tyred Gantry) cranes and trucks. The system allows the allocation of resources required to achieve an optimal working capacity according to technical, operative and economic factors. The operating of these resources is interrelated in the same way as a set of friction wheels (see Fig. 32.1).

32.3 Logistic Optimization Model: Simplified

The planning aim of every operating process is to optimize the number of STS cranes which guarantees a quick ship turnaround, by making the most of the working time with the crane and minimizing the operating cost. Regarding the

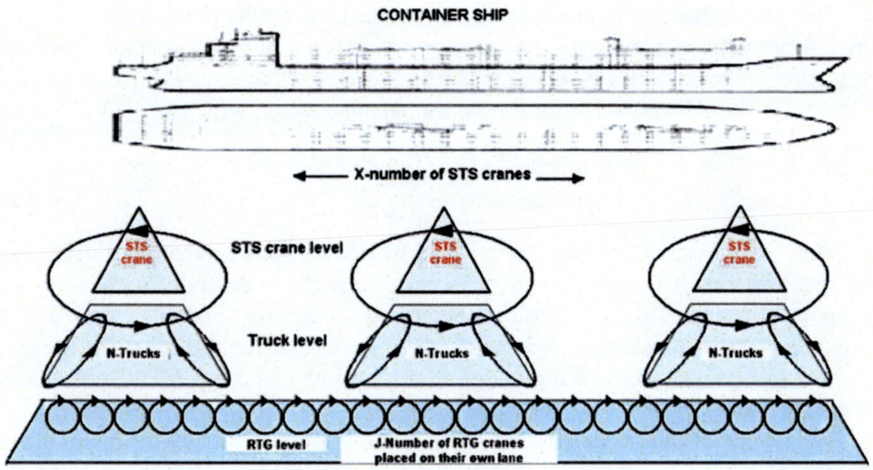

Fig. 32.1 Correlation between the elements (STS, RTG cranes and trucks) in a port operation

gauging of the rest of operating elements, waiting times of the STS crane must be avoided, since it is the most valuable element, based on acceptable costs.

The excellence of the resources allocation of a terminal and the aim of this research is to specify the number of trucks (for each STS crane) and cranes (for each lane) needed to assist the STS crane properly. This study could be carried out through the development of some software to simulate in great detail the cycle of every STS crane, RTG crane and truck. It would be complex but definitely interesting, due to its power as a tool, even if we would obviate important and random factors which could provide some uncertainty to the developed model. Therefore, this paper is focused on dealing with the problem of each independent element, by achieving a suitable performance level for each of them without causing waiting times in the rest. Our process of resources assignation has been denominated Logistic Optimization Model (simplified).

32.3.1 Productivity Optimization for Ship-to-Shore Cranes

The research about the productivity of the STS crane according to the number of assigned trucks shows that there is one point from which productivity does not increase significantly when the number of trucks does. The productivity of the crane might be modelled through a simple function, such as:

$$\text{Productivity}\ (n) = K^*(1 - (e - A^*n) \tag{32.1}$$

where "K" and "A" would be constant parameters, and "n" would be the variable which represents the number of trucks.

The "K" parameter is the upper limit for machine productivity. This is limited by a theoretical maximum of productivity which is given by the manufacturer but it is also influenced by several factors which might reduce this limit, such as the operator's skilfulness, the ship design, the weather conditions, etc.

$$
\begin{aligned}
K &= \text{Theor. productivity} - (\text{koperator} - \text{kship} - \ldots) \\
&= \text{Theor. productivity} - \sum \text{ki}
\end{aligned}
\tag{32.2}
$$

The "A" parameter shows the speed at which the peak productivity value can be reached, and it is very important to reach a high productivity with a limited number of trucks. This parameter is influenced by different variables, such as the container distribution in the yard, the speed of the trucks, the skilfulness of dockers and controllers, the work configuration in the yard, the amount of RTG cranes and their attributes.

The results of this modelling are applied in an intuitive way by the operators in charge at container terminals. It would be useful to establish and quantify the different values for "K" and "A" through statistic studies which might be the base for decision-making.

32.3.2 Simulation of the STS Crane Performance Bay After Bay

For each bay, we calculate the time needed to carry out the loading and unloading of all the containers and their compatibility to load and unload in different orders. Once we have gathered the data corresponding to each bay, we elaborate a graph similar to a Gant chart (Fig. 32.2) to observe which order is the most appropriate to carry out the unloading according to the working day in the port and the limited movement of STS cranes due to the fact that they cannot cross each other's path.

32.3.3 Optimization of RTG Cranes in Each Lane

The assignation of RTG crane resources in each lane needs detailed information about the number of containers in each lane, the chronological order to be loaded, the occupation level of the yard and the performance data of the RTG crane. The elaboration of decision-making tools in this situation is complex, due to the fact that the workload is piled up in one or various locations around the yard.

In Fig. 32.3 we observe the results given by the graphic method which has been used in this calculation. It represents workload density by using a colour code for similar time periods, in order to be able to appreciate in a qualitative way the number of RTG cranes needed in the operation.

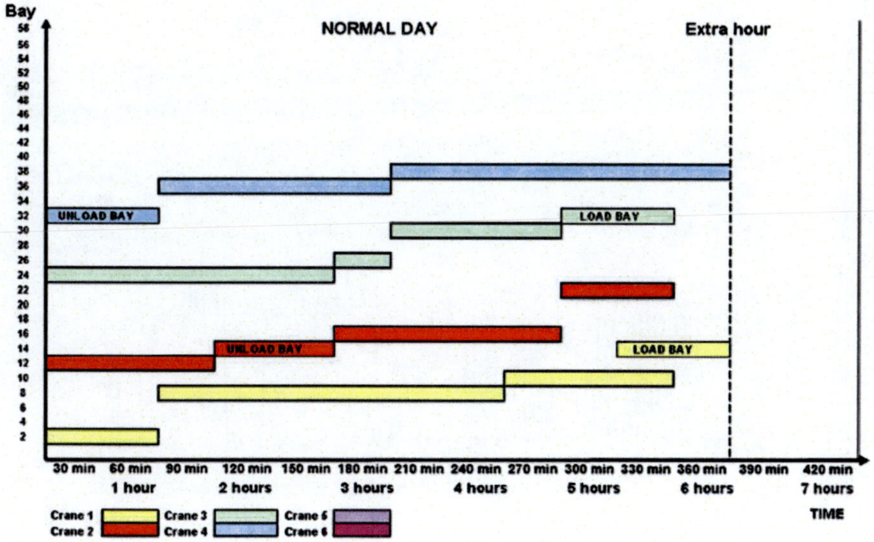

Fig. 32.2 Assignation of the number of STS cranes through bay-to-bay simulation

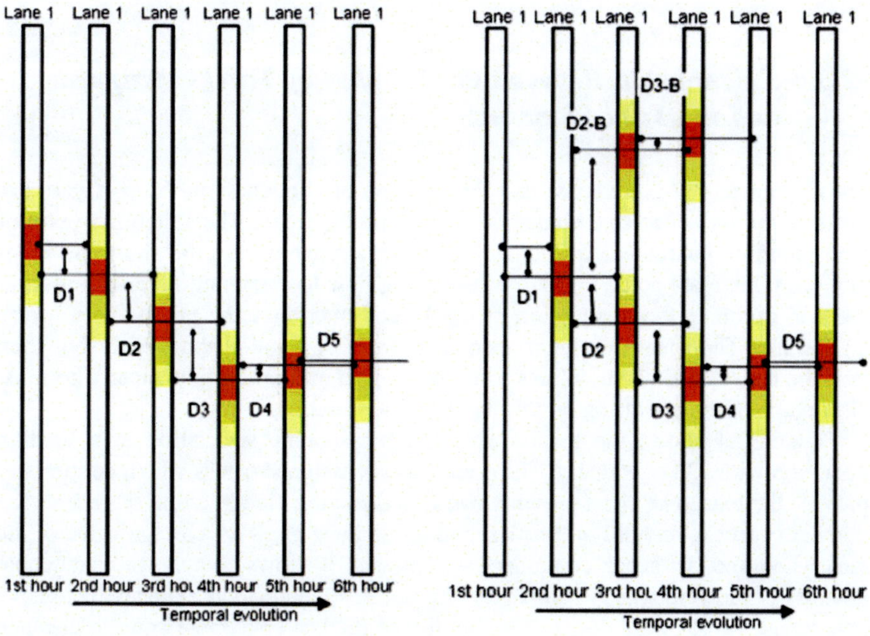

Fig. 32.3 Examples of assignation of RTG cranes

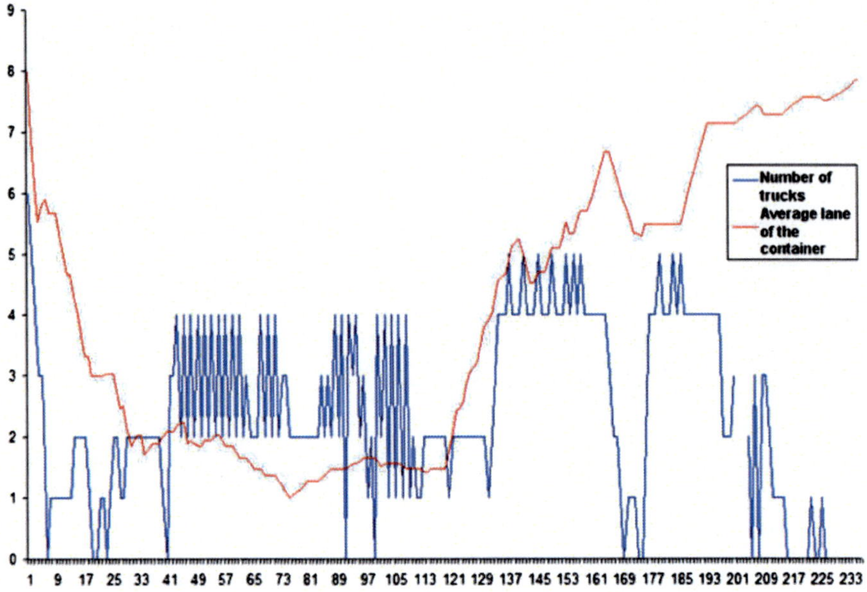

Fig. 32.4 Relationship between the distance to the yard and the trucks on hold

32.3.4 Relationship Between the Number of Trucks Assigned and the Yard Planning

The distribution of boundaries and the location of containers in the port yard are some of the most decisive elements for port productivity. The suitability level of their planning guarantees or jeopardizes truck exploitation in the short and medium term. Here are some measures describing how to carry out the distribution in order to optimize the assignation of trucks and maximize the productivity of the STS crane. The measures mentioned might not be applied thoroughly but they must be considered as guidelines to be taken into account by planners, who must also evaluate the operating conditions in each situation.

Figure 32.4 shows the results obtained for a crane simulation in a loading operation with 233 containers. The crane has six assigned trucks which continually provide the load. The chart presents two functions: the blue function represents the number of trucks on hold in the pier, and the red one gives an average value of the lane where the last seven containers are situated. It shows an average changeable quantity of the yard lanes in which are the last seven containers. The aim is to have an average value of the distance to the pier of the last containers and the distance they have to cover. It is specifically calculated through:

$$Average\ Distance(ADi) = (1/h) * \sum i\, i - h(Container\ lane)i \qquad (32.3)$$

where "h" is the number of trucks selected to calculate the average distance and "i", the container value.

The chart indicates that the average distance (average lane) to the pier is inversely proportional to the number of trucks on hold in the pier. In an operation where trucks are distributed throughout the yard, the container loading must follow a certain order to alternate containers located far from the pier with those nearby. However, the work order of the containers cannot be changed in the short term during each operating, since it is established in the first place by the location of boundaries in the previous operation. According to these data, we can conclude that one of the main factors for the suitable assignation of trucks to the cranes is the optimization of the container distribution in the yard through the boundaries and spaces. Due to the difficulties that this task presents, a short- and medium-term planning must be carried out in each operation to determine the load and unload positions in a short period.

32.3.5 Vehicle Assignation for Each STS Crane

As previously indicated, the number of containers needed in a operation not only depends on the way they are distributed but also, and more importantly, on the average distance, so that each assignation of trucks can give a quick response to a certain container distribution. The average response, AR(j), is the fictional lane in which every position in the yard would be in AR(j) with "j" trucks for the operating and there would not be any STS cranes or trucks on hold at the pier.

These two concepts—Average Distance (AD_i) and Average Response (ARj)—provide a simple way to carry out a near-optimal allocation of the number of trucks for each STS.

The response to a certain assignation depends on the productivity of the crane. The average response of the trucks has to be calculated. Such calculation is done through field tests in the yard. Each trial consists of obtaining the average time on hold of the crane and the average number of trucks on hold in operations where all the movements are concentrated in a single lane with a certain assignation of trucks. The test is run for every useful combination of yard lanes and assigned trucks. With the aim of carrying out the assignation of the number of trucks, we represent in a chart the average distance of the planned operation, we analyze the response that each assignation of trucks would give and we evaluate which assignation is the most suitable.

32.4 Discussion

The system developed for the assignation of the number of STS cranes is intended to optimize their use. Such operation is currently carried out with approximate calculations which might often prevent the operation from fulfilling the work on time, or provoke excessive idle times. Moreover, each STS crane is associated to eight operators, whose time must be efficiently spent because of the high costs involved. A conservative calculation is that through the application of the assignation of STS cranes we can improve their productivity by 5–10%.

The simplified method to assign the number of trucks for each STS crane is a really innovative procedure, since most of the Spanish ports surprisingly lack in any equivalent or similar method. The concepts of *average distance*, *truck response* and the whole procedure (simplified method) to assign the trucks derived from them enable, in a fast and simple way, to assess the number of trucks for each STS crane. This allows avoiding "*undergauging*" or "*overgauging*" of trucks regarding the STS cranes. Undergauging is the worst of the two situations, since it causes the stopping of the STS crane during a period of the operating time, which means that the most valuable resource in the terminal is not properly profited. On the other hand, overgauging is the most common occurrence, since the number of trucks needed is usually unknown, so they are overgauged in order to avoid idle times, which would incur the highest expenses. The mentioned method allows some increase of the productivity of STS cranes by eliminating idle times and some reduction of the operating cost by eliminating the overgauged trucks.

Even with a very conservative view on the benefits of this method we can estimate that one out of every three operating terminals is overgauged by at least one truck, which could be avoided with this method. This means, in approximative data, a reduction of three trucks per day in a terminal which moves 400,000 containers (with three STS cranes working 24 h a day in three shifts) and implies a reduction of costs of approximately 300 € per day, or 100,000 € per year. Moreover, we consider that one out of every six operating facilities is overgauged, so this method could also increase the global productivity of the terminal by approximately 5%.

Finally, it would be interesting to recognise that the number of trucks cannot be optimized in the short term. However, it is also important to underline that those concepts and methods here developed are relatively easy to understand and even universal, since they can be applied to any terminal.

References

Drewry SCL (2006) Annual report container market review and forecast 2006/2007. Drewry Shipping Consulting Ltd, London

Gómez MA, Camarero A (2010) Grandes buques Vs equipamiento en muelle. IX Congreso de Ingeniería del Transporte, Madrid

Palacio P (2001) Transporte Marítimo de Contenedores: Organización y Gestión. Instituto Portuario de Estudios y Cooperación

UNCTAD (2008) UNCTAD Handbook of Statistics. General Information. United Nations, New York

Chapter 33
A Simulation-Based Solution for Optimal Logistics of Heavy and Variable-Size Items

M. Victoria de-la-Fuente, Eloy Hontoria and Lorenzo Ros-McDonnell

Abstract The Vehicle Routing Problem seeks to attain the optimal determination of the routes to be used by a fleet of vehicles stations at one or several depots. When investigating acceptable solutions in this context, researchers favour the optimization technique based on simulation models, which permits descriptions of complex systems without too many assumptions. This paper presents a new solution to add onto the mathematical model developed by Ros et al. (2010), where orders are grouped by route with the objective of improving the efficiency of the transported load and the reduction of associated costs. This new work changes the previous situation by introducing additional constraints. Fleet variations are modeled for the different cases studied by using the Monte Carlo simulation technique. The aim is to find a variety of operative solutions since they have been validated by Google Maps web application.

33.1 Introduction

The current tough competition in the market is forcing companies to study and rethink all their processes in order to make them more efficient. Improved transport-related logistic behaviours further help to serve clients by offering total satisfaction in terms of time and cost. In the case of the ceramic companies, this

M. Victoria de-la-Fuente (✉) · E. Hontoria · L. Ros-McDonnell
ETSII-Universidad Politécnica de Cartagena, C/Dr. Fleming, s/n,
30202 Cartagena, Spain
e-mail: marivi.fuente@upct.es

S. P. Sethi et al. (eds.), *Industrial Engineering: Innovative Networks*,
DOI: 10.1007/978-1-4471-2321-7_33, © Springer-Verlag London Limited 2012

study is oriented toward the processes connected to the *grouping* of merchandises and customers orders. The lack of appropriate tools to support optimal logistics of such items has led to the development of different techniques to facilitate both the calculation and implementation of these tasks.

Our plan is to improve on what Ros et al. (2010) have contributed for the efficient assignment of load to a fleet of vehicles. This is why, with the objective of picking up all orders from all clients on a certain date, we will run a Monte Carlo simulation aiming to establish a new, more well-organized vehicle assignment solution, with lower costs than those of the mathematical model initially proposed by the said authors.

This paper is organized as follows: Sect. 33.2 presents a brief review of the literature related to the vehicle routing problem (VRP), and lays out the group of constraints associated with the description of the VRP and the discussion on the solutions contributed by mathematical models and simulation models. Sections 33.3 and 33.4 describe the real problem of a logistics operator, the objectives pursued, and the constraints that must be confronted, as well as the calculation method used to simulate the new solutions. Section 33.5 describes the analysis of the results obtained, by comparing them with the original solution, and the validation of the new solution. Finally, Sect. 33.6 presents the conclusions drawn after the analysis of the results and the future lines of research to be carried out.

33.2 The Vehicle Routing Problem

The Vehicle Routing Problem (VRP) seeks to attain the optimal determination of the routes to be used by a fleet of vehicles stationed at one or several depots. These vehicles serve a group of clients. The problem has been thoroughly studied (Christofides 1985; Laporte 1992; Powell et al. 1995; Psaraftis 1995; Bertsimas and Simchi-Levi 1996; Laporte et al. 2000; Desaulniers et al. 2001; Toth and Vigo 2002; Cordeau et al. 2004; Bachelet and Yon 2007).

Many of the requirements and operative constraints, such as route specifications and length, number of vehicles in the fleet and their loading capacities, composition of orders, type of demands, number of warehouses, etc., may be imposed on the practical applications of the VRP problem (Ralphs et al. 2003).

When investigating acceptable solutions in this context, researchers favour the optimization techniques based on mathematical models, which imply simplifications of the constraints. These methods have important difficulties in their practical application, mainly when changes in the definition of the problem appear, such as the addition of constraints to the problem. At this stage the problem could cease to be a robust one, and not have a solution via the previous optimization techniques. In this situation, the application of simulation models permits descriptions of complex systems without too many assumptions, obtaining near-optimal solutions and guaranteeing their feasibility (Bachelet and Yon 2007).

This paper puts forward a new solution to add onto the mathematical model of Ros et al. (2010). In doing so, the basis is the solution offered by the approximate algorithm, based on the Branch & Bound technique (Carpaneto 1980; Fischetti 1994) and the application of decision rules for the assignment of loads in a fleet of vehicles. Orders are grouped by route with the objective of improving the efficiency of the transported load, with the consequent reduction of associated costs.

For this reason, the basis is a model that must be limited by a number of constraints and meet the objective of picking up all the orders of all the clients on the agreed date. This new work changes the previous situation by introducing additional constraints, which must be checked after the assignment of vehicles is carried out by means of a simulation process. A subsequent process of validation of the solutions found will ascertain their compatibility with real constraints, so the appropriateness of assignments could be queried.

33.3 Problem Description

The client is a logistics operator receiving daily orders from distributors. These orders must be picked up from manufacturers and grouped by companies and routes in order to eventually assign a means of transport for their collection. This strategy grants the logistics operator a better use of the logistical resources, and minimizes the associated costs of these.

However, there are a number of limitations that the company must bear in mind when it comes to providing their services:

- Type and capacity of road transport vehicles.
- Pick-up routes defined by the logistics operator.
- Number of suppliers that can be visited per vehicle per route.
- Cost of each vehicle.

There are multiple combinations of these variables and constraints in order to find solutions. In an urgent context the calculations of the composition of the fleet are difficult, because at the end of the working day all orders to be collected per manufacturer per route must be grouped, the orders must be assigned to the various vehicles, and these must be recruited.

For these reasons, it is necessary to use a quick formula to find solutions, which is easy to use and reduces costs when compared with the current work method of the logistics operator.

33.4 Solution for the Logistics Operator Problem

Taking the solution proposed by Ros et al. (2010) as a beginning, with the approximate algorithm developed by these authors and their set of decision-making rules which carried out an efficient load assignment in a vehicle fleet based on the routes determined in advance by the logistics operator, this improved version, in which all the objectives are met in terms of order collection, aims to find better solutions (based on the cost) than those supplied by the mathematical model. To this end, fleet variations are modelled for the different cases studied by using the Monte Carlo simulation technique. With this in mind, the aim is to find a variety of solutions that can be used under the defined constraints (see Ros et al. 2010).

The work method for the search of solutions is as follows:

1. Calculation of the load capacity to be hired on a daily basis. This capacity of the fleet must be greater than the total load to be picked up on any particular day.
2. Calculation of the possible solutions, by means of a Monte Carlo simulation, for the fleet to be used to collect the daily orders.
3. Cost study of each solution found in the previous step. Those solutions granting lower costs than those of the approximate algorithm must be selected.
4. Choice of the cheapest solution and route definition for each vehicle, along several alternative routes, on condition that the constraints are observed and the collections of all received orders are carried out satisfactorily.
5. Calculation of the working hours for of each vehicle, verifying that the daily limit of 8 h is not surpassed.

The last point has been included in the calculation procedure with the purpose of offering a practical, realistic solution to the logistics operator. Next, we will explain the manner in which the working hours of the vehicles are calculated.

Taking the definition of the VRP as a starting point, the vehicle leaves from one single depot and must travel one or several routes, with the restriction of a limitation of routes. As an example, Fig. 33.1 shows the route for a vehicle that must visit two suppliers in a working day. The terms included in Fig. 33.1 are defined as follows:

- TDep O_i: time elapsed from the depot to the start of the first route (“i”).
- $Troute_i$: travelling time of the complete “i” route.
- Tpr_i: time needed to collect all orders from all the suppliers in the “i” route (calculated service time at each supplier’s facilities is 30 min).
- $TDiO_j$: time from the end of the “i” route to the start of the “j” route.
- $Troute_j$: travelling time of the “j” route.
- Tpr_j: time needed to collect all orders from all the suppliers in the “j” route (calculated service time at each supplier’s facilities is 30 min).
- TD_jDep: return time from the end of the last route to the depot.

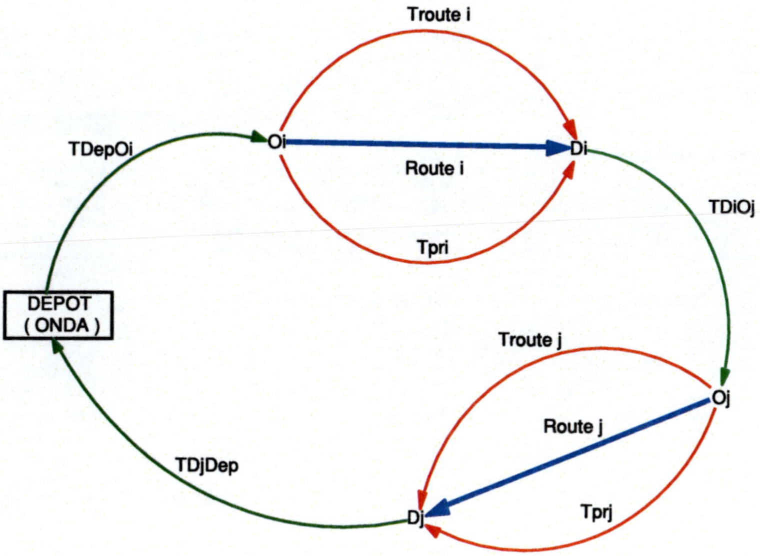

Fig. 33.1 Diagram for the calculation of vehicle times

Service time is defined as the time needed for the vehicle to leave the route, enter the supplier's facilities, pick up the order and resume travel along the day's route. This has been calculated to be 30 min for each supplier. This is additional to the time needed for the vehicle to travel the whole route, which has been calculated by using Google Maps web application.

The time that a vehicle would take to collect all the orders in the routes that have been assigned to it is calculated by the formula:

$$\mathrm{Ttotal} = \mathrm{TDep\ Oi} + \mathrm{Troutei} + \mathrm{Tpri} + \mathrm{TDiOj} + \mathrm{Troutej} + \mathrm{Tprj} + \mathrm{TDjDep} \tag{33.1}$$

33.5 Discussion

With the solution contributed by Ros et al. (2010), this information is taken as a start in order to proceed with the simulation of the transport fleet composition, in accordance with the procedure described in Sect. 33.2 above.

Firstly, among all possible simulations, those vehicle arrangements in which the total load capacity is more than the total weight of the daily orders to be collected must be found. The total load to be collected is 108,185 kg. The fleet must therefore have enough capacity to collect this load, always operating within the

DAY:19/1							
	ROUTE A	ROUTE B	ROUTE C	ROUTE D	ROUTE E	ROUTE F	TOTAL
KGS	40,906	24,402	4,577	12,517	1,885	23,898	108,185
No OF SUPPLIERS	14	22	4	2	1	18	61
VEHICLES	T1	T2	T3	T4	T5	T6	T7
CAPACITY	25,000	16,000	8,000	5,500	4,500	4,000	1,100
COST	300	220	170	160	160	160	125
SIMULATION OF FLEET VEHICLES & COSTS							
FLEET COMPOSITION	4T1+T3+T7	3T1+2T2+T7	2T1+3T2+T3+2T7	2T1+2T2+3T3+2T7	2T1+2T2+2T3+T4+T6+T7	2T1+2T2+T3+3T4+T7	
MAX CAPACITY (KGS)	109,100	108,100	108,200	108,200	108,600	108,700	
COST(€)	1,495	1,465	1,680	1,800	1,826	1,815	
		SOLUTIONS WITH A BETTER RESULT THAN THE MATHEMATICAL MODEL					
		SOLUTIONS WITH A WORSE RESULT THAN THE MATHEMATICAL MODEL					

Fig. 33.2 Simulation of vehicle fleets and calculation costs figures

DAY 9/1: 3T1+2T2+T7									
	START	ROUTE	ALCORA	No OF SUPPLIERS	SUPPLY TIME (MIN)	ROUTE TIME	RETURN	TOTAL	
FIRST T	ONDA	A	13	10	300	30	21	6.07	
25,000			KGS					24,713	OK
			SUPPLIERS					10	OK
			TIME					6.07	OK
SECOND T1	START	ROUTE	ALCORA	SUPPLIERS	TIME	ROUTE TIME	RETURN	TOTAL	
25,000	ONDA	A	13	4	120	30		2.72	
	START	ROUTE		SUPPLIERS	TIME	ROUTE TIME	RETURN	TOTAL	
	CASTELLON	B	10	5	150	36	18	3.27	
			KGS						OK
			SUPPLIERS					10	OK
			TIME					5.98	OK

Fig. 33.3 Validation of the selected solution

tolerance margin (±3%) of the weigh bridge. The cost calculation of each possible result must also be done.

Data regarding orders and load calculations are shown in Fig. 33.2. Among all the simulations obtained, the one with the lowest cost is chosen and the selected vehicles are assigned the various orders and routes, as laid out by the mathematical model, with routes organized from larger to smaller load, and vehicles from more to less capacity.

Once the assignment of routes to vehicles is completed, the defined solution must be validated to verify that the fleet can in fact carry out the collection of all the merchandise in no longer time than the usual 8-hour working day, as shown in Fig. 33.3.

Figure 33.4 shows the connections between routes and depot. All times are calculated by using Google Maps web application.

33.6 Conclusions

Decision-making in terms of fleet composition and vehicle sourcing is an important job which must be done quickly, and with relevant economic implications for a company. Orders being collected on one particular day will determine the size of

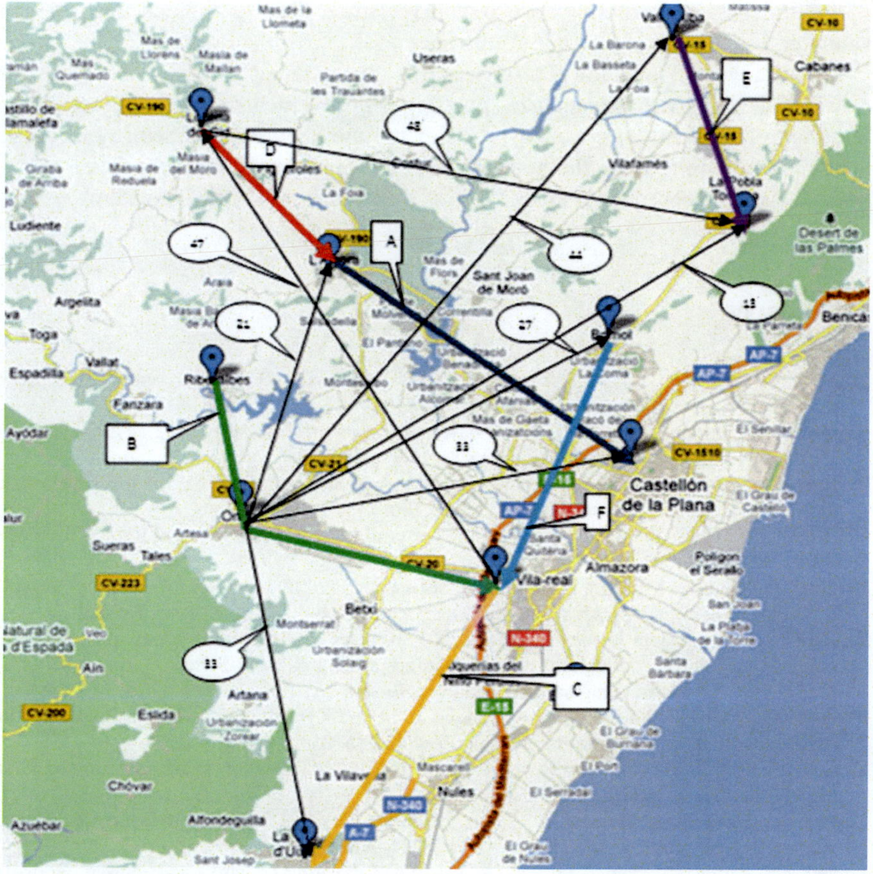

Fig. 33.4 Map of routes and time connection between routes and depot, by using Google Maps web application

the fleet for the following day. This situation, together with the lack of an appropriate method, may result in solutions that are quick and effective, but not efficient.

The Monte Carlo-based calculation method improves the results obtained by the approximate algorithm of Ros et al. (2010) for the resolution of the problem of arranging the vehicle fleet composition of a logistics operator. The improvements obtained by means of the simulation are reflected in a reduction of costs close to 20%, and in the increase of the variable efficiency of the load for each one of the vehicles in the fleet.

The advantages offered by this method as opposed to other existing methods are: ease and speed of calculation and use, and the possibility of improving solutions depending on both cost and fleet size.

All the obtained solutions are operative, since they have been verified and validated by Google Maps time & distance tool.

As a suggested line for future research, the detailed study of the routes is considered, in order to analyze possible improvements by means of route re-drawing. In addition, systematization of the methodology defined in this paper is being contemplated by the authors, with the purpose of enabling faster calculations, for the subsequent explanation of a new mathematical algorithm.

References

Bachelet B, Yon L (2007) Model enhancement: improving theoretical optimization with simulation. Simul Model Pract Theor 15:703–715

Bertsimas DJ, Simchi-Levi D (1996) A new generation of vehicle routing research: robust algorithms, addressing uncertainty. Oper Res 44(2):286–304

Carpaneto G, Toth P (1980) Some new branching and bounding criteria for the asymmetric travelling salesman problem. Manage Sci 26:736–743

Christofides N (1985) Vehicle Routing In: Lawler et al (eds) The Traveling Salesman Problem. Wiley, Chichester

Cordeau JF, Gendreau M, Hertz A, Laporte G, Sormany JS (2004) New Heuristics for the Vehicle Routing Problem, in Logistics Systems: Design and Optimization. Kluwer, Norwell

Desaulniers G, Desrosiers J, Solomon MM (2001) Accelerating strategies in column generation methods for vehicle routing and crew scheduling problems. In: Ribeiro CC, Hansen P (eds) Essays and surveys in metaheuristics. Kluwer, Boston

Fischetti M, Toth P, Vigo D (1994) A branch-and-bound algorithm for the capacitated vehicle routing problem on directed graphs. Oper Res 42:846–859

Laporte G (1992) The vehicle Routing Problem: an overview of exact and approximate algorithms. Eur J Oper Res 59:345–358

Laporte G, Gendreau M, Potvin JY, Semet F (2000) Classical and modern heuristics for the vehicle routing problem. Int Tra Oper Res 7:285–300

Powell WB, Jaillet P, Odoni A (1995) Stochastic and dynamic networks and routing. Network routing, handbooks in Operations Research and Management Science 8:141–295. Elsevier, Amsterdam

Psaraftis H (1995) Dynamic vehicle routing: status and prospects. Ann Oper Res 61:143–164

Toth P, Vigo D (2002) Models, relaxations and exact approaches for the capacitated vehicle routing problem. Discrete Appl Math 123:487–512

Ralphs TK, Kopman L, Pulleyblank WR, Trotter LE (2003) On the capacitated Vehicle Routing Problem. Math Program 94:343–359

Ros L, de-la-Fuente MV, Bogataj M (2010) An approximate algorithm for optimal logistics of heavy and variable size items. Cent Eur J Oper Res. doi: 10.1007/s10100-010-0155-6

Chapter 34
Analysis of Processes for a Collaborative Network

M. Victoria de-la-Fuente, Lorenzo Ros-McDonnell and Eloy Hontoria

Abstract The work presented in this paper focuses on the analysis of processes the fruit-and-vegetable supply chain. It is in a context of innovation and new technologies where the project is defined and initiated due to the need for improvements in logistics and transport of perishable goods. The paper presents the operational model of enterprises currently working within the fruit-and-vegetable supply chain, and an analysis of the collaborative network will be carried out throughout the enterprises or agents integrating it, as well as on the processes performed by each of these agents.

34.1 Introduction

The fruit-and-vegetable sector is extremely dynamic. Due to heavy domestic competition, it is continuously incorporating new methods and modern technologies and especially because of the development that is occurring in other regions and countries. It is in this context of innovation and new technologies where the PROCONET Research Project is defined and initiated due to the need for improvements in logistics and transport of perishable goods (De la Fuente and Ros 2010).

In line with the methodology chosen for the development of this project (ERE-GIO methodology, by Ros et al. 2009), the research team presents here the operational model of enterprises currently working within the fruit-and-vegetable supply chain. This model will make it possible to detect existing problems and

M. Victoriade-la-Fuente (✉) · L. Ros-McDonnell · E. Hontoria
ETSII-Universidad Politécnica de Cartagena,
c/Dr. Fleming, s/n, 30202 Cartagena, Spain
e-mail: marivi.fuente@upct.es

S. P. Sethi et al. (eds.), *Industrial Engineering: Innovative Networks*,
DOI: 10.1007/978-1-4471-2321-7_34, © Springer-Verlag London Limited 2012

needs, as a previous step to the development of the As-Is model of the collaborative network.

This paper is organized as follows: Sect. 34.2 presents a description of the PROCONET project and discusses its main objective and the design of the fruit-and-vegetable collaborative network to be developed within the project. Section 34.3 describes the current working manner of the fruit-and-vegetable supply chain and details its agents, the relations among them and the improving processes taking place within the collaborative network. Finally, Sect. 34.4 presents the conclusions drawn after the analysis of the processes in the fruit-and-vegetable collaborative network.

34.2 The PROCONET Project

The pursuit of the PROCONET Project research team may be defined as "the comprehensive traceability of agri-food road transport at controlled temperature", which, on the basis of logistic processes (either full loads or *groupages*) may grant access to experimenting with electronic CMRs by means of controlling technological variables involved in the said logistical process.

In these processes, relations are established among the agents involved in the food supply chain: producers, growers, exporters, importers, hauliers, etc., which must be adapted and managed by a collaborative business model.

The organizational model "Collaborative Network" is composed of a variety of entities (organizations and individuals) largely autonomous, geographically distributed, all collaborating to better achieve common/compatible goals, with the interactions among them supported by computer networks; and with common patterns, such as autonomous entities from varied locations, driven by common goals/intentions to be achieved by collaboration, and operationally based on agreed principles and inter-operable infrastructures which allow them to cope with their heterogeneity (Camarinha-Matos and Afsarmanesh 2005, Camarinha-Matos et al. 2009).

The PROCONET Project focuses on the construction of a prototype for strategic design of nodes in a logistics network. Simultaneously, it will develop a platform for knowledge management in a collaborative fruit-and-vegetable network, which supports the communication processes within that network. This network incorporates the necessary concepts, models and supporting tools related to the design of the process flow and the choice of technology:

- The operation of the prototype will be validated with global supply chain processes for perishable products.
- Global supply chain processes will consider both full loads and *groupages*.
- Special attention will be given to the impact of INCOTERM agreements.
- The prototype will support intermodal logistics processes as an added value.

34.3 The Fruit-and-Vegetable Collaborative Network

The development of a fruit-and-vegetable collaborative network calls for an appropriate paradigm and a relevant approach for different enterprise information systems of cold chain integration and collaboration. It is for that reason that the research team are working with the ERE-GIO methodology (Ros et al. 2009), which will allow the establishment of a modelling framework for CNs by integrating multiple perspectives. This is because this methodology presents a life cycle approach based on the “As-Is” model as well as on the “To-Be” model.

This double-model methodology will be used to develop a manner of approaching the cold supply chain from an endogenous CN perspective, and also to integrate the GPS system and the platform that will support cold chain processes, whilst analyzing the performance and constraints of the traceability and safety prototype and its further implementation within the collaborative network.

In this sense, in Reverse Engineering phase of ERE-GIO methodology, the research team tackles the identification and description of the business entity subject to study, which is the fruit-and-vegetable collaborative network of the various agents integrating it and of the processes to be carried out in order to eventually define the model of collaborative processes taking place in this CN.

34.3.1 The Fruit-and-Vegetable CN Framework

The current goal of companies in the fruit-and-vegetable CN is to foster the ideas of top-quality products and customer satisfaction by means of thorough and efficient production-marketing programming, the result of which will be the ability to give their products an added value, while offering a wide range of fruits and vegetables with a strict level of food safety.

In order to do so, CN companies perform over a surface of approximately 10,000 hectares located in the south-east of Spain. These farms yield more than 40,000 tonnes a year.

Production is divided into a wide range of fruits (melon, watermelon, citrus, stone fruits) and vegetables (tomato, pepper, cucumber, lettuce, artichoke, bean, etc.), all of them of first quality and marketed through 15 distribution chains, not only at a domestic level but also internationally: mainly middle- and eastern-Europe markets, as well as the newly-opened Asian and American markets.

In parallel, the philosophy of maintaining a high standard of quality is reflected by the various certifications held by several members of the CN: ISO 9001, BRC (*British Retail Consortium*), EUREPGAP, IFS (*International Food Standard*), which accounts for the opening of European markets and permits having control over such aspects of the CN as food safety, suppliers and farming products, the food processing industry and distribution.

It is convenient to note the difference between both the supply chain and the collaborative network, as defined by the PROCONET project research team. The fruit-and-vegetable supply chain includes the final consumer of perishable goods as an agent participating in the chain, whereas the CN considered in this project, although mindful of final objectives of traceability and food safety in all its logistical processes, starts with the farmer and ends with the distributor as a customer of the network, thus regarding the final consumer as an integrating part of the distribution chain, but not of the CN itself.

34.3.2 Relations Among Companies in the CN

As contemplated in the ERE-GIO methodology (de la Fuente and Ros 2010), in these early stages of the PROCONET project, an analysis of the CN has been carried out throughout the enterprises or agents integrating it, as well as on the processes performed by each of these agents. Four kinds of agents have been typified:

- *Producers*: farmers or agrarian associations dedicated to the production of fruits and vegetables, as well as on the packing operations for subsequent marketing of their products.
- *Marketers*: companies exporting these products and coordinating sales campaigns.
- *Hauliers*: companies committed to national and/or international transport of perishable goods.
- *Distributors*: acting as a customer in the collaborative network. These are mainly logistical platforms and large food chains.

In the fruit-and-vegetable collaborative network, where the various agents are geographically separated, the study of the relations among these agents in both quantitative and qualitative terms becomes especially relevant.

The quantitative and qualitative network analysis will be based on data acquisition, representation and examination, but also on identifying the types of interactions among the agents, analysing the collaborative relationships or interactions and intensity of the different co-operation tasks as the basis for an interaction value chain within the collaborative network. This kind of analysis of the collaborative network will provide a detailed way to better understand and forecast collaborative relationships and processes.

The first analysis carried out is that of the relationships among the agents. There are direct relations among the four groups of agents defined, although not all relations are comparable. Based on a study by Eschenbächer et al. (2009), three levels on intensity have been specified for these relations: low, medium and high, depending on the quantity and quality of contacts among agents.

For instance, a CN producer will usually have a main marketer, with whom work is regular, although contacts may exist with other marketing companies with whom the producer will choose on certain occasions (sales of specific products and/or in specific markets).

34.3.3 Process Analysis in the CN

Once the relations among the agents in the fruit-and-vegetable collaborative network have been scrutinized, the subsequent analysis of all processes completed by the agents has resulted in the following classification:

1. *Producer's processes:* producers (either individually or by means of some type of association) generate the fresh produce, namely fruits and vegetables. Just-harvested produce is then taken to storage facilities (either self-owned or belonging to another producer in the CN) for washing, calibrating and sorting. Upon receipt of the produce at the storehouse, the producer will contact a marketer, who will need to commission one or several distributors in order to bring the products to the market. It will be the distributor who will determine packaging specifications of the various products (Fig. 34.1).
2. *Marketer's processes:* once the existence of an order (batch of fresh produce) has been reported by any producer in the CN, marketers will seek the business areas within their scope for potential distributors. When one or several distributors have been located, both agents will negotiate the terms of the order (product, quantity, price, delivery time, delivery system, etc.). Load and delivery details will be agreed upon with the distributor and noted on the CMR or bill of payment. The marketer will then select a haulier who is then capable of fulfilling the task (national/international dispatch; refrigerated lorry; full load/*groupage*, etc.). The marketer will finally notify both the transport company and the producer about the time when the load must be collected.
3. *Haulier's processes*: after the load details have been exacted between marketer and distributor, the haulier will move the fresh produce from the designated storehouse by the producer. The producer is the agent who will monitor product condition during the loading process and will issue and stamp the copies of the CMR which will accompany the load in the lorry. Transport will have to be performed in the terms stated in the CMR and will terminate with the delivery of the load to the distributor. The distributor storehouse will receipt the load as long as it is in perfect condition (quantity, quality, freshness, food safety) and will stamp the corresponding copy of the CMR, which will be handed to the haulier as a confirmation of delivery. Should the load be out of the agreed terms and specifications, it will be rejected by the distributor, who will inform the haulier so that the corresponding agents may make the appropriate decisions (Fig. 34.2).

After the description of all processes put into practice by each of the agents in the CN, in Fig. 34.3 all processes are represented jointly, and those processes

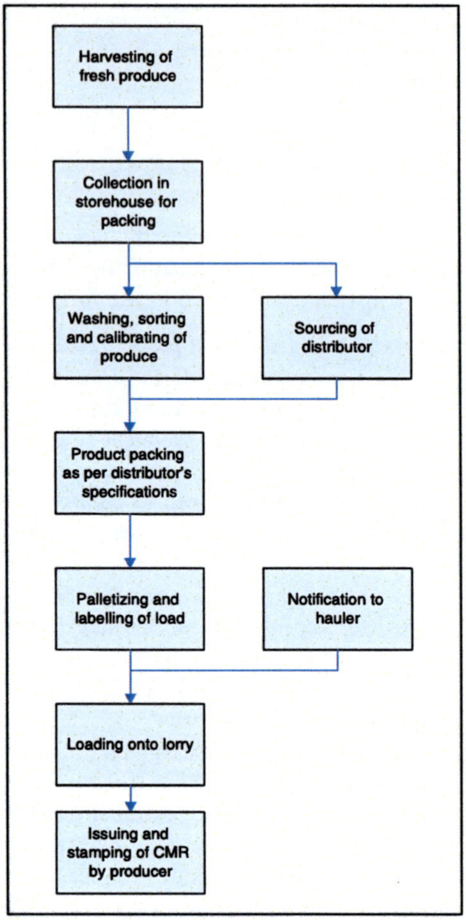

Fig. 34.1 Producer's processes diagram

related to producers, marketers and hauliers are individualized. Collaborative processes (producer-marketer; producer-haulier; marketer-distributor; marketer-haulier; haulier-distributor) currently taking place in the CN have also been differentiated.

34.3.4 Discussion Process

After analysing the processes acted on by the agents in the CN, the working model for the collaborative network (Fig. 34.3) presented shows a limited number of collaborative processes, as opposed to the amount of processes performed by each agent individually. In fact, agents' behaviour is more similar to that of a supply chain, whereas CN models produce collaborative work approaches.

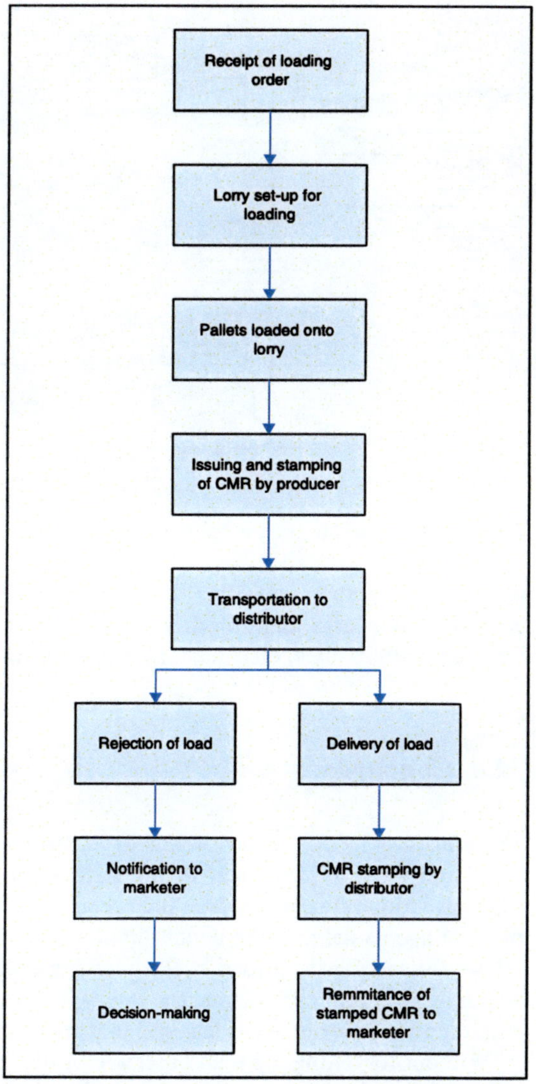

Fig. 34.2 Haulier's processes diagram

From a traceability point of view, it has been ascertained that each agent (producer, marketer and distributor) uses their own traceability systems. Contrarily, it has been detected that there is a lack of control in the processes performed by hauliers, given the fact that they do not possess a specific traceability system.

It is worth noting that the process of delivery of loads to a distributor stands out as a problematic issue in the CN. There is an undesirable percentage of rejections (due to non-fulfillment of specifications or CMRs, poor quality in production or transportation processes, etc.); CMRs left unstamped by distributors, which make

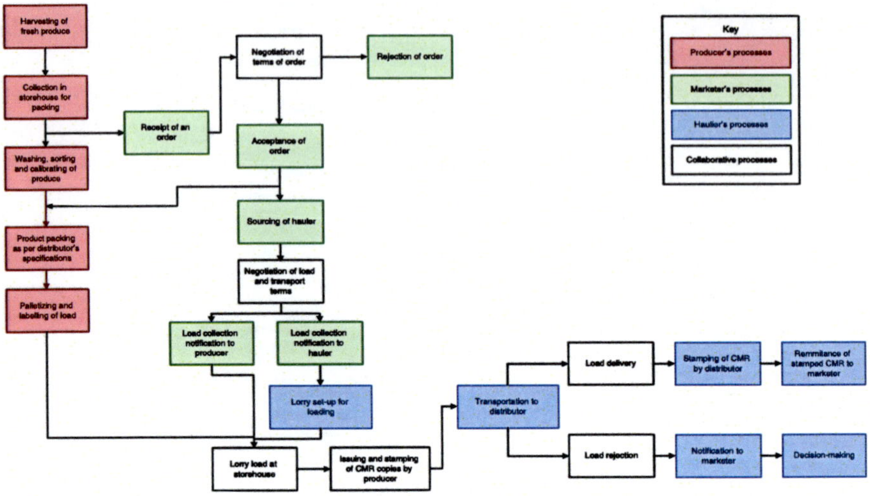

Fig. 34.3 Processes diagram in the Fruit-and-Vegetable Collaborative Network

hauliers unable to assure other agents that the load that has been delivered (and consequently charge them for their services) and the non-existence of return flows (rejected load, pallets, etc.) have also been detected.

34.4 Conclusions

The processes model defined in this paper for the collaborative network requires a collaborative work approach in the light of the following detected needs: the transformation of current individual processes into collaborative processes among the four agents defined; the definition of a collaborative traceability system which allows for monitoring a load at all times, especially during transport processes, in which there is zero control, and the 2-way functioning of the collaborative network where return flows of materials and information are defined.

Without forgetting the final objective of the PROCONET Project: to develop a prototype of collaborative network which guarantees traceability and food safety in all logistic processes of the defined network.

References

Camarinha-Matos L, Afsarmanesh H (2005) Collaborative networks: a new scientific discipline. J Intell Manuf 16:439–452

Camarinha-Matos L, Afsarmanesh H, Galeano N, Molina A (2009) Collaborative networked organizations–Concepts and practice in manufacturing enterprises. Comput Ind Eng 57:46–60

De la Fuente MV, Ros L (2010) Cold supply chain processes in a fruit-and-vegetable collaborative network. In: Proceedings of the 9th IFIP International Conference on Information Technology for Balanced Automation Systems (BASYS 2010)

Eschenbächer J, Seifert M, Thoben KD (2009) Managing distributed innovation processes in virtual organizations by applying the collaborative network relationship analysis. In: Proceeding of the 10th IFIP WG5.5 Working Conference on Virtual Enterprises, PRO-VE 2009

Ros L, de la Fuente MV, Ortiz A (2009) Enterprise engineering versus cycle re-engineering methods. In: Proceeding of the 13th IFAC Symposium on Information Control Problems in Manufacturing (INCOM'09)

Chapter 35
A Supply Chain Operations Lot-Sizing and Scheduling Model with Alternative Operations

Julien Maheut, José Pedro Garcia-Sabater and Josefa Mula

Abstract The aim of this paper is to propose an mixed integer linear programming (MILP) model for operations lot-sizing and scheduling (assignment and sequencing) in the supply chain of an international company which produces and delivers customized products through several geographically distributed assembly plants. The model schedules the purchase of raw materials in the various plants considered, the transshipments, shipments to customers and the various operations to assemble the product. The model considers different alternative production operations such as product substitution (upgrading), alternative procurement and transport operations. It also addresses the different lead times associated with these operations. Specific constraints such as space availability on each plant and workforces are contemplated. A novel approach based on the stroke concept is applied to the MILP model to model alternatives.

35.1 Introduction

A supply chain (SC) is a network of organisations involved through upstream and downstream relations where different processes and activities are carried out to produce value in the form of products and/or services for the end customer. In order to face increasingly demand, SCs must offer a product or service with a

J. Maheut (✉) · J. P. Garcia-Sabater
Reengineering Operations GroupWork Logistics Excellence (ROGLE), Universitat Politècnica de València, Camino de Vera S/N, 46022 Valencia, Spain
e-mail: juma2@upvnet.upv.es

J. Mula
Research Center on Production Management and Engineering (CIGIP), Universitat Politècnica de València, Plaza Ferrándiz y Carbonell 2, 03801 Alcoy, Spain

S. P. Sethi et al. (eds.), *Industrial Engineering: Innovative Networks*,
DOI: 10.1007/978-1-4471-2321-7_35, © Springer-Verlag London Limited 2012

minimum cost and a short lead time. To do so, Stadtler (2005) considers that SC management must be based on two pillars: SC integration and coordination.

For SCs to be able to coordinate efficiently, the literature contemplates two phases at the strategic level: SC design (Mohammadi Bidhandi et al. 2009) or redesign (Nagurney 2010) and SC configuration. Graves and Willems (2003) were the first to introduce the SC configuration problem. In general, this problem contemplates different possible configurations because, for instance, raw material can be purchased from different suppliers, products can be produced or assembled on different machines, or products can be delivered by different forms of transport (Li and Womer 2008). Selecting a configuration implies reaching a compromise between the costs involved and the service levels to be offered to the customer. The literature includes a large amount of mathematical models which address the SC configuration problem. We refer readers to the following review (Mula et al. 2011). The literature includes some cases such as the work of Li and Womer (2008) in which problem deal with the configuration problem at the same time with considerations at the tactical and/or operational level.

In relation to lot-sizing and scheduling problems, one of the first proposed models was that of Wagner-Whitin (1958) which proposes lot-sizing for a single product. Afterwards, work was done on the capacitated lot-sizing problem (Karimi et al. 2003). Later, another concepts were introduced: sequence-dependent setup times (Haase 1996), lead time concepts (Hnaien et al. 2008), multi-stage production (Tempelmeier and Buschkühl 2008), products substitution (Lang 2010) or other multi-site scheduling problems into scheduling models (Lloret et al. 2008). Nevertheless, to the best of our knowledge, the multi-site, multi-stage, capacitated lot-sizing and scheduling problem with lead times has not been studied by considering alternative operations for purchasing, transport (replenishment, transshipments and distribution) and production.

This article proposes an MILP model to optimise lot-sizing and scheduling (assignment and sequencing) of SC operations with the arrival of new firm orders. The model proposed is based on the stroke concept (Garcia-Sabater et al. 2009a, b) (a similar concept to the Resource-Task Networks (Pantelides 1994)) to consider purchase, transport and production alternatives in the SC. The model's objective is to minimise total costs by fulfilling lead times.

The structure of the paper is as follows. Section 2 describes the aims of the model. Section 3 presents the basic assumptions of the model. Section 4 formulates the MILP model for supply chain lot sizing and scheduling. Finally, the last section draws conclusions and provides future research lines.

35.2 The Supply Chain Operations Lot-Sizing and Scheduling Model with Alternative Operations

The SC operations lot-sizing and scheduling model with alternative operations that considers lead times (SCOLSS-AO) is a multi-site, multi-level and multi-period problem with transport among plants (transshipments) that considers alternative production routes and products substitution. It contemplates intermediate (or half-finished) items and finished goods, as well as different resources.

When new firm orders arrive, SC configuration and multi-plant scheduling are performed to deliver the product to customers. SC scheduling must consider restrictions in all the plants. Furthermore, setup times, costs and all the possible alternatives, i.e., replenishment, production/assembling, and transport alternatives, must be studied.

Since it is assumed that already sequenced products cannot be amended, and as resources have been assigned and scheduled with a defined sequence, the available resources capacity considers an assignment prior to these operations.

In general, the objective of the model is to determine operations scheduling which minimises costs by fulfilling customers′ expectations in terms of the characteristics of the product to be delivered and due dates. In detail, the model provides: (i) the different products and SC configurations that respond to the strategy selected; (ii) the supplier that best responds to each strategy efficiently; (iii) the costs of each configuration; and (iv) the due date of the product ordered for each configuration. In short, when a new order arrives, the model must assign the production of the various modules to each plant and resource, generate the raw materials purchase order to suppliers, and manage the transshipments among plants.

35.3 Assumptions

The SCOLSS-AO is a deterministic model that considers a P set of product, a W set of sites, a Z set of operations (known as strokes) and an R set of resources.

Each product are defined with both their packaging and site. A stock-keeping unit (SKU) corresponds to each product. Consequently, two products in different sites are considered two different items. Two products in the same site but in different packagings are considered two different SKUs. Demand refers to an SKU with specific characteristics, which is accomplished with the assembly of several modules that are produced basically with a raw material purchased from several suppliers. Demand has to be served with a due date at a given site without backlogs. Substitution of SKUs is considered (through upgrading or because of the suppliers' capacity to supply similar components).

Each stroke corresponds to a determined located operation. It is characterised by the use of located resources. Two technically identical operations executed in two different sites are considered as two different strokes since they are using different SKU and creating different SKU. Moreover, two identical operations executed in the same site but using different resources are considered as two different strokes. A set of SKUs is assigned to each stroke which is consumed when a stroke unit is executed. This set (known as "stroke input") can be null, unitary or multiple, while its coefficients (the Gozinto factor) can be above a unit. A set of SKUs is assigned to each stroke which is produced when a stroke unit is executed. This set (called a "stroke output") can consist in several different items, a single item or none, and its

Table 35.1 Sets and indexes

Symbol	Definition
Sets and indexes	
$i \in P$	SKUs
$r \in R$	Resources
$k \in Z$	Strokes including dummy stroke 0
$0 \in Z$	Dummy stroke for modeling time during which a resource is not set up for any stroke
$j \in J$	Sites
$t = 1, \ldots, T$	Periods
Z_r	Set of strokes that are executed in resource r ($Z_r \subseteq Z$)
Z_j	Set of strokes that are executed in site j ($Z_j \subseteq Z$)
L_X	Set of strokes whose lead time is lower than $X \in \{0, 1, .., l\}$
P_j	Set of SKUs in site j ($P_j \subseteq P$)

coefficients (amount of each item produced) can be above one unit. Lead times are considered and assigned to each stroke. Setup times and the associated costs are invariable in time, but depend on the sequence of the stroke considered (consequently, on the resources set employed). The scheduling (assignment and sequencing) of strokes is executed in each period. There can be different setups in one same period. The existence of alternative operations is considered. Some modules can be produced in different resources and/or sites.

Resources are localised, immovable and heterogeneous. In resources, strokes have been assigned and sequenced in accordance with previous orders. Thus for each resource, the production capacity corresponding to the different strokes needed for these orders has been reduced. Resources have different limits of the capacity available in each period. The consumption of the production capacity in each resource varies in terms of both the strokes executed in it in each period and the setup changes contemplated. Setups must be complete during each period.

35.4 Formulation

Table 35.1 contains the notations for the constants, sets and indexes used for formulating the SCOLSS-AO.

Table 35.2 contains the notations for parameters.

Table 35.3 contains the notations for variables.

By assuming, for instance, that the maximum lead time is 3, then $X \in \{0, 1, 2, 3\}$. Thus, there are three lists $L_X \in \{L_0, L_1, L_2, L_3\}$. The SCOLSS-AO can be formulated as shown below:

$$Minimise\, F(x, y, \delta, \theta) = \sum_{i \in P} \sum_{t=1}^{T} (H_i \cdot y_{i,t}) + \sum_{k1 \in Z} \sum_{t=1}^{T} \left(CO_{k1,t} \cdot x_{k1,t} + \sum_{k2 \in Z} CS_{k1,k2} \cdot \delta_{k1,k2,t} \right) \quad (35.1)$$

Table 35.2 Parameter notation

Symbol	Definition
Parameters	
D_{it}	Demand for SKU i in period t (due date)
X_{it}^{rec}	Planned reception for SKUs i in period t
H_i	Non-negative holding cost per period for storing one unit of SKU i
$Y_i^{max} \backslash Y_i^{min}$	Upper\Lower inventory limit for SKU i
Y_i^0	Initial inventory of SKU i
A_i	Space consummation for storing one unit of SKU i
A_k	Workforce utilisation for executing one unit of stroke k
K_j^{sp}	Space capacity of site j (in space unit)
Γ_k	Workforce utilisation for executing one unit of stroke k
K_{rt}^{lab}	Workforce capacity available of resource r in period t (in the workforce unit)
CO_{kt}	Cost of stroke k in period t
$CS_{k1,k2}$	Setup cost that is incurred when the setup state changes form stroke 1 to $k2$
$\theta_{k,1} = \delta_{k,k,1}$	Binary parameter that indicates whether stroke k is set up at the beginning of the first period
SO_{ik}	Number of SKU i units produced by making one unit of stroke k (stroke output)
SI_{ik}	Number of SKU i units required for making one unit of stroke k (stroke input)
$LT(k)$	Lead time of stroke k
M_r	Number of strokes executed in resource r

Table 35.3 Variable notation

Symbol	Definition
Variables	
x_{kt}	Quantity of strokes k starting in period t
y_{it}	Inventory of SKU i at the end of period t
v_{kt}	Auxiliary variable: the larger it is, the later the stroke is scheduled in period t
$\delta_{k1,k2,t}$	Binary variable which indicates whether stroke $k2$ is set up immediately after stroke $k1$ in period t
θ_{kt}	Binary variable which indicates whether stroke k is set up at the beginning of period t

subject to

$$y_{it} = y_{i,t-1} + X_{it}^{rec} + \sum_{k \in Z_i^+} \left(SO_{ik} \cdot x_{k,t-LT(k)} \right) - D_{it} - \sum_{k \in Z_i^-} \left(SI_{ik} \cdot x_{kt} \right) \quad i \in P, t = 4, \ldots, T \tag{35.2}$$

$$y_{i,X} = y_{i,X-1} + X_{i,X}^{rec} + \sum_{k \in L_{X-1}} \left(SO_{ik} \cdot x_{k,t-LT(k)} \right) - D_{iX} - \sum_{k} \left(SI_{ik} \cdot x_{k,X} \right) \quad i \in P, X = 1, \ldots, 3 \tag{35.3}$$

$$Y_i^{\min} \leq y_{i,t} \leq Y_i^{\max} \quad i \in P, t = 1, \ldots, T \tag{35.4}$$

$$\sum_{k \in Z_j} \sum_{\tau=t-LT(k)}^{\tau=t} A_k \cdot x_{k\tau} + \sum_{i \in P_j} A_i \cdot y_{it} \leq K_j^{sp} \quad i \in P, t = 4, \ldots, T \tag{35.5}$$

$$\sum_{k \in Z_j \wedge k \in L_{X-1}} \sum_{\tau=X-LT(k)}^{\tau=t} A_k \cdot x_{k\tau} + \sum_{i \in P_j} A_i \cdot y_{i,X} \leq K_j^{sp} \quad j \in J, X = 1, \ldots, 3 \tag{35.6}$$

$$\sum_{k \in Z_r} \Gamma_k \cdot x_{kt} \leq K_{rt}^{lab} \quad r \in R, t = 1, \ldots, T \tag{35.7}$$

$$\Gamma_{k1} \cdot x_{k1,t} \leq K_{rt}^{lab} \left(\theta_{k1,t} + \sum_{k2 \in Z_r} \delta_{k1,k2,t} \right) \quad r \in R, k1 \in Z_r, t = 1, \ldots, T \tag{35.8}$$

$$1 = \sum_{k \in Z_r} \theta_{kt} \quad Z_r | r \in R, t = 1, \ldots, T \tag{35.9}$$

$$\theta_{k,t} + \sum_{k1 \in Z} \delta_{k1,k,t} = \theta_{k,t+1} + \sum_{k2 \in Z} \delta_{k,k2,t} \quad k \in Z, t = 1, \ldots, T \tag{35.10}$$

$$v_{k2,t} \geq v_{k1,t} + 1 - |M_r|(1 - \delta_{k1,k2,t}) \quad k1, k2 \in Z_r,\ k1 \neq k2,\ t = 1, \ldots, T \tag{35.11}$$

$$x_{kt}, v_{vt} \in \mathbb{Z}_0^+ \quad k \in Z, t = 1, \ldots, T \tag{35.12}$$

$$y_{it} \geq 0 \quad r \in R, t = 1, \ldots, T \tag{35.13}$$

$$\delta_{k1,k2,t} \in \{0, 1\} \quad k1, k2 \in Z, t = 2, \ldots, T \tag{35.14}$$

$$\theta_{kt} \in \{0, 1\} \quad k \in Z, t = 2, \ldots, T \tag{35.15}$$

Objective (35.1) consists in minimising the sum of the storage costs, the stroke execution costs and those costs associated with sequencing strokes. (35.2) provides the continuity equation of the inventory levels. The inventory level at the end of a period considers the inventory level at the end of the previous period, planned receptions, product demand and the execution of those strokes with associated lead times. (35.3) presents the continuity equation for the first periods given the lead times (in this case, the maximum assumes 3 periods). (35.4) offers a limit for the maximum and minimum inventory levels for each item. (35.5)–(35.6) present the limits of the space resource. These limits imply that the sum of the space consumed by executing strokes k (a similar concept to the WIP stock) in each plant, plus the space consumed by the inventory levels of items i in the considered plant, cannot exceed a limit. (35.7) offers availability limit of the workforce resource as a result of executing the different strokes in each resource. Equation (35.8) ensures that the execution of stroke k in the considered resource occurs only in period t if stroke k is a setup in the resource at the end of period t-1, or a change in the setup state is made in period t. Equation (35.9) implies that only one stroke is in the setup at the end of period t. Equation (35.10) conserves the setup state (Lang 2010). Constraint (35.11) enables the creation of a sequence of the strokes for each resource throughout each period thanks to the use of variable $v_{k,t}$. If $v_{k1,t} > v_{k2,t}$, then stroke $k1$ will be sequenced after stroke $k2$ during period t. Equations (35.12)–(35.15) define the domains of the considered variables.

We have solved the SCOLSS-AO problem employing the Gurobi optimizer 4.5. We tested the performance on a real size case study problem with 2 sites, about 60 products taking into account alternative operations, under various space and workforce resources constraints and including transshipments between the 2 sites for different instances. The results showed an average running time of 1 min per instance with an Intel Core i7 3.22Ghz processor, 24 GB RAM and Windows 7 as OS.

35.5 Conclusions

The model presented in this work has been created to plan and schedule the supply chain operations activities in an international company with distributed plants. The novelty of the MILP model presented in this paper is its capacity to schedule alternative operations in a multi-site context, like transshipments, shipments to customers, product substitution and alternative operations considering lead times. Moreover, in order to model alternatives, a novel approach based on the stroke concept is introduced and applied to the MILP model.

Although the work presented herein lacks a practical application given the limited extension of a paper, a real application will be presented in an extended paper.

Further research has been identified throughout this work as follows: (i) extending the model by including backorders; (ii) designing specific heuristics for the problem considered herein, and incorporating the central stroke concept for modelling and solving distributed problems; and (iii) incorporating variants such as uncertainty (whether it is stochastic or uses fuzzy methods) is another future research line.

Acknowledgments The research leading to these results has received funding from the European Community's Seventh Framework Programme (FP7/2007-2013) under grant agreement no. NMP2-SL-2009- 229333. Julien Maheut holds a Val I+D grant funded by the Generalitad Valenciana (Regional Valencian Government, Spain) (Ref. ACIF/2010).

References

Garcia-Sabater JP, Maheut J, Garcia-Sabater JJ (2009a) A capacited material requierements planning model considering delivery constraints. In: 3rd international conference on industrial engineering and industrial management. pp 793–803

Garcia-Sabater JP, Maheut J, Garcia-Sabater JJ (2009b) A capacited material requierements planning model considering delivery constraints: a case study from the automotive industry. In: 39th international conference on computers and industrial engineering. pp 378–383

Graves SC, Willems SP (2003) Supply chain design: safety stock placement and supply chain configuration. Handb Oper Res Manag Sci 11:95–132
Haase K (1996) Capacitated lot-sizing with sequence dependent setup costs. OR Spectr 18:51–59
Hnaien F, Dolgui A, Louly MA (2008) Planned lead time optimization in material requirement planning environment for multilevel production systems. J Syst Sci Sys Eng 17:132–155
Karimi B, Fatemi Ghomi SMT, Wilson JM (2003) The capacitated lot sizing problem: a review of models and algorithms. Omega-Int J Manage S 31:365–378
Lang JC (2010) Production and inventory management with substitutions. Springer, Germany
Li H, Womer K (2008) Modeling the supply chain configuration problem with resource constraints. Int J Proj Manag 26:646–654
Lloret J, Garcia-Sabater JP, Marin-Garcia J (2008) Cooperative multisite production re-scheduling
Mohammadi Bidhandi H, Yusuff R, Megat Ahmad MMH et al (2009) Development of a new approach for deterministic supply chain network design. Eur J Oper Res 198:121–128
Mula J, Maheut J, Garcia-Sabater JP (2011) Mathematical modelling for supply chain configuration. In: Brennan. Christopher R (ed) Mathematical modelling. Nova Science Publisher, New York
Nagurney A (2010) Optimal supply chain network design and redesign at minimal total cost and with demand satisfaction. Int J Prod Econ 128:200–208
Pantelides CC (1994) Unified frameworks for the optimal process planning and scheduling. In: Proceedings of the 2nd Conference on the Foundations of Computer Aided Operations. 253–274
Stadtler H (2005) Supply chain management and advanced planning–basics, overview and challenges. Eur J Oper Res 163:575–588
Tempelmeier H, Buschkühl L (2008) Dynamic multi-machine lotsizing and sequencing with simultaneous scheduling of a common setup resource. Int J Prod Econ 113:401–412
Wagner HM, Whitin TM (1958) Dynamic version of the economic lot size problem. Manage Sci 5:89–96

Part VI
Operations Research

Chapter 36
The Power of ECOTOOL MATLAB Toolbox

Diego J. Pedregal and Juan R. Trapero

Abstract This paper presents the ECOnometrics TOOLbox (ECOTOOL), a new MATLAB forecasting toolbox that embodies several tools for identification, validation and forecasting of dynamic models based on time series analysis. Tools to perform a wide range of exploratory and statistical tests with visual counterparts are included, designed in easy-to-use front ends. The models implemented so far are classical in essence (in the very first version of the toolbox), among them, ARIMA, Exponential Smoothing, Unobserved Components, ARX, ARMAX, Transfer Function, Dynamic Regression and Distributed Lag models. The main idea of this development is providing a tool that, with a few code lines, performs a great number of tasks. The toolbox is presented in all its potentiality on the Spanish Industrial Production Index.

36.1 Introduction

MATLAB is a very powerful and flexible tool in engineering at many levels, but also in many other disciplines. This paper concentrates on a toolbox developed recently for signal processing, analysis of time series and econometrics that is intended to overcome some of the rigidities found in other pieces of software (some MATLAB toolboxes included).

D. J. Pedregal (✉)
ETSI Industriales e IMACI, Universidad de Castilla-La Mancha,
13071 Ciudad Real, Spain
e-mail: Diego.Pedregal@uclm.es

J. R. Trapero
Facultad de Ciencias Químicas, Universidad de Castilla-La Mancha,
13071 Ciudad Real, Spain

S. P. Sethi et al. (eds.), *Industrial Engineering: Innovative Networks*,
DOI: 10.1007/978-1-4471-2321-7_36,

The user often finds that some of the official toolboxes in MATLAB do not offer the flexibility usually required by real applications in specific contexts. For example, missing observations are sometimes a problem, there is no easy treatment for outliers, no quick way of imposing constraints among parameters in models, models are defined in a narrow way, there are some difficulties to code repetitive routines, etc. Due to these limitations, we developed a MATLAB toolbox (ECOTOOL,[1] Pedregal et al. 2011) that overcomes such difficulties. Its visual capabilities and user-friendly interface can be very effective for students, professors and forecasting professionals dealing with problems in a wide range of different contexts. Among the salient properties of the toolbox we could list i) it is user-oriented, i.e. a full time series analysis may be performed with just a few MATLAB functions, so the user should keep in mind just a few function names; ii) thorough help is available for the toolbox as a whole and for each function in particular, as it happens with standard MATLAB toolboxes; iii) specification of models is rather simple and flexible, e.g. in the case of TF models, they are written in MATLAB code in just the same way as they would be written analytically on paper; iv) imposing constraints on the parameters is straightforward; v) estimation methods for some models include both conditional and exact Maximum Likelihood; vi) automatic detection and estimation of four types of outliers is implemented for some models: additive, innovative, level shift and transitory change functions (see Tsay 1986); vii) to deal with moveable holidays, trading day, Easter, leap year, and other calendar effects are also included; viii) additional identification and diagnostic tools are made available by means of the `toolTEST` function, as well as a forecast function to compute forecast errors and other statistics related to the appropriateness of the models (`toolFORECAST`); ix) step-by-step demo files based on real data are included to show how the toolbox works.

36.2 Modelling Options

The toolbox offers the possibility of estimating and forecasting time series by using models with different degrees of complexity, either Single-Input–Single-Output (SISO), Multiple-Input–Single-Output (MISO) or Multiple-Input—Multiple-Output (MIMO) systems.

36.2.1 SISO Models

Three main families of SISO systems (or univariate models) are available in ECOTOOL, namely ARIMA, Exponential Smoothing (ES) and Unobserved Components models (UC).

[1] The toolbox is available upon request to the authors via email.

ARIMA Models

Following the standard ARIMA jargon (Box et al. 1994), the models implemented on the toolbox are of the $ARIMA(p_0, d_0, q_0) \times (p_1, d_1, q_1)_{s_1} \times \ldots \times (p_k, d_k, q_k)_{s_k}$ class. It is clear that such a model is very general because it includes as many seasonal multiplicative polynomials as necessary. It also includes multiple ARIMA models as particular cases. Estimation of this type of models may be performed by a number of methods, the ones implemented in ECOTOOL being the conditional sum of squares and exact maximum likelihood (Box et al. 1994).

Exponential Smoothing Models

Six different cases of additive Exponential Smoothing (ES) models may be used, depending on the number of components included. Each of the six models are coded with two letters (NN, AN, DN, NA, AA, DA). The first letter in the code indicates the model for the trend, while the second letter is reserved for the seasonal model. 'N' indicates that the component is not present; 'A' implies an additive component; 'D' implies a damped component, reserved solely for the trend.

Any ES model may be expressed in a constrained ARIMA model as a *reduced form* (see Hyndman et al. 2008). ES models are observationally equivalent to their reduced counterparts, and produce the same forecasts and prediction intervals. Therefore, ES estimation in the toolbox is worked out from the reduced form, based on the MATLAB code developed for ARIMA estimation. The main advantages are: i) either exact or conditional likelihood is possible, ii) further procedures for outlier automatic detection and estimation would be applicable (see next section), and iii) further extensions of the ES models with inputs may be used (see MISO models below).

Unobserved Components Models

In a univariate Unobserved Components model (UC), the time series are assumed to be the addition of several components, usually a trend, seasonal and irregular, each one with its own physical interpretation (in a similar way to Exponential Smoothing models).

The SS representations of each of them used in this paper are typical of the so-called Basic Structural Model (BSM) of Harvey (1989). Given the full SS system, the Kalman Filter (KF) (1960) and the Fixed Interval Smoothing algorithms (FIS) (Bryson and Ho 1969) provide the optimal estimation of the first and second order moments of the state vector in the sense of minimizing the Mean Squared Error.

36.2.2 MISO Models

The basic MISO models implemented are a natural generalization of the SISO counterparts. The main model is a Linear Transfer Function (Box et al. 1994) with coloured noise. The important point here is that the noise may be modelled as any of the previous SISO models, i.e. ARIMA, ES or UC.

36.2.3 MIMO Models

VARX models are also included in the toolbox. Estimation of unrestricted VARX models is particularly simple, since least squares of each equation separately produce consistent and efficient estimates. However, when constraints are imposed onto some of the elements of the coefficients matrices, iterated generalized least squares may be used, in order to avoid inefficient estimation (Lütkepohl 1991).

36.3 ECOTOOL in Use

Due to space constraints it is not possible to fully explain each one of the functions, but in the remainder of this section the capabilities of the two GUI included and the use of the most important modelling functions are briefly described.

One key function in the toolbox is the `toolTEST` GUI, in which, as well as the standard figure menus, the tool offers three additional possibilities: i) *Tests* to perform a long list of statistical tests (see below); ii) *Series* to select the time series (or all available) on which the text will be executed; and iii) *Options* to choose particular options for the current test (the second GUI, `toolFORECAST`, is built in a similar way). The options available in the *Tests* menu are:

- Descriptive information: time plots, box plots to detect seasonal behaviour, scatter plots, descriptive statistics, histograms with hypothetical Gaussian distribution, quantile plots, gaussianity standard tests.
- Identification tools: Univariate Autocorrelation and Partial Autocorrelation Functions, Vector Autocorrelation and Partial Autocorrelation Functions for multivariate time series, several information criteria applied to increasing VAR models, Granger causality tests based on VAR models.
- Heteroscedasticity tests: CUSUM and CUSUMSQ tests, range-mean plot (standard deviation mean or median plots), variance ratio tests, estimation of univariate or multivariate Box-Cox homoscedasticity transformation.
- Unit root and co integration tests: Dickey-Fuller, Phillips Perron and Johansen tests.
- Non-linearity tests: Tsay, Schwarz criterion on squares, Brock-Dechert-Scheinkman test, etc.
- Spectral tools: cumulative periodogram, smoothed or raw periodogram, AR-spectrum.

The main MATLAB functions in ECOTOOL are `modelES`, `modelTF`, `modelUC`, `modelVARX`. These are the functions that estimate the different types of models. All of them have been written using a common syntax in order to make their use easier. Next section shows a hands-on example in which one code makes it possible to perform an analysis of a time series by ECOTOOL and is shown in great detail.

Then the usual procedure for a full analysis of a time series would pass, basically, through the use of `toolTEST` for the identification of tentative models, `model??` for the estimation and once more `toolTEST` for model validation. These stages are shown in practice within the case study in the next section.

36.4 Case Study

The case study analyzed to show the toolbox working in practice is the Spanish monthly Industrial Production Index (IPI) from January 1975 to January 2011 (433 observations). The case is not intended as a comprehensive analysis of the IPI time series, but only as an example of the use of ECOTOOL.

Provided the data are already loaded onto the MATLAB environment coded as a vector with name `ipi`, a possible MATLAB code to analyze the series might be:

```
Line 1: toolTEST(ipi)
Line 2: z = log(ipi);
Line 3: toolTEST(vdif(z, [1 12], [1 1]))
Line 4: model = '(1 + a1*B)(1 + a12*B12 + a24*B24)/(1-B)
(1-B12)
                  (1 + b1*B)';
Line 5: uu = easter([1975 1], 433);
Line 6: u = uu(1 : end-12); uf = u(end-11 : end);
Line 7: [e1, pz1, stdpz1] = modelTF(z(1 : end-12), u,...
        {model, 'EASTER'}, uf, 'EML', 4);
Line 8: [e2, pz2, stdpz2] = modelES(z(1 : end-12), u,...
        {'AA12', 'EASTER'}, uf, 'EML', 4);
Line 9: [e3, pz3, stdpz3] = modelUC(z(1 : end-12), u,...
        {'LLT12', 'EASTER'}, uf, 'EML', 4);
Line 10: toolTEST([e1 e2 e3])
Line 11: toolFORECAST(ipi(end-11 : end), exp
        ([pz1 pz2 pz3]));
```

Line 1 provides a first exploratory analysis of the time series in many graphical and formal ways, as it was briefly shown in the previous section. The series is shown in Fig. 36.1 where the effect of the recent economic crisis is clearly visible, since the index has fallen to levels of 1996 or even 1990.

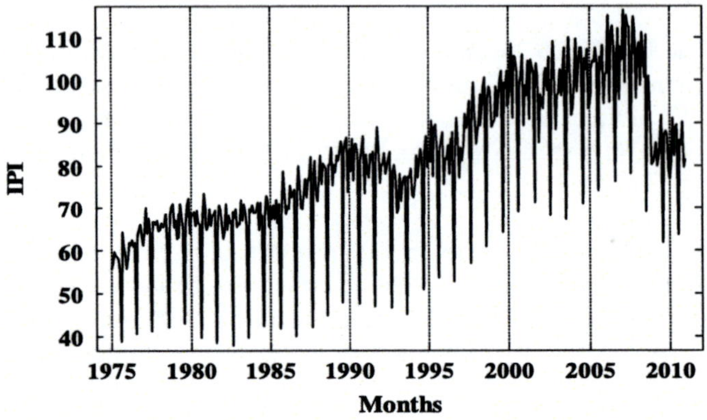

Fig. 36.1 The Spanish industrial production index from January 1975 to January 2011

Line 2 transforms the series into natural logarithms to avoid heteroscedasticity problems. This could be further analyzed with `toolTEST` in Line 1, since specific homoscedasticity tests are implemented and Maximum Likelihood estimation of the transformation parameter in the Box-Cox family is possible.

Line 3 provides the code necessary to perform a number of tests on the stationary time series. Be aware that differentiation is done on the log-transformed time series by the function `vdif`. One of the most important tests here is the simple and partial autocorrelation functions, shown in Fig. 36.2.

Line 4 sets up an ARIMA model for the series. The model is passed to the estimation function `modelTF` in Line 5 in variable `model`. That variable is a ratio of polynomials as they are written analytically on a paper, reserving the letter `'B'` for the backshift operator and using any names (formed just by letters mixed with numbers) to identify the parameters.

Lines 5 and 6 sets up a dummy EASTER variable by using the function `Easter`, and divides it into the variable in the estimation period and the forecast period. Additional variables for trading effects and leap year effects may also be included by running `trading` and `leap year` functions, but they have not been included here to keep the example simple enough.

Lines 7 to 9 show typical calls to the estimation functions `modelTF`, `modelES` and `modelUC`. Only three outputs of the estimation functions are shown in the examples, namely the residuals for each model, the forecasts and their standard deviation necessary to build confidence intervals (not shown in the paper). The inputs to the function are the output series (`z`, reserving the last 12 months for forecasting checking), input series (`u`, empty in this case because there are not any inputs), the model, estimation method (`'EML'` stands for Exact Maximum Likelihood), and a number that activates the outlier automatic detection routine. The model in the ES model is `'AA12'`, implying that the model will contain an additive trend and a seasonal component of period 12 observations. The model in the UC call is similar; `'LLT12'` stands for a model with a Local Linear

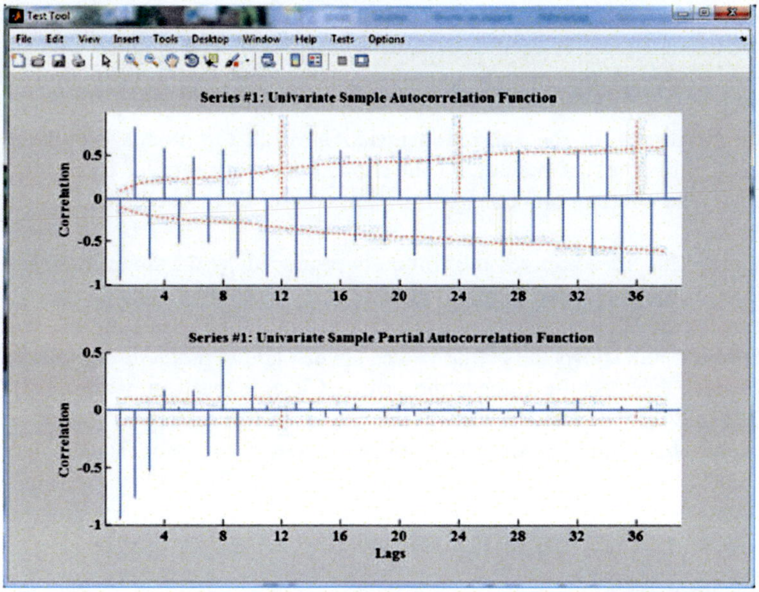

Fig. 36.2 Simple and partial autocorrelation functions for the stationary IPI

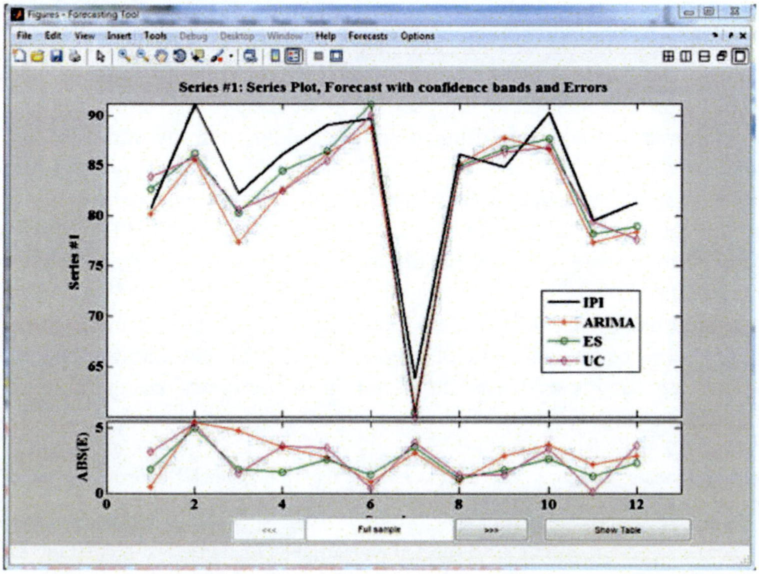

Fig. 36.3 Estimation of the ARIMA model on the Spanish IPI with automatic outlier detection

Trend and a seasonal component of period 12 observations. The EASTER estimated effect is about 7% drop in industrial production. The automatic detection of outliers by any of the models unambiguously show that sudden drops in the industrial production where found in March and November 2008.

Line 10 allows for the validation residuals of all the models simultaneously. Statistical properties of residuals (not shown) are very similar among them, as expected. The contemporaneous correlation among them oscillates between 0.90 and 0.92.

Finally, Line 11 compares the forecasts produced by the three models against the actual values (see Fig. 36.3). Be aware that the comparisons are made on the true units of the IPI variable, hence the exponential function in Line 11. The upper panel of Fig. 36.3 shows the actual values against all the predictions, and the lower panel shows the absolute forecasting errors. Close inspection of the lower panel does not show a model that is clearly the winner in forecasting terms, but with ES it produces marginal improvements over the other two models, UC being the second best.

36.5 Conclusions

This paper presents very briefly the power of the new general MATLAB ECOnometrics TOOLbox (ECOTOOL). The toolbox is intended mainly for professional practitioners, academic researchers, students, and anyone involved in the analysis of time series, forecasting or signal processing.

ECOTOOL is composed of a number of powerful functions to estimate a wide range of SISO, MISO and MIMO models (ARIMA, Exponential Smoothing, Unobserved Components, VARX); user-friendly tools of identification, validation and graphical representation of results, and other general functions for showing results, building moving holidays, dummy variables and other calendar effects, etc.

The case study included is a forecasting exercise based on monthly Spanish Industrial Production Index from January 1975 up to January 2011. It is shown and compared with the models available highlighting the diversity of possible forecasts that may be produced.

As a summary, ECOTOOL is a powerful, easy-to-use toolbox intended to satisfy the needs of a wide range of users, all with different requirements. Its main advantages are and user-friendliness and it is perfectly comparable to other existing commercial tools.

Acknowledgements Diego J. Pedregal acknowledges the financial support by the Junta de Comunidades de Castilla—La Mancha, grant PII1I09-0209-6050.

References

Box GEP, Jenkins GM, Reinsel GC (1994) Time series analysis forecasting and control, 3rd edn. Prentice Hall, Englewood Cliffs

Bryson AE, Ho YC (1969) Applied optimal control optimization estimation and control. Blaisdell Publishing, Waltham

Harvey AC (1989) Forecasting structural time series models and the Kalman filter. Cambridge University Press, Cambridge

Hyndman RJ, Koehler AB, Ord JK, Snyder RD (2008) Forecasting with exponential smoothing. Springer, Berlin

Kalman RE (1960) A new approach to linear filtering and prediction problems. ASME Trans J Basic Eng 83-D:95–108

Lütkepohl H (1991) Introduction to multiple time series analysis. Springer, Heidelberg

Pedregal DJ, Contreras J, Sánchez A (2011) ECOTOOL: a general MATLAB forecasting toolbox with applications to electricity markets. In: Pardalos PM, Pereira MVF, Iliadis NA Rebennack, Sorokin A (eds) Handbook of networks in power systems, Springer, Berlin

Tsay RS (1986) Time series model specification in the presence of outliers. J Amer Stat Assoc 81:132–141

Chapter 37
Solving Flow Shop Problems with Bounded Dynamic Programming

Joaquín Bautista, Alberto Cano, Ramon Companys and Imma Ribas

Abstract We present some results attained with the bounded dynamic programming algorithms to solve the Fm|*prmu*|C_{max} and the Fm|*block*|C_{max} problems using the well-known Taillard instances as experimental data. We have improved four of the best-known solutions of the Taillard's instances for the Fm|*block*|C_{max} problem and we have confirmed the optimality of six solutions for the Fm|*prmu*|C_{max} case.

37.1 Introduction

The permutation flow-shop problem (PFSP) consists of sequencing a set of jobs in a set of machines. Each job must go through all the machines in exactly the same order and the job order must be the same on each machine. This problem has been studied since Johnson (1954)'s seminal paper. Various reviews can be found in the literature of flow shop scheduling, (e.g. Reisman et al. 1994; Hejazi and Saghafian 2005;

J. Bautista (✉) · A. Cano
Nissan Chair, Universitat Politècnica de Catalunya,
Avda. Diagonal 647, 7th Floor, 08028 Barcelona, Spain
e-mail: joaquin.bautista@nissanchair.com
URL: http://www.nissanchair.com

A. Cano
e-mail: alberto.cano-perez@upc.edu

R. Companys · I. Ribas
Departament D'Organització d'Empreses, ETSEIB,
Universitat Politècnica de Catalunya, Avda. Diagonal 647, 7th Floor,
08028 Barcelona, Spain
e-mail: ramon.companys@upc.edu

I. Ribas
e-mail: imma.ribas@upc.edu

S. P. Sethi et al. (eds.), *Industrial Engineering: Innovative Networks*,
DOI: 10.1007/978-1-4471-2321-7_37,

Ruiz and Maroto 2005), where a complete survey of flow shop scheduling problems with a *makespan* criterion is presented.

For $m = 2$ an optimal schedule can be attained in $O(n \log n)$ time using Johnson's algorithm. For $m \geq 3$, Garey et al. (1976) show that the problem is NP-hard. In this paper the permutation flow shop problem is considered with and without storage space between machines. If there is no space between two consecutive machines, a job completed on one machine may block that machine until the next machine downstream is free. This problem is known as a blocking flow shop problem (BFSP). A good review on flow shop with blocking and no-wait in-process can be found in Hall and Sriskandarajah (1996). If the number of machines is two, Reddi and Ramamoorthy (1972) presented a polynomial algorithm, which gives an exact solution. The problem F2|*block*|C_{max} can be reduced to a travelling salesman problem (TSP). In Gilmore et al. (1985), a polynomial algorithm to solve this problem in $O(n \log n)$ time is proposed.

Since both problems are NP-hard, heuristic procedures are used to find good solutions. In this paper we propose a procedure based on a bounded dynamic programming (BDP) to solve the PFSP and BFSP problems, whose principles can be found in Bautista (1993) and Bautista et al. (1996).

37.2 Problem Definition

At time zero, n jobs must be processed, in the same order, on each of m machines. Each job goes from machine 1 to machine m. The processing time for each operation is $p_{i,k}$, where $k \in K = \{1, 2, \ldots, m\}$ denotes a machine and $i \in I = \{1, 2, \ldots, n\}$ a job. These times are fixed, known in advance and positive. The objective function considered is the minimisation of the *makespan* (C_{max}).

Given a permutation, π, of the n jobs, $[t]$ indicates the job that occupies position t in the sequence. For example, in $\pi = (3, 1, 2)$ $[3] = 2$. In a feasible schedule associated to a permutation, let $s_{k,t}$ be the beginning of the time destined in machine k to the job that occupies position t and $e_{k,t}$ the time when the job that occupies position t releases machine k. The PFSP, denoted as Fm|*prmu*|C_{max}, can be formalised as follows:

$$s_{k,t} + p_{[t],k} \leq e_{k,t} \quad k = 1, 2, \ldots, m; \ t = 1, 2, \ldots, n \tag{37.1}$$

$$s_{k,t} \geq e_{k,t-1} \quad k = 1, 2, \ldots, m; \ t = 1, 2, \ldots, n \tag{37.2}$$

$$s_{k,t} \geq e_{k-1,t} \quad k = 1, 2, \ldots, m; \ t = 1, 2, \ldots, n \tag{37.3}$$

$$C_{max} = e_{m.n} \tag{37.4}$$

Being $e_{k,0} = 0 \ \forall k$, $e_{0,t} = 0 \ \forall t$, the initial conditions.

The schedule is semi-active if equation (37.1) is written as $s_{k,t} + p_{[t],k} = e_{k,t}$ and equations (37.2) and (37.3) are summarised as $s_{k,t} = \max\{e_{k,t-1}, e_{k-1,t}\}$.

When there is no storage space between stages, Fm|*block*|C_{max} problem, if a job i finishes its operation on a machine k and if the next machine, $k+1$, is still busy on the previous job, the completed job i has to remain on the machine k blocking it. This condition requires an additional equation (37.5) in the formulation of the problem.

$$e_{k,t} \geq e_{k+1,t-1} \quad k = 1,2,\ldots,m; \; t = 1,2,\ldots,n \tag{37.5}$$

The initial condition $e_{m+1,t} = 0,\; t = 1,2,\ldots,n$ must be added.

The schedule obtained is semi-active if equations (37.1) and (37.5) are summarised as (37.6):

$$e_{k,t} = \max\{s_{k,t} + p_{[t],k}, e_{k+1,t-1}\} \quad \forall k, \forall t \tag{37.6}$$

Consequently, the Fm|*prmu*|C_{max} problem can be seen as a relaxation of the Fm|*block*|C_{max} problem.

37.3 Graph Associated with the Problem

Similarly to Bautista et al. (1996) and Bautista and Cano (2011), we can build a linked graph without loops or direct cycles of $T+1$ stages (T=n). The set of vertices in level t ($t = 0,\ldots,T$) will be noted as $J(t)$. $J(t,j)(j = 1,\ldots,|J(t)|)$ being a vertex of level t, which is defined by the triad $(\vec{q}(t,j),\; \vec{e}(t,j),\; C_{max}(t,j))$, where:

- $\vec{q}(t,j) = (q_1(t,j),\ldots,q_n(t,j))$, where $q_i(t,j)\; \forall i$ takes the value 1 if the job i has been completed and the value 0 in the opposite case.
- $\vec{e}(t,j) = (e_1(t,j),\ldots,e_m(t,j))$ represents the vector of completion times of the operations at the stations.
- $C_{max}(t,j)$ represents the completion time of the last programmed job, at stage t and vertex j.

The vertex $J(t,j)$ has the following properties:

$$\sum_{i=1}^{n} q_i(t,j) = t\,; \quad q_i(t,j) \in \{0,1\}\; \forall i\,; \quad C_{max}(t,j) = e_m(t,j) \tag{37.7}$$

In short, a vertex $J(t,j)$ will be represented as $J(t,j) = \{(t,j), \vec{q}(t,j), \vec{e}(t,j)\}$.

At level 0 of the graph, there is only one $J(0)$ vertex. Initially, we may consider that at level t, $J(t)$ contains the vertices associated with all of the sub-sequences that can be built with t jobs and which satisfy properties (37.7). However, it is easy to reduce the cardinal that $J(t)$ may present a priori, establishing the following relationship of dominance and equivalence:

$$J(t,j) \prec J(t,j') \Leftrightarrow [\vec{q}(t,j) = \vec{q}(t,j')] \wedge [\vec{e}(t,j) \prec \vec{e}(t,j')] \tag{37.8}$$

$$J(t,j) \equiv J(t,j') \Leftrightarrow [\vec{q}(t,j) = \vec{q}(t,j')] \wedge [\vec{e}(t,j) = \vec{e}(t,j')] \tag{37.9}$$

With these relationships, we can reduce the search space for solutions in the graph. Therefore, at level t of the graph, $J(t)$ will contain the vertices associated with non-dominated and non-equivalent sub-sequences, and at level T, $J(T)$ will contain all the vertices associated with non-equivalent and non-dominanted completed sequences.

A transition arc through the type of job i exists between the vertex $J(t,j)$ and the vertex $J(t+1,j_i)$ if $\vec{q}(t,j) \prec \vec{q}(t+1,j_i)$. For vertex $J(t+1,j_i)$ to be completely defined through the transition from $J(t,j)$, it is necessary to determine $J(t+1,j_i) = \{(t+1,j_i), \vec{q}(t+1,j_i), \vec{e}(t+1,j_i)\}$. For the PFSP problem this is done as follows:

$$q_i(t+1,j_i) = 1; \quad q_h(t+1,j_i) = q_h(t,j) \quad \forall h : h \neq i \tag{37.10}$$

$$e_0(t+1,j_i) = 0; \quad e_k(t+1,j_i) = \max\{e_k(t,j) + p_{i,k}, e_{k-1}(t+1,j_i)\} \quad \forall k \in K \tag{37.11}$$

Instead, for the BFSP case, it is necessary to replace equation (37.11) with a new equation which has an additional element:

$$e_0(t+1,j_i) = 0; \quad e_k(t+1,j_i) = \max\{e_k(t,j) + p_{i,k}, e_{k-1}(t+1,j_i), e_{k+1}(t,j)\} \quad \forall k \in K \tag{37.12}$$

Under these conditions, finding a sequence that optimizes the total $C_{\max}$ is equivalent to finding an optimum path from vertex $J(0)$ to the set of vertices $J(T)$ of stage T.

Therefore, any algorithm of extreme paths in the graphs is valid for finding solutions to the proposed problem. However, realistic industrial problems where n and m are large give rise to graphs with a large number of vertices. Therefore, we recommend resorting to procedures that do not explicitly require the presence of all of the vertices for calculation.

37.4 Bounds for $C_{\max}$

First, we establish overall bounds for $C_{\max}$ (defined in (37.4)). If we account for the machines independently, we can write a bound for $C_{\max}$ for each machine k ($k = 1,\ldots,|K|$) and therefore, considering all machines:

$$LB1(k) = \sum_{i=1}^{n} p_{i,k} + \min_{(i,h)\in I:\, i\neq h}\left\{\sum_{k'=1}^{k-1} p_{i,k'} + \sum_{k'=k+1}^{m} p_{h,k'}\right\} \forall k \in K$$
$$\Rightarrow LB1 = \max_{k\in K}\{LB1(k)\} \tag{37.13}$$

In the same manner, we can also consider a bound for $C_{\max}$ through the jobs:

$$LB2(i) = \sum_{k=1}^{m} p_{i,k} + \sum_{h \in I: h \neq i} \min_{k \in K}\{p_{h,k}\} \; \forall i \in I \Rightarrow LB2 = \max_{i \in I}\{LB2(i)\} \quad (37.14)$$

Both bounds are valid for the PFSP and BFSP problems.

Moreover, let us assume that we have built a path from $J(0)$ to vertex $J(t,j)$, and thus we have the information $\vec{q}(t,j)$ and $\vec{e}(t,j)$. To complete a sequence up to stage T, we will need to link with $J(t,j)$, T-t vertices, each of them associated with a different unscheduled job. Under these conditions, we can delimit $C_{\max}$ through the vertex $J(t,j)$ adapting the overall bound *LB1* and *LB2*.

$$LB1(t,j) = \max_{k \in K}\left\{ e_k(t,j) + \sum_{i \in I: q_i(t,j)=0} p_{i,k} + \min_{i \in I: q_i(t,j)=0}\left\{ \sum_{k'=k+1}^{m} p_{i,k'} \right\}\right\} \quad (37.15)$$

$$LB2(t,j) = e_1(t,j) + \max_{i \in I: q_i(t,j)=0}\left\{ \sum_{k=1}^{m} p_{i,k} + \sum_{h \in I-\{i\}: q_h(t,j)=0} \min_{k \in K}\{p_{h,k}\} \right\} \quad (37.16)$$

37.5 The Use of Bounded Dynamic Programming

For this study, we used a procedure based on BDP. This procedure combines features of dynamic programming (determination of extreme paths in graphs) with features of branch and bound algorithms. The principles of Bounded Dynamic Programming have been described in Bautista (1993) and Bautista et al. (1996). The procedure is described below (details can be seen in Bautista and Cano (2011):

BDP—P/B: FSP
Input: T, I, K, $d_i(\forall i \in I)$, $p_{i,k}$ $(\forall i \in I, \forall k \in K)$, Z_0, H
Output: list of sequences obtained by BDP
0 Initialization: $t = 0; LBZ_{\min} = \infty$
1 While $(t < T)$ *do*
2 $t = t + 1$
3 While (list of consolidated vertices in stage *t-1* **not empty**) *do*
4 Select_vertex (t)
5 Develop_vertex (t)
6 Filter_vertices (Z_0, H, $LBZ_{\min}$)
7 end while
8 End_stage ()
9 end while
end BDP—P/B: FSP

The procedure is formed by the following functions: (1) Select_vertex (t): selects in order one of the vertices consolidated in stage *t-1*; (2) Develop_vertex (t): develops the selected vertex in previous function adding a new product unit;

(3) Filter_vertices (Z_0, H, LBZ_{min}): choose, from all vertices, the most promising vertices, according to Z_0 and H and LBZ_{min} (lower bound for C_{max} from non-developed vertices); and (4) End_stage (): consolidates the most promising vertices in stage t.

The bounding scheme implemented in BDP, for both problems (PFSP and BFSP) is $LB1(t,j)$ because a previous test to compare both showed that BDP had better performance with LB1 than with LB2.

For the function filter_vertices (Z_0, H, LBZ_{min}), we have developed two variants for the BFSP (see Bautista et al. 2011):

- Variant 1 (V1). A partial solution is more promising than another when it has a best bound for C_{max} ($LB1(t,j)$). In case of a tie between two partial solutions (equal $LB1(t,j)$), the partial solution with less $e_{|K|}(t,j)$ is chosen.
- Variant 2 (V2). A partial solution is more promising than another when it has less value for his partial C_{max} $\left(e_{|K|}(t,j)\right)$. In case of a tie between two partial solutions (equal $e_{|K|}(t,j)$), the partial solution with less bound $LB1(t,j)$ is chosen.

For the PFSP we have only tested V1.

37.6 Computational Experiment

We have performed an operation test with the first 11 sets from Taillard's benchmark instances (Taillard (1993)). The Taillard's benchmark instances consist of 120 instances, grouped in 12 sets. Each set has 10 instances, each of them with the same number of jobs and machines. In our computational experiment, we have tested the first 110 instances from the 120 supplied. The number of jobs goes from 20 (set 1) to 200 (set 11) and the number of machines goes from 5 (set 1) to 20 (set 11).

To obtain solutions, we have used two variants of BDP programmed in C ++, compiled with gcc v. 4.01, running on an Apple Macintosh iMac computer with an Intel Core i7 2.93 GHz processor and 8 GB RAM using MAC OS X 10.6.4. Neither the implementation nor the compiler used threads or any type of parallel code; therefore, the computer can be considered a single 2.93 GHz processor. The 110 instances were solved by using seven correlative window widths, H_1 to H_7 with values 1, 10, 50, 100, 250, 500 and 750, respectively.

For the initial solution Z_0, we used the value of the solution obtained with the previous width $H_{\alpha-1}$ for each window width H_α ($\alpha = 1, .., 7$), except for the case with width $H_1 = 1$ in which Z_0 was fixed at infinity.

To analyse the experimental results, we used the relative percentage deviation (RPD) calculated as follows:

Table 37.1 Results obtained for the BFSP (V1 and V2) for each window width

	H_1	H_2	H_3	H_4	H_5	H_6	H_7
Average RPD-V1	12.92	6.93	4.82	3.92	3.36	2.92	2.69
Average CPU-V1	0.49	5.02	24.41	55.46	176.07	475.36	905.06
Average RPD-V2	20.55	12.51	9.39	7.72	6.86	6.00	5.52
Average CPU-V2	0.44	4.44	25.45	58.39	200.19	571.17	1117.27

Table 37.2 Results obtained for the PFSP (V1) for each window width

	H_1	H_2	H_3	H_4	H_5	H_6	H_7
Average RPD-V1	18.43	9.08	6.27	5.38	4.64	4.03	3.68
CPU-V1	0.43	4.44	24.67	54.14	173.23	465.92	871.81

Table 37.3 RPD per set obtained by BDP with V1 and $H = 750$ for the PFSP and best solution between V1 and V2 and $H = 750$ for the *(BFSP-B)*

	20x5	*20x10*	*20x20*	*50x5*	*50x10*	*50x20*	*100x5*	*100x10*	*100x20*	*200x10*	*200x20*	*All*
PFSP	0.64	1.91	4.75	0.60	3.81	8.89	0.84	1.82	9.87	1.26	6.08	3.68
BFSP-B	1.02	2.2	4.4	2.1	3.32	6.05	0.41	2.1	5.11	0.47	2.04	2.66

$$RPD = \frac{BDP_{Best} - Best_{solution}}{Best_{solution}} 100 \tag{37.17}$$

Table 37.1 shows the average RPD and CPU time for the first 11 sets of Taillard, per each window width and each variant of the BDP, with bound $LB1(t,j)$, for the BFSP.

The main results are: (1) CPU times are similar for both variants of filter_vertices; (2) the V1 always offered better overall average results than V2 (taking the index RPD); and (3) with $H = 750$ we are, on average, at 2.69% of the best known solutions. The found better solutions to instances 68, 70, 92 and 95 than those reported in current literature.

Given the best performance of V1 for BFSP, we have adapted this algorithm to solve the PFSP. The results are shown in Table 37.2. These results placed us, with $H = 750$, at 3.68% on average of the best solutions known for the PFSP. In addition, we have obtained 8 of the best solutions found in the literature and we have confirmed their optimality in 6 instances.

To analyze the behavior of the BDP, with window width $H = 750$, we determined the average values of RPD per each set of Taillard's instances; the results shown in Table 37.3 and Fig. 37.1, testify that sets of instances with 20 machines have worse solutions than the other sets, regardless of the number of jobs.

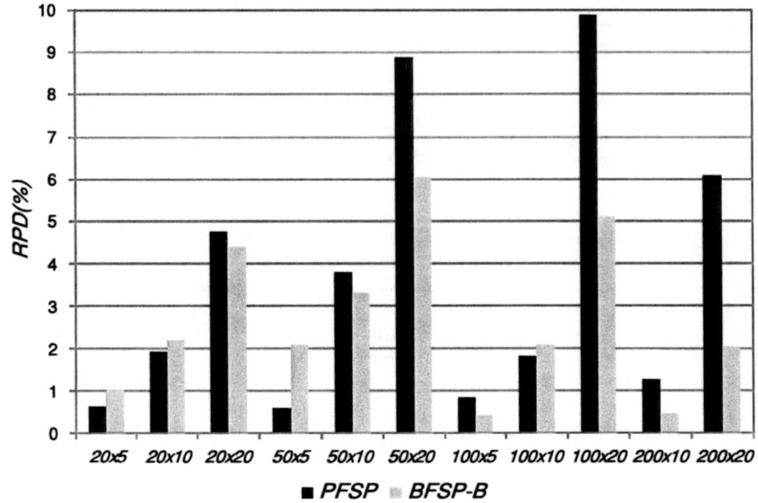

Fig. 37.1 RPD values obtained with BDP ($H = 750$) solving the 11 sets of Taillard for the PFSP and BFSP problems

37.7 Conclusions

We have analyzed the behaviour of a BDP-based procedure for the PFSP and BFSP problems, using the first 11 sets from Taillard.

The results provided by the BDP are satisfactory for the problem BFSP, where we have improved four best-known solutions. However, for the PFSP we have only confirmed their optimality in six instances.

We have noticed that the BDP performs better on sets of instances with $m < 20$ probably because the bound used is worse when the number of machines is bigger. Therefore, the next step will be to try to improve the performance of the BDP with a tight lower bound.

Acknowledgements This work is supported by the UPC Nissan Chair and the Spanish Ministerio de Educación y Ciencia under project DPI2010-16759 (PROTHIUS-III) including EDRF fundings.

References

Bautista J (1993) Procedimientos heurísticos y exactos para la secuenciación en sistemas productivos de unidades homogéneas (contexto J.I.T.). Doctoral Thesis, DOE, ETSEIB-UPC

Bautista J, Companys R, Corominas A (1996) Heuristics and exact algorithms for solving the Monden problem. Eur J Oper Res 88:101–113

Bautista J, Cano A (2011) Solving mixed model sequencing problem in assembly lines with serial workstations with work overload minimization and interruption rules. Eur J Oper Res 210:495–513

Bautista J, Cano A, Companys R, Ribas I (2011) Solving the Fm|block|Cmax problem using bounded dynamic programming. Engineering Applications of Artificial Intelligence, Corrected Proof (DOI: 10.1016/j.engappai.2011.09.001) in press

Garey MR, Johnson DS, Sethi R (1976) Complexity of flowshop and jobshop scheduling. Math Oper Res 1(2):117–129

Gilmore PC, Lawler EL, Shmoys DB (1985) Well-solved special cases. In: Lawler EL, Lenstra KL, Rinooy Kan AHG, Shmoys DB (eds) The traveling salesman problem: a guided tour of combinatorial optimization, Wiley, New York

Hall NG, Sriskandarajah C (1996) A survey of machine scheduling problems with blocking and no wait in process. Oper Res 44(3):510–525

Hejazi RS, Saghafian S (2005) Flowshop-scheduling problems with makespan criterion: a review. Int J Prod Res 43(14):2895–2929

Johnson SM (1954) Optimal two-and three-stage production schedules with set up times included. Naval Res Logist Quart 1:61–68

Reddi SS, Ramamoorthy B (1972) On the flow-shop sequencing problem with no wait in process. Oper Res Quart 23(3):323–331

Reisman A, Kumar A, Motwani J (1994) 'Flowshop scheduling/sequencing research: A statistical review of the literature, 1952–1994'. IEEE Transact Eng Manag 44(3):316–329

Ruiz R, Maroto C (2005) A comprehensive review and evaluation of permutation flowshop heuristics. Eur J Oper Res 165(2):479–494

Taillard E (1993) Benchmarks for basic scheduling problems. Eur J Oper Res 64(2):278–285

Chapter 38
Performance of Bounded Dynamic Programming Applied to a Variant of the MMSP-W Problem

Joaquín Bautista, Alberto Cano and Rocío Alfaro

Abstract In this study we present a formulation for the MMSP-W (Mixed Model Sequencing Problem with Work overload Minimization) for productions lines with both serial workstations and rules for interrupting operations. We propose a combination of the models M5 and M6 proposed by Bautista and Cano (Eur J Oper Res 210:495–513, 2011) and analyze the results obtained, using the BDP (Bounded Dynamic Programming), through a case study of the powertrain line of the Nissan plant in Barcelona.

38.1 Introduction

In mixed-products manufacturing lines, very common in JIT (Just In Time) and DS (Douki Seisan) environments, it is possible to handle different variants of products. This flexibility determines the sequence by which the units on the line should be handled to achieve the following goals: (1) a drastic reduction in component stocks and intermediate semi-manufactured parts and (2) efficient use of the available manufacturing time.

J. Bautista (✉) · A. Cano · R. Alfaro
Nissan Chair, Universitat Politècnica de Catalunya, Avda. Diagonal 647, 7th floor, 08028 Barcelona, Spain
e-mail: joaquin.bautista@nissanchair.com
URL: http://www.nissanchair.com

A. Cano
e-mail: alberto.cano-perez@upc.edu

R. Alfaro
e-mail: rocio.alfaro@upc.edu

S. P. Sethi et al. (eds.), *Industrial Engineering: Innovative Networks*,
DOI: 10.1007/978-1-4471-2321-7_38,

This study falls under the second goal established by Boysen et al. (2009).

Among the works dedicated to this goal are Yano and Rachamadugu (1991), Tsai (1995), Scholl et al. (1998) and Bautista and Cano (2008, 2011), who focus on sequences with a minimum work overload (*MMSP-W*).

In this work we synthesize the models M5 and M6 proposed in Bautista and Cano (2011) in a new model. In addition, we analyze the performance of the BDP applied to a case study linked to the Nissan Powertrain plant in Barcelona.

38.2 Reference Models

To minimize work overload or maximize the total work completed in the problem of mixed product sequencing in a production line with multiple linked stations, we begin with two models proposed by Bautista and Cano (2011):

- M5: that focuses on maximizing the total work completed in the assembly line.
- M6: that focuses on minimizing work overload or additional effort.

These models comply with the following interruption rule: an operation will only be interrupted at the maximum finishing instant established by the time window if this instant is reached during the execution of the operation.

The combination of these models and taking into account several processors on every workstation, allowed us to establish a new model, M5∪6, for this variant of the MMSP-W problem. The parameters and variables of the model are described below:

Parameters	
K	Set of workstations ($k = 1,\ldots,\lvert K\rvert$)
b_k	Number of equivalent processors in station k
I	Set of product types ($i = 1,\ldots,\lvert I\rvert$)
d_i	Programmed demand of the i type of product
$p_{i,k}$	Processing time required by a unit of the i type of product at station k for each processor (at normal activity)
T	Total demand. Obviously: $\sum_{i=1}^{\lvert I\rvert} d_i = T$
t	Position index in the sequence ($t = 1, \ldots, T$)
c	Cycle time (standard time assigned to each processors in the workstations to process any product unit)
l_k	Time window (i.e. the maximum time that each processor assigned to the workstation k is allowed to work on any product unit), where $l_k - c > 0$ is the maximum time that the WIP (Work In Process) is held at workstation k, once the cycle time has finished
Variables	
$x_{i,t}$	Binary variable that is equal to 1 if a product unit i ($i = 1,\ldots,\lvert I\rvert$) is assigned to the position t ($t = 1, \ldots, T$) of the sequence, and 0 if otherwise

(continued)

(continued)

$s_{k,t}$	Start instant for the *t*th unit of the sequence of products in station *k* ($k = 1,\ldots,\|K\|$)
$v_{k,t}$	Processing time applied to the *t*th unit of the product sequence in workstation *k* for each processor (at normal activity)
$w_{k,t}$	Overload generated for the *t*th unit of the product sequence in workstation *k* for each processor (at normal activity) measured in time.

Under these conditions, we can define the following mathematical model M5U6:

$$Max \quad V = \sum\nolimits_{k=1}^{|K|}\left(b_k\sum_{t=1}^{T} v_{k,t}\right) \Leftrightarrow Min \quad W = \sum\nolimits_{k=1}^{|K|}\left(b_k\sum_{t=1}^{T} w_{k,t}\right) \tag{38.1}$$

subject to:

$$\sum\nolimits_{t=1}^{T} x_{i,t} = d_i \quad i = 1,\ldots,|\mathrm{I}| \tag{38.2}$$

$$\sum\nolimits_{i=1}^{|I|} x_{i,t} = 1 \quad t = 1,\ldots,T \tag{38.3}$$

$$v_{k,t} + w_{k,t} = \sum\nolimits_{i=1}^{|I|} p_{i,k}x_{i,t} \quad k = 1,\ldots,|K|;\ t = 1,\ldots,T \tag{38.4}$$

$$s_{k,t} \geq (t + k - 2)c \quad k = 1,\ldots,|K|;\ t = 1,\ldots,T \tag{38.5}$$

$$s_{k,t} \geq s_{k,t-1} + v_{k,t-1} \quad k = 1,\ldots,|K|;\ t = 2,\ldots,T \tag{38.6}$$

$$s_{k,t} \geq s_{k-1,t} + v_{k-1,t} \quad k = 2,\ldots,|K|;\ t = 1,\ldots,T \tag{38.7}$$

$$s_{k,t} + v_{k,t} \leq (t + k - 2)c + l_k \quad k = 1,\ldots,|K|;\ t = 1,\ldots,T \tag{38.8}$$

$$v_{k,t} \geq \sum\nolimits_{i=1}^{|I|} p_{i,k}x_{i,t} - M(1 - y_{k,t}) \quad k = 1,\ldots,|K|;\ t = 1,\ldots,T \tag{38.9}$$

$$s_{k,t} + v_{k,t} \geq (t + k - 2)c + l_k - My_{k,t} \quad k = 1,\ldots,|K|;\ t = 1,\ldots,T \tag{38.10}$$

$$v_{k,t} \geq 0 \quad k = 1,\ldots,|K|;\ t = 1,\ldots,T \tag{38.11}$$

$$w_{k,t} \geq 0 \quad k = 1,\ldots,|K|;\ t = 1,\ldots,T \tag{38.12}$$

$$x_{i,t} \in \{0,1\} \quad i = 1,\ldots,|I|;\ t = 1,\ldots,T \tag{38.13}$$

$$y_{k,t} \in \{0,1\} \quad k = 1,\ldots,|K|;\ t = 1,\ldots,T \tag{38.14}$$

In the model, *V* and *W* represents the value of the total work performed and total overload, respectively; constraint (38.2) requires that the programmed demand be satisfied; constraint (38.3) indicates that only one product unit can be assigned in each position of the sequence; constraint (38.4) establishes the equivalence between the processing time applied to each unit in each workstation and the overload generated to each unit in each workstation; constraints (38.5), (38.6) and (38.7) establish that the earliest time at which the tth product of the sequence at workstation k can be operated is the maximum of the following three—the expected arrival time of the product at the beginning of the workstation ($s_{k,t}^{\min}$), the instant at which workstation k is released from operating on the previous product (t-1), or the instant at which the previous workstation (k-1) releases the tth product; constraint (38.8) limits the maximum working time that can be applied by a processor to the tth product unit of the sequence at workstation k; constraints (38.9) and (38.10) enable the value of the completed work to be fixed, $v_{k,t}$, differentiating between situations with and without interruption of the operation according to the proposed rule (considering $M = (T + |K|)c$ and $y_{k,t}$ a binary variable equal to 1 if the work required from the tth operation at station k is completed and 0 if there is an interruption); constraints (38.11) and (38.12) indicate that the processing times applied and the overload generated are not negative; and, finally, constraints (38.13) and (38.14) necessitate that the assigned variables be binary. Let $s_{1,1} = 0$ as the start instants for the operations.

38.3 Overall Bounds

We establish overall bounds for function of the overload defined in (38.1). Let V_0 the total work required. Obviously, a bound of W is:

$$LB0 = \left[V_0 - \left((T \cdot |K| - 1)c + l_{|K|}\right)\right]^+ \tag{38.15}$$

Considering the workstations independently, we can write a bound of the work that cannot be concluded at station k, and, therefore, considering all stations.

$$LB1(k) = b_k\left[\sum\nolimits_{i=1}^{|I|} d_i p_{i,k} - (T-1)c - l_k\right]^+, \quad \forall k \in K \Rightarrow LB1 = \sum\nolimits_{k=1}^{|K|} LB1(k) \tag{38.16}$$

We can also consider the work required for each type of product independently, and considering the demands of the products, we have the following:

$$LB2(i) = \left[\sum_{k=1}^{|K|} b_k p_{i,k} - c\sum_{k=1}^{|K|} b_k - b_{|K|}\left(l_{|K|} - c\right)\right]^+, \quad \forall i \in I \Rightarrow LB2 = \sum_{i=1}^{|I|} d_i LB2(i) \tag{38.17}$$

Table 38.1 Processing times ($p_{i,k}$) for each processor at normal activity. *A*, *B* and *C*, corresponds to products types and m_1, m_2, m_3 corresponds to workstations

	A	*B*	*C*
m_1	5	4	3
m_2	5	4	4
m_3	4	3	5

38.4 An Example

There are six product units ($T = 6$), of which three are type *A*, one is type *B*, and two are type *C*. The product units are processed in three workstations ($|K| = 3$), and the processing time of each unit of type of product at each station is indicated in Table 38.1.

The following conditions are also taken into consideration: $c = 4$ (*cycle time*) and $l_k = 6$, for all $k = 1,\ldots,3$ (*time windows*).

Figure 38.1 shows the optimum solution given by the mixed linear program associated with model M5∪6, considering $b_k = 1$ for all $k = 1,\ldots,3$. The sequence of products that presents the maximum total work completed is *C–A–A–B–C–A*. In Fig. 38.1 we can see: (1) the total work completed is $V = 74$ out of a total of $V_0 = 77$ units required; (2) the total overload is $W = 3$ (the grey area in Fig. 38.1), which is generated at workstation m_2 on products A_2 and A_3 at periods 19 and 31 and at workstation m_3 product C_2 at period 31; (3) workers will interrupt the operation on units A_2, C_2 and A_3 at the precise moment at which the limit imposed by the time window is reached.

38.5 The Use of Bounded Dynamic Programming

For this study, we used a procedure based on BDP. This procedure combines features of dynamic programming (determination of extreme paths in graphs) with features of branch and bound algorithms. The principles of BDP have been described by Bautista (1993) and Bautista et al. (1996). The procedure is described below (see details on Bautista and Cano (2011)):

BDP – MMSPW

Input: $T, I, K, d_i(\forall i \in I), p_{i,k}, b_k, l_k(\forall i \in I, \forall k \in K), c, Z_0, H$

Output: list of sequences obtained by BDP

Initialization: $t = 0$; $LBZ_{min} = \infty$

While ($t < T$) *do*

  $t = t + 1$

  While (list of consolidated vertices in stage *t-1* **not empty**) *do*

    Select_vertex (t)

(continued)

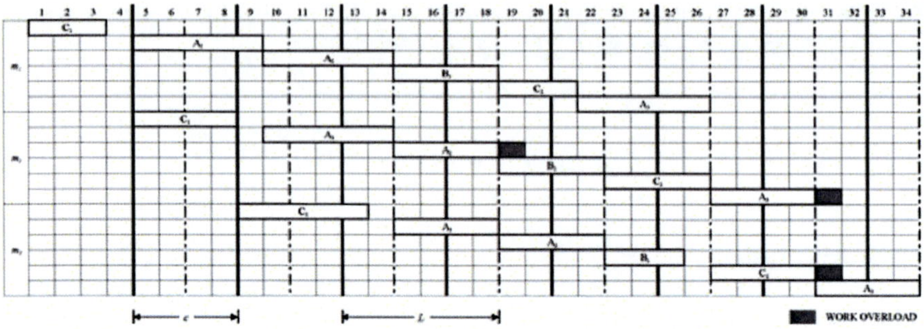

Fig. 38.1 Gantt chart for the optimum solution for the example provided by M5∪6

(continued)

```
            Develop_vertex (t)
            Filter_vertices (Z0, H, LBZmin)
        end while
        End_stage ()
    end while
end BDP – MMSPW
```

In the procedure appear the following functions: (1) Select_vertex (t): selects, in order, one of the vertices consolidated in stage $t-1$; (2) Develop_vertex (t): develops the selected vertex in previous function adding a new product unit; (3) Filter_vertices (Z_0, H, LBZ_{min}): chooses, from all vertices, the vertices most promising, according to Z_0 and H; and (4) End_stage (): consolidates the most promising vertices in stage t.

38.6 Computational Experiment

The Nissan powertrain plant in Barcelona has an assembly line with twenty-one workstations (m_1, …, m_{21}) assembling nine types of engines (p_1, …, p_9) that are grouped into three families (4 × 4, vans and trucks).

For the experiment, we considered the processing times of each engine type at each station and a set E of 46 instances ($\varepsilon = 1, \ldots, 46$) which are associated to a demand plan, for the nine types of engines, and grouped them into two blocks (I: $T = 270$; II: $T = 540$). These data are shown in Bautista and Cano (2011) (see www.nissanchair.com). In addition, we considered an effective cycle time $c = 175$ s and an identical time window for all stations $l_k = 195$ s ($k = 1,\ldots,21$).

The solutions were obtained using the following two resources: (1) Solver CPLEX v11.0 (single-processor license) and (2) BDP programmed in C++ (see Bautista and Cano (2011)). The 46 instances were solved using a set A of 8 large window widths ($\alpha = 1,\ldots,8$) associated with values 1, 10, 50, 100, 250, 500,

Table 38.2 Results obtained for each block of instances and each window width

		$H_1 = 1$	$H_2 = 10$	$H_3 = 50$	$H_4 = 100$	$H_5 = 250$	$H_6 = 500$	$H_7 = 750$	$H_8 = 1000$
Block-I	*RPD*	3.82	1.27	0.94	0.88	0.77	0.74	0.70	0.69
	CPU	0.15	1.71	13.19	37.55	178.11	617.19	1319.43	2261.18
Block-II	*RPD*	3.15	1.29	0.86	0.78	0.74	0.70	0.67	0.66
	CPU	0.97	10.13	60.19	143.53	525.95	1605.25	3233.05	5320.96

750 and 1000, for H_1 to H_8 respectively. The initial solution Z_0 for each window width was the solution ($W_{\varepsilon,\alpha}$) obtained by BDP with the previous window width ($H_{\alpha-1}$), except in the case $H_1 = 1$, where Z_0 was established as ∞.

To estimate the quality of the experimental results, we use two indicators:

1. The relative percentage deviation using the best lower bound found (LB_ε) for an instance ε. In addition to use of the lower bounds defined above, we also consider the bounds found by CPLEX ($LBLP_\varepsilon$) and the BDP ($LBDP_\varepsilon$). These bounds correspond to the bound offered by the solver CPLEX and value LBZ_{min} from BDP [minor value $LB(t)$ between the discarded or replaced vertices during the procedure], respectively. In these conditions, we can write

$$LB_\varepsilon = \max\{LB0_\varepsilon, LB1_\varepsilon, LB2_\varepsilon, LBLP_\varepsilon, LBDP_\varepsilon\} \quad \forall \varepsilon \in \mathrm{E} \tag{38.18}$$

$$RPD_{\varepsilon,\alpha} = \frac{W_{\varepsilon,\alpha} - LB_\varepsilon}{LB_\varepsilon} \quad \forall \varepsilon \in \mathrm{E}, \quad \forall \alpha \in \mathrm{A} \tag{38.19}$$

2. The CPU time required, $CPU_{\varepsilon,\alpha}$, to obtain a solution for an instance ε and a window width H_α.

Therefore, we can obtain average values for each block of instances and each window width H_α as following:

$$RPD_\alpha = \frac{1}{|\mathrm{E}|}\sum_{\varepsilon \in \mathrm{E}} RPD_{\varepsilon,\alpha} \ \forall \alpha \in \mathrm{A}; \quad CPU_\alpha = \frac{1}{|\mathrm{E}|}\sum_{\varepsilon \in \mathrm{E}} CPU_{\varepsilon,\alpha} \quad \forall \alpha \in \mathrm{A} \tag{38.20}$$

According to the results obtained (see Table 38.2), we can comment the following: (1) *RPD* decreases asymptotically with the increment of *H* and (2) CPU_α and H_α are related through the function $CPU_\alpha = 0.0866 \cdot H_\alpha^{1.4113}$ with $R^2 = 0.986$ for Block-I, and $CPU_\alpha = 0.6557 \cdot H_\alpha^{1.2481}$ with $R^2 = 0.989$ for Block-II. Moreover, for all instances, the best lower bound was the obtained by the BDP (LB_ε).

Finally, to evaluate the computational performance of the BDP procedure applied to this variant of the MMSP-W problem, we define the rate of *RPD* for each window width as following:

$$\rho_\alpha = \frac{RPD_{\alpha-1} - RPD_\alpha}{RPD_{\alpha-1} \cdot CPU_\alpha} \quad \forall \alpha \in \mathrm{A} \wedge \alpha \geq 2 \tag{38.21}$$

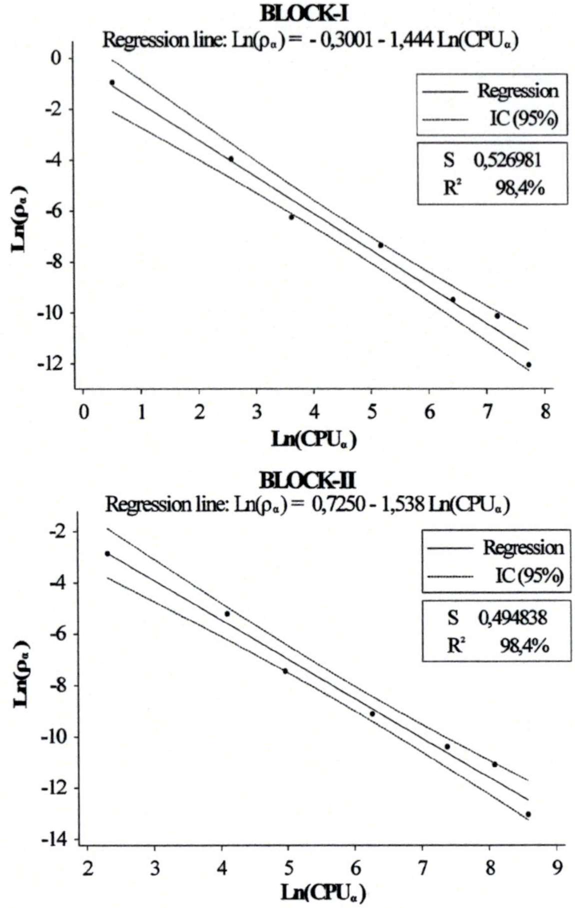

Fig. 38.2 Graphic of the real (points in the figure) and adjusted values of $Ln(\rho_\alpha)$ versus $Ln(CPU_\alpha)$ for the Block-I of instances (*up*) and the Block-II (*down*)

From the definition of the rate of *RPD* (38.21), we performed a linear regression between $Ln(\rho_\alpha)$ and $Ln(CPU_\alpha)$. The results are shown in Fig 38.2.

The quality of the adjustment obtained is $R^2 = 0.984$ for both blocks of instances. From these results it is easy to establish a function to estimate the average value of *RPD*(*t*) (by blocks of instances). This function depends on the average of CPU time taken to solve an instance; in our case, we obtain, for Block-I [$RPD_{\mathrm{I}}(t)$] and Block-II [$RPD_{\mathrm{II}}(t)$], the following result:

$$RPD(t) = RPD_\infty f(t);\ with\, f(t) = e^{\int_t^\infty \rho(u)du} : \left\{ \begin{array}{l} RPD_{\mathrm{I}}(t) = 0.593e^{1.669t^{-0.444}} \\ RPD_{\mathrm{II}}(t) = 0.593e^{3.838t^{-0.538}} \end{array} \right\} \tag{38.22}$$

In view of the estimation functions (38.22) we can conclude that the estimated values of *RPD*(*t*), when $t \to \infty$, tend asymptotically to value 0.593, for both blocks of instances. These values have been obtained through a constant linear regression on the points $RPD_\alpha / f(CPU_\alpha) \forall \alpha \in A$. These functions can be used to: (1) estimate the average improvement that can be produced in a block of instances, given a CPU time or a window width; and (2) to make a prediction of the CPU time (or window width) that should be used to achieve a required average improvement in a block of instances.

As an example of using expressions (38.22) we can conclude that if we do not change the bounds employed, we can achieve a minimum improvement of 6% in the solution set of Block-I and 4% in Block-II.

38.7 Conclusions

We have presented a model, M5∪6, on the sequencing of products to minimize the work overload or maximize the value of work completed in an assembly line with serial workstations.

Afterwards, we have studied a case related to the Nissan Powertrain line in Barcelona through 46 possible demand plans, using the solver CPLEX and the BDP implementation (Bautista and Cano 2011).

Then, after the analysis of the obtained results we can conclude: (1) the CPU times used by the BDP grow potentially with the increase of window width; (2) the logarithms of the rate of *RPD*, for a set of solutions, and the CPU times are linearly related; (3) the expressions of *RPD* as a function of time, allow a prediction of the average improvement of the solutions, as minimum, by 6% in instances of Block-I and 4% in Block-II.

Acknowledgements This work is supported by the UPC Nissan Chair and the Spanish Ministerio de Educación y Ciencia under project DPI2010-16759 (PROTHIUS-III) including EDRF fundings.

References

Bautista J, (1993) Procedimientos heurísticos y exactos para la secuenciación en sistemas productivos de unidades homogéneas (contexto J.I.T.). Doctoral Thesis, DOE, ETSEIB-UPC

Bautista J, Companys R, Corominas A (1996) Heuristics and exact algorithms for solving the Monden problem. Eur J Oper Res 88(1):101–113

Bautista J, Cano A (2011) Solving mixed model sequencing problem in assembly lines with serial workstations with work overload minimization and interruption rules. Eur J Oper Res 210(3):495–513

Bautista J, Cano J (2008) Minimizing work overload in mixed-model assembly lines. Int J Prod Econ 112(1):177–191

Boysen N, Fliedner M, Scholl A (2009) Sequencing mixed-model assembly lines: survey, classification and model critique. Eur J Oper Res 192(2):349–373

Scholl A, Klein R, Domschke W (1998) Pattern based vocabulary building for effectively sequencing mixed-model assembly lines. J Heuristics 4:359–381

Tsai LH (1995) Mixed-model sequencing to minimize utility work and the risk of conveyor stoppage. Manage Sci 41:485–495

Yano CA, Rachamadugu R (1991) Sequencing to minimize work overload in assembly lines with product options. Manage Sci 37(5):572–586

Chapter 39
Solution Approaches for Material Requirement Planning* with Fuzzy Costs

Josefa Mula and Manuel Díaz-Madroñero

Abstract This paper valuates different solution approaches of a fuzzy linear programming model to be applied to an MRP (Material Requirement Planning) problem with capacity constraints and uncertainty in costs data. The proposed model represents the possible lack of knowledge or epistemic knowledge in the inventories maintenance cost coefficients, delayed demand and idle time, and considers all constraints to be deterministic. In order to obtain the solution to this model, various solution approaches are analysed in terms of modelling complexity and computational efficiency, and the accomplished solution is presented.

39.1 Introduction

Several models have been developed under uncertainty in the production planning context (Mula et al. 2006a, b; Peidro et al. 2009), which include fuzzy mathematical programming approaches. The work of Mula et al. (2006c) models lack of knowledge or epistemic uncertainty of information about costs likely to be present in an MRP. This model considers delayed demand costs, inventory costs and undertime capacity costs to be ill-known parameters, represented by triangular possibility distributions, and all constraints to be deterministic. In general, there is no ideal solution available to solve linear programming problems with fuzzy coefficients in the objective function with non-fuzzy constraints, and several authors propose different approaches. This article uses the fuzzy number ranking

J. Mula (✉) · M. Díaz-Madroñero
Research Centre on Production Management and Engineering (CIGIP),
Universitat Politècnica de València, Plaza Ferrándiz y Carbonell, 2,
03801 Alcoy (Alicante), Spain
e-mail: fmula@cigip.upv.es

S. P. Sethi et al. (eds.), *Industrial Engineering: Innovative Networks*,
DOI: 10.1007/978-1-4471-2321-7_39,

methods of the first and third indexes of Yager (1978, 1981), as well as Lai and Hwang's approach (1992), which transform the fuzzy objective function into three deterministic objectives through a triangular possibility distribution. The main contribution of this work is to apply these solution approaches to the MRP model with the considered fuzzy costs and to evaluate these approaches in terms of modelling complexity, computational efficiency and the accomplished solution.

The structure of this article is as follows: Sect. 39.2 presents the problem as a fuzzy linear programming model. Section 39.3 explains the development of the equivalent crisp models by means of different solution approaches. Section 39.4 evaluates these approaches using a numerical example. Finally, Sect. 39.5 provides the conclusions drawn throughout this work.

39.2 Fuzzy Model Formulation

The model of Mula et al. (2006c) is a formulation to optimise a mid-term planning problem in an MRP manufacturing setting with multi-product, multi-level and multi-period capacity constraints and fuzzy costs. Table 39.1 defines the decision variables and parameters in the fuzzy model.

$$\begin{aligned}\text{Minimise } z = &\sum_{i=1}^{I}\sum_{t=1}^{T}\left(cp_{it}P_{it} + \tilde{ci}_{it}\, INVT_{it} + \tilde{crd}_{it}\, Rd_{it}\right)\\ &+\sum_{r=1}^{R}\sum_{t=1}^{T}\left(\tilde{ctoc}_{rt}\, Toc_{rt} + ctex_{rt}Tex_{rt}\right)\end{aligned} \tag{39.1}$$

Subject to

$$\begin{aligned}&INVT_{i,t-1} + P_{i,t-TS_i} + RP_{it} - INVT_{i,t} - Rd_{i,t-1}\\ &-\sum_{j=1}^{I} a_{ij}(P_{jt} + RP_{jt}) + Rd_{it} = d_{it} \quad i = 1\ldots I, t = 1\ldots T\end{aligned} \tag{39.2}$$

$$\sum_{i=1}^{I} AR_{ir}P_{it} + Toc_{rt} - Tex_{rt} = CAP_{rt} \quad r = 1\ldots R, t = 1\ldots T \tag{39.3}$$

$$Rd_{iT} = 0 \quad i = 1\ldots I \tag{39.4}$$

$$P_{it}, INVT_{it}, Rd_{it}, Toc_{rt}, Tex_{rt} \geq 0 \quad i = 1\ldots I, r = 1\ldots R, t = 1\ldots T \tag{39.5}$$

Objective function (39.1) attempts to minimise those costs related with production, maintaining inventories, delayed demand, undertime capacity and required overtime capacity. A triangular possibility distribution represents inventory costs, $\tilde{ci}_{it} = \left(ci_{itm}, ci_{ito}, ci_{itp}\right)$, delayed demand, $\tilde{crd}_{it} = \left(crd_{itm}, crd_{ito}, crd_{itp}\right)$, and undertime

Table 39.1 Decision variables and model parameters

Sets			
T	Set of periods in the planning horizon ($t = 1\ldots T$)		
I	Set of products ($i = 1\ldots I$)		
J	Set of parent products in the bill of materials ($j = 1\ldots J$)		
R	Set of resources ($r = 1\ldots R$)		
Decision Variables		***Data***	
P_{it}	Quantity to produce product i in period t	d_{it}	Demand of product i in period t
$INVT_{it}$	Inventory of product i at the end of period t	α_{ij}	Required quantity of i to produce one unit of product j
Rd_{it}	Delayed demand of product i at the end of period t	TS_i	Lead time of product i
Toc_{rt}	Undertime hours for resource r in period t	$INVT_{i0}$	Inventory of product i in period 0
Tex_{rt}	Overtime hours for resource r in period t	Rd_{i0}	Delayed demand of product i in period 0
Objective Function Cost Coefficients		RP_{it}	Programmed receptions of product i in period t
cp_{it}	Variable cost of production of one unit of product i in period t	***Technological Coefficients***	
$\underset{it}{\tilde{ci}}$	Fuzzy inventory cost of one unit of product i in period t	AR_{ir}	Required time of resource r for one production unit of product i
$\underset{it}{\tilde{crd}}$	Fuzzy delayed demand cost of one unit of product i in period t	CAP_{rt}	Available capacity of resource r in period t
$\underset{rt}{\tilde{ctoc}}$	Fuzzy undertime hour cost for resource r in period t		
$ctex_{rt}$	Overtime hour cost for resource r in period t		

capacity, $\underset{rt}{\tilde{ctoc}} = (ctoc_{rtm}, ctoc_{rto}, ctoc_{rtp})$. Thus, inventory maintenance costs contain those costs relating to fixed capital, insurances, price variations, etc. The delayed demand costs for finished goods imply explicit costs due to loss of profit, but also include the costs incurred through a damaged image or loss of customers. An excessive supply of key machines in relation to the current production volume in each period affects undertime capacity costs. These three costs are fuzzy in nature, and those constraints relating to production activities are considered deterministic. The balance equations for inventory are given by the group of constraints (39.2). The availability of a group of shared resources limits production in each period, while Eq. 39.3 considers these resources' capacity limits. A constraint has also been added (39.4) to avoid the delays in the last period (T) of the planning horizon. The model also contemplates the non-negativity constraints (39.5) for the decision variables.

39.3 Solution Approaches

39.3.1 First Index of Yager

According to the first index of Yager (1978, 1981), the following equivalent model is obtained:

$$\text{Minimise}\, z=\sum_{i=1}^{I}\sum_{t=1}^{T}\left(cp_{it}P_{it}+\frac{ci_{itm}+ci_{ito}+ci_{itp}}{3}INVT_{it}+\frac{crd_{itm}+crd_{ito}+crd_{itp}}{3}Rd_{it}\right)$$
$$+\sum_{r=1}^{R}\sum_{t=1}^{T}\left(\frac{ctoc_{rtm}+ctoc_{rto}+ctoc_{rtp}}{3}Toc_{rt}+ctex_{rt}Tex_{rt}\right) \tag{39.6}$$

Constraints (39.2)–(39.5) remain inalterable in this model.

39.3.2 Third Index of Yager

The use of the third index of Yager (1978, 1981) for ranking fuzzy numbers provides an auxiliary model as follows:

$$\text{Minimise z;} =$$
$$\sum_{i=1}^{I}\sum_{t=1}^{T}\left(cp_{it}P_{it}+\frac{ci_{itm}+2ci_{ito}+ci_{itp}}{4}INVT_{it}+\frac{crd_{itm}+2crd_{ito}+crd_{itp}}{4}Rd_{it}\right)$$
$$+\sum_{r=1}^{R}\sum_{t=1}^{T}\left(\frac{ctoc_{rtm}+2ctoc_{rto}+ctoc_{rtp}}{4}Toc_{rt}+ctex_{rt}Tex_{rt}\right) \tag{39.7}$$

Constraints (39.2)–(39.5) remain inalterable in this model.

39.3.3 The Lai and Hwang Approach

This section applies the approach put forward by Lai and Hwang (1992) to transform a fuzzy objective function modelled through triangular possibility distributions into three objective functions

$$\text{Minimise } z_1 = \sum_{i=1}^{I}\sum_{t=1}^{T}\left(cp_{it}P_{it} + ci_{itm}\,INVT_{it} + crd_{itm}\,Rd_{it}\right) + \sum_{r=1}^{R}\sum_{t=1}^{T}\left(ctoc_{rtm}\,Toc_{rt} + ctex_{rt}Tex_{rt}\right) \tag{39.8}$$

$$\text{Maximise } z_2 = \sum_{i=1}^{I}\sum_{t=1}^{T}\left((ci_{itm} - ci_{ito})\,INVT_{it} + (crd_{itm} - crd_{ito})Rd_{it}\right) + \sum_{r=1}^{R}\sum_{t=1}^{T}\left((ctoc_{rtm} - ctoc_{rto})Toc_{rt}\right) \tag{39.9}$$

$$\text{Minimise } z_3 = \sum_{i=1}^{I}\sum_{t=1}^{T}\left((ci_{itp} - ci_{itm})\,INVT_{it} + (crd_{itp} - crd_{itm})Rd_{it}\right) + \sum_{r=1}^{R}\sum_{t=1}^{T}\left((ctoc_{rtp} - ctoc_{rtm})Toc_{rt}\right) \tag{39.10}$$

This model minimises the most possible value of the uncertain costs (39.8), whereas it maximises the possibility of obtaining the lowest costs (39.9) and minimises the possibility of obtaining the highest costs (39.108). The rest of the model constraints (39.2), (39.3), (39.4) and (39.5) remain the same.

Next, the model in this section is solved with the approaches of Zimmermann (1978) and Werners (1988). We formulate the corresponding continuous linear membership functions for each objective function as follows (Bellman and Zadeh 1970):

$$\mu_1 = \begin{cases} 1 & z_1 < z_1^l \\ \frac{z_1^u - z_1}{z_1^u - z_1^l} & z_1^l < z_1 < z_1^u \\ 0 & z_1 > z_1^u \end{cases} \tag{39.11}$$

$$\mu_2 = \begin{cases} 1 & z_2 > z_2^u \\ \frac{z_2 - z_2^l}{z_2^u - z_2^l} & z_2^l < z_2 < z_1^u \\ 0 & z_2 < z_2^l \end{cases} \tag{39.12}$$

μ_3 is defined similarly to μ_1 where μ_1, μ_2, μ_3 are the membership functions of z_1, z_2, z_3. z_1^l, z_2^l, z_3^l, while z_1^u, z_2^u, z_3^u are the lower and upper bounds of the objective functions. We can determine each membership function by asking the decision maker to specify the fuzzy objective value interval (39.11)–(39.12). Besides, we can obtain these bounds from the optimisation values of each objective function.

39.3.3.1 The Zimmermann Solution Method

According to the approach of Zimmermann (1978), based on the minimum operator of Bellman and Zadeh (1970), the former multi-objective linear programming model can be transformed into an equivalent model with a single objective by maximising an auxiliary variable $\lambda_0 \in [0, 1]$, which represents the minimum degree of fulfilment of all the objectives:

Maximise λ_0

Subject to

$$z_1 \leq z_1^u - \lambda_0 \left(z_1^u - z_1^l \right) \tag{39.13}$$

$$z_2 \geq z_2^l + \lambda_0 \left(z_2^u - z_2^l \right) \tag{39.14}$$

$$z_3 \leq z_3^u - \lambda_0 \left(z_3^u - z_3^l \right) \tag{39.15}$$

This model also considers Constraints (39.2)–(39.5).

39.3.3.2 The Werners Solution Method

According to Werners (1988), the multi-objective model can be formulated as follows:

$$\text{Maximise } \lambda(x) = \gamma \lambda_0 + (1 - \gamma) \frac{1}{3} (\lambda_1 + \lambda_2 + \lambda_3) \tag{39.16}$$

Subject to

$$z_1 \leq z_1^u - (\lambda_0 + \lambda_1) \left(z_1^u - z_1^l \right) \tag{39.17}$$

$$z_2 \geq z_2^l + (\lambda_0 + \lambda_2) \left(z_2^u - z_2^l \right) \tag{39.18}$$

$$z_3 \leq z_3^u - (\lambda_0 + \lambda_3) \left(z_3^u - z_3^l \right) \tag{30.19}$$

This model also considers Constraints (39.2)–(39.5). Besides, λ_0 has an equivalent meaning to that of Zimmermann's solution method. However, parameter γ corresponds to the compensation coefficient among the objectives in accordance with the decision maker's preferences.

39.4 Numerical Example

The proposed model is applied in a first-tier supplier of an automobile assembler. The hypotheses to carry out the computational experiment are summarised as follows: the study considers a single product; decision variables P_{it}, $INVT_{it}$, and Rd_{it} are integer;

Table 39.2 Model statistics

Model	First index of Yager	Third index of Yager	Lai and Hwang approach Zimmerman solution method	Werners solution method
Iterations	0	0	0	0
Variables	4202	4202	4203	4209
Integer	5612	5612	5612	5612
Constraints	2762	2762	2765	2768
Non-zero elements	8204	8204	16667	16676
Array density (%)	0.07	0.07	0.14	0.14
CPU time (seconds)	0.06	0.08	0.17	0.17

Table 39.3 Solutions for each model

Model	First index of Yager	Third index of Yager	Lai and Hwang's approach	
			Zimmerman solution method	Werners solution method
Total production costs	381503.742	381503.742	381503.742	381503.742
Total inventory costs	50127.629	50127.629	–	–
Most possible inventory costs	–	–	50127.785	50129.912
Most optimistic inventory costs	–	–	45115.007	45116.921
Most pessimistic inventory costs	–	–	55140.564	55142.903
Total costs (z)	431631.371	431631.371	–	–
Most possible total costs $z1$ ($\mu 1$)	–	–	431631.527 (0.967)	431633.654 (0.967)
Most optimistic total costs $z2$ ($\mu 2$)	–	–	426618.749 (0.967)	426620.663 (0.998)
Most pessimistic total costs $z3$ ($\mu 3$)	–	–	436644.306 (0.967)	436646,645 (0.997)

external demand exists only for the final product; delayed demand for the final product is considered; idle costs are considered null; only a productive resource restricts production: the assembly line; and a 6-month planning horizon with weekly period planning has been considered. The model has been implemented with the MPL modelling language (2010). Resolution has been carried out with the optimisation solver CPLEX 9. Finally, a Microsoft Access 2010 database manages the input and output data of the model. Table 39.2 shows the computational efficiency of the evaluated models, and how the Lai and Hwang approach requires a double amount of CPU time in relation to the Yager approaches. All the models can obtain an optimal mixed integer linear programming solution with a null number of iterations. On the other hand, the Werners model has a larger number of constraints, but the same number of integer variables; this does not imply additional information storage requirements, but more modelling complexity. Additionally, the Werners model requires more parameter settings for the decision maker.

Table 39.3 summarises the solutions yielded by the proposed models. All the models have provided null backorder and overtime costs. From Table 39.3, we can conclude that both of the models based on the first and third indexes of Yager provide the same level of total costs. Lai and Hwang's approach (1992) generates two different fuzzy solutions, depending on the solution method applied. From the optimistic perspective, these fuzzy solutions could improve the inventory level and the total costs generated.

39.5 Conclusions

In many manufacturing environments, it is necessary to make production planning decisions under uncertain conditions in very important parameters such as costs. In order to define objective function costs, this paper presents a fuzzy linear model for production planning under conditions of uncertainty. Different solution approaches have been applied and evaluated in terms of model complexity and computational efficiency, and the accomplished solution is presented. We conclude that Lai and Hwang's approach (1992), solved by the Werners method, implies more modelling complexity and CPU time, but provides more flexibility for the decision maker to achieve a fuzzy solution according to his/her preferences. Finally, we have identified the following further lines of research: (1) test the models with real-world problems in a rolling horizon; (2) compare the models with other solution approaches; and (3) use other membership functions patterns.

References

Bellman R, Zadeh L (1970) Decision making in a fuzzy environment management science 17: 141–164

Lai YJ, Hwang CL (1992) A new approach to some possibilistic linear programming problems. Fuzzy Sets Syst 49:121–133

Maximal Software Incorporation (2010) MPL modeling system. Release 4.2 m, U S A

Mula J, Poler R, Garcia JP (2006a) MRP with flexible constraints, Fuzzy Sets and Systems, Vol. 157, N° 1, 74–97

Mula J, Poler R, Garcia-Sabater JP, Lario FC (2006b) Models for production planning under uncertainty: A review. Int J Prod Econ Vol 103, 271–285

Mula J, Poler R, Garcia-Sabater JP (2006c) Un modelo de programación lineal multi-objetivo para la resolución del MRP con restricciones de capacidad. X Congreso de Ingeniería de Organización, Valencia, 7th and 8th September

Peidro D, Mula J, Poler R, Lario FC (2009) Quantitative models for supply chain planning under uncertainty. Int J Adv Manuf Technol 43:400–420

Werners B (1988) Aggregation models in mathematical programming. In: Mitra G, Greenberg HJ, Lootsma FA, Rijckaert MJ, Zimmermann H-J (eds) Mathematical models for decision support. Springer, Heidelberg, pp 295–305

Yager RR (1978) Ranking fuzzy subsets over the unit interval. In: Proceedings of the IEEE conference on. decision and control, pp 1435–1437
Yager RR (1981) A procedure for ordering fuzzy subsets of the unit interval. Inf Sci 24(2):143–161
Zimmermann HJ (1978) Fuzzy programming and linear programming with several objective functions. Fuzzy Sets Syst 1:45–55

Chapter 40
Forecasting Methods for Determining the Level of Safety Stock in Electronic Industry

Fauzia Dianawati, Isti Surjandari and Rainy Nafitri

Abstract The competition in industries force many companies to reduce production costs in order to gain more profit. This decision requires a good planning for production schedule. In production planning, forecasting theory to estimate future demand is an important aspect. The problem in this article is to choose the best forecasting method based on the characteristics of historical data in seasonal demand patterns. Results forecast error is then used to calculate the safety stock as a form of anticipatory strategies shortage of material. The results indicate that the best forecasting method is a combination method of Holt-Winters Exponential Smoothing and Naïve, it can decrease Mean Absolute Percentage Error (MAPE) up to 63%.

40.1 Introduction

The increasing competition among the industry in achieving its primary goal of all companies that obtain maximum benefit and with the minimum of production factors resulted a tight competition because of the many similar industries or companies engaged in the same industry while the market share of consumers are tend to be limited. One industry that increasingly compete with each other is the manufacturing industry. These companies have the ideal manufacturing production planning strategies that are effective in adjusting the production target of the available capacity. Lack of production capacity will lead to failure to meet

F. Dianawati (✉) · I. Surjandari · R. Nafitri
Industrial Engineering Department, University of Indonesia,
Kampus UI Depok, Depok 16424, Indonesia
e-mail: fauzia@ie.ui.ac.id

S. P. Sethi et al. (eds.), *Industrial Engineering: Innovative Networks*,
DOI: 10.1007/978-1-4471-2321-7_40,

production targets, delays in shipments to customers and the loss of customer confidence (Chase and Jacobs 2004). Conversely excess production capacity will lead to lower utilization rates, cost increases, prices become less competitive products, loss of market share, declining profits, and others. Therefore, all the factors of production must be managed through good management company which are planning, organizing, actuating, and controlling (Montgomery and Johnson 1998).

In production planning, forecasting is an important tool in making decisions (Montgomery and Johnson 1998). Forecasting, in various methods is done by using information on past and present to identify expected future conditions (Arnold and Chapman 2004). In the manufacturing industries that produce electronics hardware, the demand shows a seasonal demand pattern.

Standard forecasting methods that are designed to cope with seasonal demand often are no longer applicable in practice. Due to growing assortments and shorter product life cycles, demand data may show too high variation or may be insufficient to construct reliable forecast models at the individual item level (Dekker et al. 2004). Therefore, we need to develop alternative methods of forecasting by using the information demand from a higher aggregation level and by combining several forecasting methods.

Furthermore, more understanding is needed in determining production planning in order to generate maximum profit with minimum cost. Problems related to safety stock become interesting to discuss. By making the prediction using the best methods and make a calculation of safety stock, production costs will be minimized.

40.2 Methodology

Research methodology for conducting this research consists of several stages. At first, what to do is collect the demand data of the printer for 2 years and forecast data by the company for 6 months. In order to be easy to identify, the data is presented per period in the form of graphs. Based on the graph, it shows that the data appertain to seasonal demand pattern with a decrease trend.

After obtaining the required data related to forecasting, data processing steps undertaken were:

1. Calculate the MSE, MAD, and MAPE

At this stage, data processing, demand forecasting is based on several methods, namely Naïve, Exponential Smoothing, Winter, Winter and the combination of Naïve, Product aggregation, combination of product aggregation and Naïve.

According to Dekker et al. (2004), there were some of quantitative forecasting methods are often used for seasonal demand patterns.

- *Naïve Methods*

$$F_t = X_{t-1} \tag{40.1}$$

- *Single Exponential Smoothing*

$$Ft = Ft - 1 + \alpha(Xt - 1 - Ft - 1) \tag{40.2}$$

- *Holt-Winters Exponential Smoothing*

$$Ft + m = (St + Ttm)\ It - L + m \tag{40.3}$$

- *Holt-Winters Exponential Smoothing combined with Naïve*
- *Product Aggregation*

$$Xj,t,t + 1 = \alpha j,t.\ \ fNt{+}1 - P \tag{40.4}$$

After we got the result of individual item forecasting, we need to do forecasting again at the product family level, using the formula:

$$Xt = \sum Xj,t \tag{40.5}$$

- *Product Aggregation combined with Naïve*

This method produces forecasts based on the combining the two forecasting methods, namely Product aggregation with Naïve. Forecasting results obtained by averaging the two of them.

Forecasting is never perfect. Forecasting of future conditions is generally not able to exactly the same as the actual reality in the future. In any forecast made are always generated bias. Forecast error is defined as the difference between the forecast values with the actual condition. The amount of demand forecast errors is the amount of actual demand minus by the forecasted value of that current period. Forecast error value indicates whether the forecast is done well enough.

Best forecasting method is the one that produces the smallest forecast error values. In order to get the forecast that has a good level of accuracy it is necessary to test the accuracy by finding the smallest error in calculating the value:

- Mean Squared Error (*MSE*)
- Mean Absolute Deviation (*MAD*)
- Mean Absolute Percentage Error (*MAPE*)

To compare the forecasting accuracy between one methods with each other, calculate MAPE is most appropriate way in finding the forecast error which are reviewed as a percentage of errors compared to the actual circumstances.

Results from the mean squared error (MSE) and mean absolute error (MAD) is less accurate in a direct comparison when reviewing some forecasting methods and data. Therefore, the forecast error of each method should be first converted into a percentage of the actual, in other words, MAPE appropriate for used in comparing the accuracy of some forecasting methods which are tested. The smaller value of MAPE, the forecasting method is more accurate.

2. Estimating the forecast for next period

At this stage, what to do is calculating the forecast for the next period with the best forecasting method that has been previously selected.

3. Conducting performance measurement

At this stage, the comparison will be made based on the performance assessment calculations for service levels before and after implementation of the chosen forecasting method.

4. Calculate safety stock

Safety stock is prepared as a buffer stock in anticipation of the differences between forecast and actual demand, the delivery time that have planned before and actual, and matters other unexpected. Total safety stock required to meet demand levels. Specific needs of safety stock can be determined through computer simulation or statistical methods.

In relation with the inaccuracies of forecasting, there is a formula in determining the safety stock (SS) (Chockalingam 2010):

$$\mathrm{SS} = \mathrm{SL} \times \mathrm{FE} \times \sqrt{\mathrm{LT}} \tag{40.6}$$

Service level (SL) used in the calculation of this formula is the value of z (safety factor) of the percentage of the expected level of service. Forecast error (FE) that used is the Root Mean Square Error value, while the Lead Time (LT) which is used is the range of time starting from when ordering.

40.3 Result and Discussion

Based on calculations performed with various forecasting methods above, can be seen that when compared one method with each other, with a view of the period, it can be concluded that the MSE and MAPE is directly proportional. Based on data processing by the various forecasting methods above, the MAPE obtained as follows.

Seen from the Table 40.1, the methods proposed (new forecasting methods), having a result of MAPE values which are smaller than the forecasting that has been done. Therefore, new methods are feasible to be applied to the company. Of the total new method proposed, the smallest MAPE values found when forecasting is done using Holt-Winters Exponential Smoothing method combined with Naïve. MAPE generated by this combination forecasting method is 20%.

Table 40.1 Result of mean absolute percentage error (MAPE)

Methods	MAPE
Forecast methods exist	83
Single exponential smoothing	46
Holt-winters exponential smoothing	26
Holt-winters exponential smoothing combined	20
With Naïve	24
Product aggregation	24
Product aggregation combined with Naive	

Table 40.2 Forecasting result with the combination method of holt-winters exponential smoothing and Naïve method

Methods	Forecast (units)
Holt-winters exponential smoothing	5,306
Naive	10,738
Holt-winters exponential smoothing combined with Na	8,022

This happens because of all the new method proposed, the method of Holt-Winters Exponential Smoothing which produces 26% MAPE is the most complex forecasting method because it involves the calculation of trend components, seasonality, and randomness in it. And apparently when combined with the Naïve method that usually known as random walk model, it produce a smaller MAPE, which is decrease by 6%. This shows that the Naïve method has a positive contribution when combined with the method of Holt-Winters Exponential Smoothing.

40.3.1 Forecasting Next Period

Next further calculations was used forecasting method that that has choosed based on the smallest MAPE value to forecast the units to produce. Holt-Winters methods combined with Exponential Smoothing Naïve method is the best method found.

After processing the data for forecasting future periods using each method, will be combine them. The datas were taken from the actual units in production. Here is the result of forecasting the next period by combining the two methods, Holt-Winters and Exponential Smoothing Naïve (Table 40.2).

Results after calculation show that the results forecast for the next period is 8,022 units. Forecasting is only possible for one following period because this forecasting method consists of two forecasting methods that combined. Holt-Winters Exponential Smoothing method can produce forecast value for some

Table 40.3 Service level comparison

	Service level forecast methods exist	Service level holt-winter exp. smoothing COMBINED with Naïve
Total backlog	16,929	5,221
Total demand that can be fulfilled	2,15,018	2,26,723
Total demand	2,31,944	2,31,944
Service level	92.7%	97.75%

period ahead, while the Naïve method can't. This happen because Na forecasting method using a single data in the previous period as the forecast period. So to get the forecast value of the combined these two methods, it needs an update routinely when actual demand comes every period. Because of these reasons, the method of Holt-Winters Exponential Smoothing combined with Na can only produce forecasts for one period ahead.

40.3.2 Comparison of Performance Measurement

Performance measurement are routinely conducted to ensure the sustainability of production companies in order to evaluate the performance of the company. Assessment of performance is calculated based on the calculation of service levels before and after implementation of the chosen forecasting method. The table shows that there is an increase of service level value as much as 5.05%. This is due to the new forecasting method accuracy is higher so that demand can be more anticipated (Table 40.3).

40.3.3 Determination of Safety Stock Strategies

Availability of products is the most important thing in ensuring the level of customer satisfaction (Narasimhan and Mcleavy 2000). Conditions where the demand is higher than the production can result backlog that causes lost sales. This represents that the company lost their sales and then loose. Therefore, the strategy is needed in determining the amount of inventory for anticipating the customer demand. This type of inventory is known as safety stock.

Calculation of safety stock requirements can be determined using the formula (40.6). Based on calculations, safety stock that must be provided based on the method of Holt-Winters Exponential Smoothing Combined with Naïve, can be seen as follows (Table 40.4).

Table 40.4 Level of safety stock requirement

Service level (%)	Safety factor	Forecast error	Lead time	Safety stock
99	2,33	3,344	1	7,778
98	2,05	3,344	1	6,867
97	1,88	3,344	1	6,288
96	1,75	3,344	1	5,853
95	1,64	3,344	1	5,500
94	1,55	3,344	1	5,198
93	1,48	3,344	1	4,934
92	1,41	3,344	1	4,698
91	1,34	3,344	1	4,483
90	1,28	3,344	1	4,285

Basically, the determination of service level is company policy because there may be other considerations that should be considered in order to gain more profit. The table above is a proposed alternative determination of the amount of safety stock that can help companies based on service level 90–99%.

Provided safety stock can also cause increased risk of warehouse costs. Therefore, the more accurate calculations in determining safety stock is an important thing in order to minimize costs. The determination of safety stock requirements is depends on the accuracy of forecasting.

Conceptually, safety stock is used to handle the demand uncertainty. With the existence of an accurate forecasting method, the amount of safety stock needs can be well controlled (Ballou 2004). The smaller error of the forecast, the smaller safety stock required. Safety stock has an important role in production management, but sometimes the number of available unused inventory when the customer demand low. Therefore, it is necessary doing forecasting as a consideration production planning for future period. Because sometimes might be occur a situations where the cost of holding inventories are much higher than that by allowing the occurrence of lost sales, or vice versa. With the help of accurate forecasting, determination of safety stock level requirement, and establishing a service level expected, the company is able to integrate all of them so that production can get the maximum profit.

40.4 Conclusion

Based on the final results of the analysis, it can be concluded that:

1. From the research, the best method wich selected based on the smallest error is a combination forecasting method of Holt-Winters Exponential Smoothing and Naïve, with MAPE of 20%.
2. Service level based on the forecast that had been applied amounted to 92.7%, while the service level based on the proposed forecasting method, namely the

combination method of Holt-Winters Exponential Smoothing and Naïve amounted to 97.75%. So, there an increase of 5.05%.

3. Estimation of safety stock requirement with a specified service level variations depands on the company can be used as a consideration in the determination of level of safety stock.

References

Arnold JRT, Chapman NS (2004) Introduction to material management. Prentice Hall Inc., Upper Saddle River

Ballou HR (2004) Business logistics management. Prentice Hall Inc., Upper Saddle River

Richard CB, Robert JF (2004) Operation management for competitive advantage. McGraw Hill Inc., New York

Chockalingam M (2010) Forecast error and safety stock strategies. Demand Planning LCC, India

Dekker M, Donselaar KV, Ouwehand Pim (2004) How to use aggregation and combined forecasting to improve seasonal demand forecasts. Int J Prod Econ 90:151–167

Montgomery CD, Lynwood Johnson A (1998) Forecasting and time series analysis. McGraw Hill Inc., New York

Narasimhan LS, McLeavey WD (2000) Production planning and inventory control. Prentice Hall Inc, Upper Saddle River

Chapter 41
On the Fill Rate Approximations in Periodic Review Systems for Discrete Demand

E. Guijarro, E. Babiloni, M. Cardós and J. M. Albarracín

Abstract One of the most common criteria used to measure the performance of a inventory system is the fill rate. In the literature we find two easy-to-compute approaches: the traditional and the Hadley-Whitin approximations, but unfortunately they only apply for normally distributed demand. In practice stock managers can use any of them, even when demand is not normal or discrete. This paper assesses the performance of these approximations in a periodic inventory system for different demand scenarios such as intermittent, erratic, lumpy and smooth. However, in the discrete context, simulation results show that both approximations present a significant bias including negative fill rate values. In keeping with these results, we re-formulate both approximations to be able to cope with any discrete demand distribution and asses their performance. In this case, simulation results show that the proposed discrete formulation of the Hadley-Whitin approximation seems to avoid the systematic bias observed before.

41.1 Introduction and Literature Review

In inventory control, the item fill rate, β, (also known as just fill rate) is often used to analyze the performance of the system and to establish the parameters of an inventory policy. Traditionally, it is defined as the fraction of total demand that can be immediately satisfied from on-hand stock and has been calculated in terms of units short. Then, the fill rate can be expressed as

E. Guijarro (✉) · E. Babiloni · M. Cardós · J. M. Albarracín
Dpto. Organización de Empresas, Universidad Politécnica de Valencia,
Camino de Vera s/n, 46022 Valencia, Spain
e-mail: esguitar@doe.upv.es

S. P. Sethi et al. (eds.), *Industrial Engineering: Innovative Networks*,
DOI: 10.1007/978-1-4471-2321-7_41,

$$\beta = 1 - \frac{E(shortage\,per\,replenishment\,cycle)}{E(total\,demand\,per\,replenishment\,cycle)} \tag{41.1}$$

Given that the expected total demand per replenishment cycle can be easily calculated, authors focus on estimating the expected shortage per replenishment cycle. However, under the traditional assumptions of the periodic, order-up-to-level, (*R*, *S*) system (see Silver et al. (1998)), only a handful of methods are available. One of the widely used ones in practical environments due to its ease of use is the method known as traditional expression, presented in classical Operations Management (OM) texts (e.g., Chase et al. (1992); Vollmann et al. (1997), or Axsäter (2000)). Its derivation is detailed by Silver et al. (1998), where demand follows a normal distribution pattern.

Hadley and Whitin (1963) derive approximations to estimate the expected shortage for Poisson and normal demand patterns. The Hadley and Whitin approach, from now on referred to as H&D, is used by de Kok (1990) to compute the fill rate as in expression (41.1) and recently Silver and Bischak (2011) present another derivation which leads to the same result as Hadley and Whitin (1963) but in a quicker and more natural way.

Johnson et al. (1995) review the traditional and the H&W approximations under the normal demand pattern and point out that they are biased and even the traditional approximation may produce negative values.

Despite discrete demand being the most usual one in distribution and maintenance environments, there is not any expression available in the literature for any discrete demand. Therefore practitioners have to apply the traditional and the H&W approximation, although they are developed for normal demand. Thus, this paper analyses the performance of the traditional (β_{Trad}) and the H&W approximation ($\beta_{H\&W}$) for different demand categories (intermittent, erratic, lumpy and smooth) when demand is discrete. The simulation results point out that this approach presents significant deviations and biases.

To provide a more accurate fill rate estimation we develop a new approach based on considering explicitly the discrete demand in both approximations and assessing their performance.

The paper has been organized as follows: first, a section is devoted to explaining the fill rate approximations, their assumption and derivation. Later, we present an experiment which shows the deviations obtained. After that, we re-formulate the classical approximations for the discrete demand context, which is assessed in the subsequent section. Finally, conclusions and further research are presented.

41.2 Approximations of the Fill Rate in Periodic Systems

A periodic review inventory system examines the status of an item every *R* fixed units of time and places a replenishment order to raise the inventory position to the order up to level *S*. The replenishment order is received *L* time periods after being launched.

Both β_{Trad} and $\beta_{H\&W}$ consider the following assumptions to derive the fill rate: (i) the lead time L is constant, (ii) the value of R is predetermined, (iii) there is a negligible chance of no demand between reviews; consequently, a replenishment order is launched at every review, (iv) all the excess demands are backlogged, (v) demands in different time periods are independent and identically distributed, (vi) demand variance is sufficiently small to ensure non-negative period demands, and (vii) the on-hand stock at the beginning of the cycle equals S.

Then, the fill rate has been traditionally calculated based on the expected shortage in $R + L$:

$$E(shortage)_{Trad} = E[D_{R+L} - S]^{+} \tag{41.2}$$

where $X^{+} = \max\{X, 0\}$ and D_t represents the accumulated demand during t consecutive periods. When demand is continuous, expression (1.2) can be rewritten as

$$E(shortage)_{Trad} = \int_{S}^{\infty} (D_{R+L} - S) \cdot f_{R+L}(D_{R+L}) \tag{41.3}$$

where $f_t(\cdot)$ is the probability density function of demand during t consecutive periods; assuming normal demand with mean μ and variance σ^2 leads to

$$E(shortage)_{Trad} = \sigma_{R+L} G(k_{R+L}(S)) \tag{41.4}$$

where σ_{R+L} is the standard deviation of demand during the review period plus lead time and $G(k_t(S))$ is a special function of the safety stock factor $k_t(S)$ and the standard normal distribution (see Appendix B of (Silver et al. 1998) for its derivation). Then, substituting the numerator of expression (41.1) by expression (41.4), the traditional fill rate is

$$\beta_{Trad} = 1 - \frac{\sigma_{R+L} G(k_{R+L}(S))}{R\mu} \tag{41.5}$$

(Hadley and Whitin 1963) improve the traditional approach by defining the expected shortage as the backorders during $R + L$ minus the backorders during L. So analogously to above

$$E(shortage)_{H\&W} = E[D_{R+L} - S]^{+} - E[D_L - S]^{+} \tag{41.6}$$

with continuous demand distribution

$$E(shortage)_{H\&W} = \int_{S}^{\infty} (D_{R+L} - S) \cdot f_{R+L}(D_{R+L}) - \int_{S}^{\infty} (D_L - S) \cdot f_L(D_L) \tag{41.7}$$

Considering normal demand distribution, expression (41.7) leads to

$$E(shortage)_{H\&W} = \sigma_{R+L} G(k_{R+L}(S)) - \sigma_L G(k_L(S)) \tag{41.8}$$

Table 41.1 Experiment set of data

Lead time, L	1, 2, 3, 4
Review period, R	1, 2, 3, 4
Order up to level, S	1, 2, 3, 4
Negative binomial, NB (r, p)	(0.25, 0.75), (0.5, 0.3), (1, 0.9), (0.85, 3)

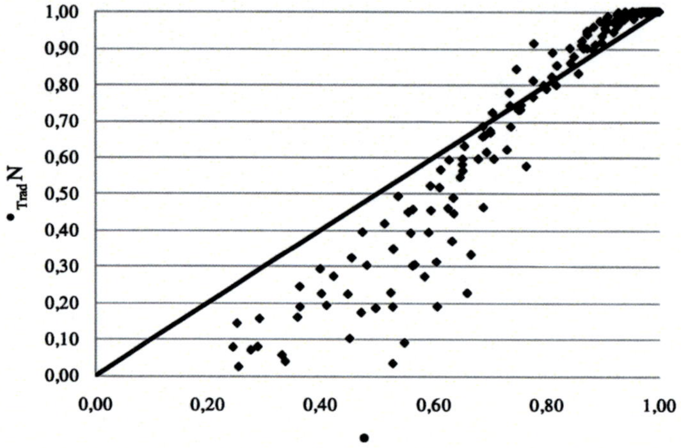

Fig. 41.1 Simulated fill rate vs. traditional approximation once negative values have been removed

and therefore, the H&W estimation of the fill rate is

$$\beta_{H\&W} = 1 - \frac{\sigma_{R+L} G(k_{R+L}(S)) - \sigma_L G(k_L(S))}{R\mu} \tag{41.9}$$

41.3 Experimental Evaluation of the Approximations

This section presents an experiment designed to explore deviations which arise while using the traditional and the H&W approximations when demand is discrete. We choose the negative binomial distribution since it covers the smooth, erratic, intermittent and lumpy categories suggested by Syntetos et al. (2005). Furthermore, this experiment uses a combination of inventory policy data (Table 41.1) which results in 256 different cases.

Figure 41.1 shows the performance of the traditional fill rate (β_{Trad}) versus the simulated fill rate (β). Not surprisingly, approximately 27% of the values are negative, so they have been removed from the figure, since a negative fill rate

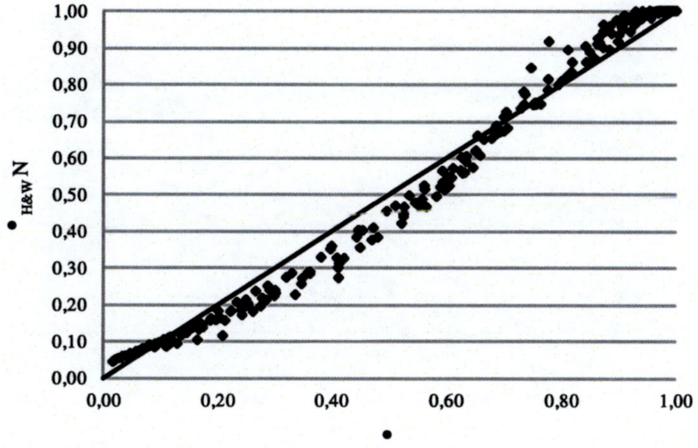

Fig. 41.2 Simulated fill rate vs. H&W approximation

would have no meaning. In this case, basically β_{Trad} seems to underestimate the fill rate for values lower than 0.70 and overestimate it for values higher than 0.80; however, for values between 0.70 and 0.80, the fill rate can be either overestimated or underestimated.

Regarding the performance of the H&W approximation, Fig. 41.2 shows that $\beta_{H\&W}$ tends to underestimate the fill rate for values below 0.7 and tends to overestimate the fill rate for values above 0.7. Both approximations show a very similar behaviour for high values of fill rate.

Obviously, $\beta_{H\&W}$ is more accurate than β_{Trad}. However, they both fall into significant deviations and biases that cannot be neglected for discrete demand. For example, a case in the experiment shows a $\beta_{H\&W} = 0.91$ whereas $\beta = 0.78$, which may lead to establish an inventory policy that does not assure the fulfilment of a target fill rate or to consider the system less protected against stock out occasions than managers expect.

41.4 A New Approach to the Fill Rate Approximations for Discrete Demand

This section re-formulates the classical approximation of the fill rate for any discrete demand distribution under the same assumptions considered above. Then, the expected shortage during $R + L$ can be rewritten for discrete demand as

$$E(shortage)_{\mathrm{Trad\_D}} = \sum_{i=S}^{\infty} (i - S)P(D_{R+L} = i) \qquad (41.10)$$

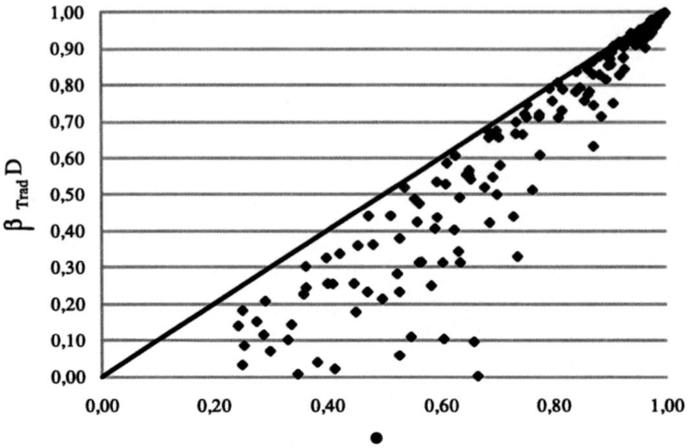

Fig. 41.3 Simulated fill rate vs. discrete traditional approximation once negative values have been removed

This can be expressed using the probability mass function, $f_t(\cdot)$, and the cumulative distribution function, $F_t(\cdot)$. Then, the discrete traditional fill rate is

$$\beta_{Trad\_D} = 1 - \frac{\sum_{i=S}^{\infty}(i - S)f_{R+L}(i)}{\sum_{k=1}^{\infty} k \cdot f_R(k)} \tag{41.11}$$

where the denominator represents the expected total demand per replenishment cycle.

On the other hand, the backorders during $R + L$ minus the backorders during L can be expressed for any discrete distribution as

$$\mathrm{E}(shortage)_{\mathrm{H\&W}} = \sum_{i=S}^{\infty}(\mathrm{i} - \mathrm{S})\mathrm{P}(\mathrm{D_{R+L}} = i) - \sum_{j=S}^{\infty}(\mathrm{j} - \mathrm{S})\mathrm{P}(\mathrm{D_L} = j) \tag{41.12}$$

and then, the discrete H&W fill rate estimation results as

$$\beta_{\mathrm{H\&W\_D}} = 1 - \frac{\sum_{\mathrm{i=S}}^{\infty}(\mathrm{i} - \mathrm{S})f_{\mathrm{R+L}}(i) - \sum_{\mathrm{j=S}}^{\infty}(\mathrm{j} - \mathrm{S})f_L(j)}{\sum_{k=1}^{\infty} k \cdot f_R(\mathrm{k})} \tag{41.13}$$

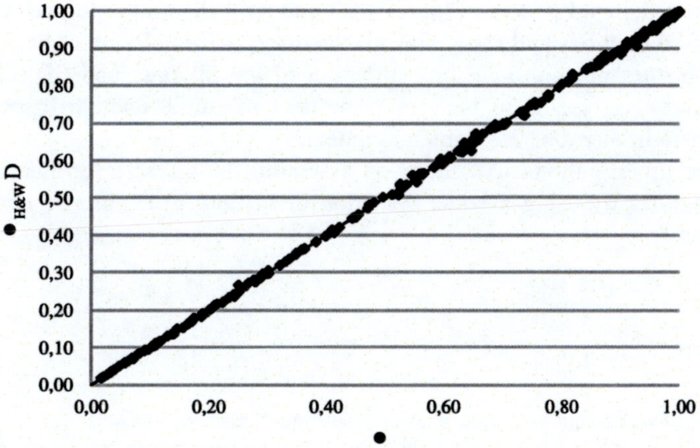

Fig. 41.4 Simulated fill rate vs. discrete H&W approximation

41.5 Experimental Evaluation of the Discrete Approximations

Using the data from Table 41.1, this section illustrates the performance of the discrete expressions derived above. Figure 41.3 compares β_{Trad_D} and the simulated fill rate β and shows that significant deviations appear, as shown in Fig. 41.1. However, the discrete formulation never overestimates the fill rate. It should be noted that even in this case, the traditional approximation provides negative values for approximately 26% of the cases. These negative cases are removed from the figure.

On the other hand, the proposed discrete formulation of the H&W approximation shows an outstanding performance. In fact, neither significant deviations nor biases can be appreciated in Fig. 41.4. This result confirms the need of this new approach which considers explicitly the discrete nature of the demand.

41.6 Summary, Conclusions and Further Research

In general, the fill rate is estimated by means of the expected shortage per replenishment cycle. When the inventory is managed by a periodic system, the traditional (β_{Trad}) and the Hadley and Whitin ($\beta_{H\&W}$) approximations are the only available choices for inventory managers. However, they both apply only for normal demand although the discrete demand is more frequent in distribution and maintenance environments. With this in mind, we test their performance when demand follows a discrete pattern. As Figs. 41.1 and 41.2 show, both approximations present significant deviations and biases, specially the traditional one, which can produce negative fill rates. In an effort to overcome this shortcoming, this paper suggests a new formulation of both approximations able to cope with any discrete demand pattern. Figures 41.3 and 41.4 illustrate the performance of the proposed methods and show that: (i) the discrete formulation of the traditional expression does not improve the estimation of the fill rate; and (ii) the discrete H&W expression seems to be very accurate but simple enough to be used by practitioners in any discrete demand context.

Further investigations will focus on assessing the discrete formulation of the approximations by using a wider selection of inventory parameters and demand distributions.

References

Axsäter S (2000) Inventory control. Kluwer Academic Publishers, Norwell

Chase RB, Aquilano NJ, Jacobs FR (1992) Production and operations management. Manufacturing and servicies. Irwin Mc Graw–Hill, Illinois

de Kok AG (1990) Hierarchical production planning for consumer goods. Eur J Oper Res 45:55–69

Hadley G, Whitin T (1963) Analysis of inventory systems. Prentice Hall, Englewood Cliffs

Johnson ME, Lee HL, Davis T, Hall R (1995) Expressions for item fill rates in periodic inventory systems. Nav Res Log 42:57–80

Silver EA, Pyke DF, Peterson R (1998) Inventory management and production planning and scheduling. Wiley, New York

Silver EA, Bischak DP (2011) The exact fill rate in a periodic review base stock system under normally distributed demand. Omega-int J Manage S 39:346–349

Syntetos AA, Boylan JE, Croston JD (2005) On the categorization of demand patterns. J Oper Res Soc 56:495–503

Vollmann TE, Berry WL, Whybark DC (1997) Manufacturing planning and control systems. McGraw Hill, NewYork

Chapter 42
A Fuzzy Approach to Forecasting the Attractiveness of Regions for Human Resources

Marija Bogataj, Samo Drobne and Danijela Tuljak Suban

Abstract Thinking flexibility and security on labour markets together is a new paradigm because business as well as labour is in need of security as well as flexibility. The flexibility will increase with accessibility of individual regions in Europe. A study of migration may show us not only how these aims can be achieved, but also may highlight other policies needed for inducing growth. Accessibility is especially increased by investments in European transportation corridors and by removing barriers on the borders (Schengen Agreement). ESPON ATTREG project aims to investigate also the motivation and behaviour of migration flows and daily commuting of human resources in gross migrations and commuting between regions.

42.1 Flexibility of the Labour Market

Chun (1996) wrote that migration is a means of achieving both economic efficiency and equity. State and local governments in faster growing EU economies have wanted to attract migration, when and where they wanted to increase employment (Anjomani, 2002). Thinking flexibility and security on labour

---

M. Bogataj (✉)
University of Ljubljana, and MEDIFAS, Mednarodni prehod 6,
Šempeter pri Gorici, Slovenia
e-mail: marija.bogataj@guest.arnes.si

S. Drobne
Faculty of Civil and Geodetic Engineering, University of Ljubljana,
Jamova cesta 2, Ljubljana, Slovenia

D. Tuljak Suban
Faculty of Maritime Studies and Transport, University of Ljubljana,
Pot pomorščakov 4, Portorož, Slovenia

S. P. Sethi et al. (eds.), *Industrial Engineering: Innovative Networks*,
DOI: 10.1007/978-1-4471-2321-7_42, © Springer-Verlag London Limited 2012

markets together is a new paradigm because business as well as labour is in need of security as well as flexibility. ESPON ATTREG project aims to investigate also the motivation and behaviour of migration flows and daily commuting of human resources in gross migrations and commuting between regions. Cohesion policy of the European Commission contributed to the development of new transport infrastructure increasing regional economic development. Extensive spending has taken place under ERDF, Cohesion Fund and ISPA to reduce disparities among regions and increase accessibility through Trans-European transport networks (TEN-T) by reducing time spending distances. How the time spending distance between EU regions influence migration of human resources is studied here. Fuzzy approach to improve the regression analysis when evaluating distance in indicators of accessibility is presented in the paper.

As listed by Janssen (2008), flexibility of labour market has advantages to both sides: while workers that are mobile and skilled have (a) a better bargaining position (b) they will not accept the 'blackmail' to cut wages in order to subsidize jobs and firms that have failed to innovate, (c) internal functional flexibility safeguards jobs through innovation and (d) flexibility and autonomy in working time could be achieved through their bargaining position. Business side is also looking at this kind of flexibility positively in favour of human resources because (a) stable jobs are good for productivity, (b) secure work force is open to innovation and change, (c) a secure work force is willing to share its (tacit) knowledge which means a competitive advantage in today's world, (d) workers will only be willing to move between jobs if the economy is built up of good and stable jobs offering secure work contracts. It means that we can expect more flexible jobs in the future and more migrations of human resource between EU regions, but this phenomenon is not negligible also in the beginning of 21st century.

Mobility of human resources is directly connected with the migration phenomenon since the motivation of job search represents one of three the main sources of human migration. There exists also intensive migration of retired population and migration of students but both have been less intensive at least in the past.

42.2 The Factors Influencing Migrations

We have to differentiate two groups of factors influencing migrations: (a) internal factors which influence the quality of places and (b) those which influence accessibility to region.

42.2.1 Internal Factors Which Influence Quality of Places

In 2010 new ESPON project ATTREG was starting, where we study the flows of human resources in EU in details. According to this study the endowment factors of regions are (Russo et al. 2010) as: (a) ***Environmental Capital*** relates to the natural environment; (b) ***Anthropic Capital*** relates to the quality of the built environment like the nature of infrastructure and different other facilities, percentage of built up land cover and availability of not expensive houses; (c) ***Economic Capital*** relates to the performance of the economy within the region. This group of factors includes salaries and wages, the generation of wealth which is often measured through GDP and employment; (d) ***Social & Cultural Capital*** captures characteristics of the people living within an area either in terms of demographic characteristics, well-being, and social capital. In a planned move, a location is preferred if migrants have a strong attachment to the community and friends residing in that location. People are more likely to move to places where relatives and friends have previously migrated so as to reduce uncertainty; (e) ***Human Capital*** is measured by the availability of labour within a region as well as the blend of skills and qualifications within the resident labour force; (f) ***Institutional Capital*** variables measured by the provision of public series of indicators that capture changes in accessibility potential resulting from labour movement restrictions. These variables are directly related to the actions of public bodies at European, national or local/regional level. Each source of territorial capital consists of a number of indicators covering the whole territory of analysis (endowment factors) which differ in places and in time. These indicators could be obtained or calculated for regions of EU separately or, combined in "synthetic indices" through clustering, often observed through proxy indicators available in EUROSTAT (2002–2010). Those people who are more skilled and educated, whose affluence permits more travel, and who are familiar with different parts of the country, tend to migrate more If we divide the flow of human resources to two groups: (a) top-workers and creative class and (b) opportunity-driven immigrants, we can assume that different endowment factors from the groups of capital listed above are differently important for this two classes.

42.2.2 Accessibility Index

Accessibility plays a significant role in European policy related to the development of territorial units. In several European policy documents as it is European Commission Green Paper on Territorial Cohesion or Territorial Agenda of the EU involving all EU Member States, accessibility is seen as key factor in improving attractiveness of Members States, their regions and cities (ESPON 2009). Accessibility is a measure of ease of access and in all of our previous research works it was assumed to be symmetrical. Converting distance to time spending or some other actual costs of access, we get a measure that may provide a changed ranking of localities by centrality. By accessibility in general we want to describe an integral view of localization qualities that result from nonlocal influences.

ESPON data of accessibility are based on measurements of airline distance, route distances and rail distance services/investment in a region. In addition to these measures there are also a travel time, from which weighted combinations of these indicators are derived. These values vary by mode or choice of routes.

42.3 Gravity Models of Migrations Flows

The growth of regions relates closely to population growth, which is mostly a result of migration, mostly migration of human resources, workers and their families. For more accurate anticipation of this movement, to be prepared for the future growth of different regions and to study policies that could increase the growth, gravity models present a good and reliable tool. Without such a tool it is difficult to forecast the future of an urban growth or population changes in some rural areas. The overbuilding of different types of construction occurring in last years are the examples of growth overestimation, contributing to the subsequent loan crisis. The analysis of residuals associated with these functions will enable to explore the regions where other indicators and mobilization activities increase or decrease the expected value of flows.

From our previous analysis (Lowry 1966; Bogataj et al. 2004, 1995; Drobne et al. 2008; Lisec et al. 2008) of Lowry-like models for migration inside Slovenian regions we found that in the model

$$GM_{i,j} = \alpha P_i^{\beta_1} P_j^{\beta_2} d_{i,j}^{\beta_3} \prod_{s \in S} K_{s,i}^{\gamma_s} K_{s,j}^{\varphi_s} \tag{42.1}$$

where P_{i} is number of inhabitants in origin; P_{j} is number of inhabitants in destination; $d_{\mathrm{i,j}}$ is time spending distance between regions and s is an indicator from the sets of economic, social, environmental and other sets S of regional capital named in the chapter 1, the following coefficients (ratios) influence GM_{ij} the most significantly:

$K_{BDPp,i}$	The ratio between GDP_p (per capita) in region of outflow iand GDP_p in total
$K_{BDPp,j}$	The ratio between GDP_p in region of inflow j and GDP_p in total area
$K_{BOD,i}$	The ratio between wages in region of outflow i and the average in total area
$K_{BOD,j}$	The ratio between wages in region of inflow j and the average in total area
$K_{ZAP,i}$	The ratio between employment in region of outflow i and the average in total area
$K_{ZAP,j}$	The ratio between employment in region of inflow j and the average in total area
$K_{OGR,i}$	The ratio between development and environmental risk in region of outflow i and the average in total area
$K_{OGR,j}$	The ratio between development and environmental risk in region of inflow j and the average in total area

Therefore we can write

$$GM_{i,j} = 0.0003 P_i^{0,8} P_j^{0,67} d_{i,j}^{-1,47} \frac{K_{BDPp,i}^{0,98} K_{BDPp,j}^{0,13} K_{BOD,i}^{-0,31} K_{BOD,j}^{4,98}}{K_{ZAP,i}^{-0,68} K_{ZAP,j}^{-3,99} K_{OGR,i}^{-0,077} K_{OGR,j}^{-0,57}} \tag{42.2}$$

From where the following regression output has been given (see Table 42.1) where at 132 observations adjusted R^2 is 0.86 and exponents has the values and significance as presented in Table 42.1, where, β_1 and γ_s are measures of

Table 42.1 The regression output for Slovenian NUTS3 regions

	Coefficients	Standard error	P-value
α	−2.5159	3.7511	0.5037
β_1	0.8093	0.1169	0.0000
β_2	0.6658	0.1167	0.0000
β_3	−1.4706	0.1075	0.0000
γ_1	0.9844	1.0686	0.3588
φ_1	0.1286	1.0688	0.9045
γ_2	−0.3072	1.6129	0.8493
φ_2	4.9838	1.6124	0.0025
γ_3	0.6766	1.3253	0.6106
φ_3	3.9891	1.3222	0.0031
γ_4	0.0768	0.1654	0.6431
φ_4	0.572	0.1644	0.0007

stickiness, while β_2 and φ_s are measures of attractiveness for each indicator of regional capital. To be focused on the fuzzy estimation of changes in β_3 **to forecasting the attractiveness through the accessibility of regions** the aim of the first stage of our research is to reduce the number of other variables being taken into consideration. Because there is a high degree of inter-correlation between Gross domestic product per capita (GDP_p), wages and unemployment, especially for this study, the other variables will be removed leaving one variable (gross domestic product per capita—GDP_p) to represent groups of highly-inter-correlated ones. Equation (42.1) was modified to:

$$GM_{i,j} = \alpha P_i^{\beta_1} P_j^{\beta_2} d_{i,j}^{\beta(t)_3}(t) K_{BDPp,i}^{\gamma_1} K_{BDPp,j}^{\varphi_1} \tag{42.3}$$

If in equation (42.3) β_3 is assumed to be constant, we get the regression output presented in Table 42.2. If we are comparing β_3 (= −0.2449) with the other exponents of endowment indicators of regions, we can see that it has the highest p-value (= 0.0016) and therefore the question arise what are this values on the shorter distances and is it dependent on distance at all. For the Slovenian case study we have got $\beta_3 = -1.4706$ and p-value very small (= 0.0000). From our study of regression outputs for interstate migrations in Member States of EU and the candidate countries on the NUTS 2 level, we found out that only Spain and Finland have higher values β_3 and that in averages this value is much **lower** than −1, as taken in accessibility indices in ESPON database at all other countries. Therefore fuzzy approach was taken to evaluate better $\beta(t)$ as a stepwise function of time spending distance when driving by car from *i* to *j*.

42.4 Fuzzy Approach to Evaluate the Accessibility of Region as the Potential for Gross Migrations

FuzzyTech 5.52 Professional Edition was used to define a fuzzy model that compute the value of $GM_{i,j}$ on the base of the linguistic variables P_i, P_j, -d_{ij}, $K_{BDPp,i}$ and $K_{BDPp,j}$. All the input linguistic variables are express by three terms

Table 42.2 The regression output for equation (4) between EU Member states, 2006

Observations:: 927	Adjusted $R^{2:}$ 55.4 %	
	Coefficients	P-value
α	0.0003	3.04E − 42
β_1	0.7633	1.55E − 82
β_2	0.8050	7.22E − 89
β_3	−0.2449	0.0016
γ_1	0.6714	3.40E − 21
φ_1	1.1745	8.96E − 56

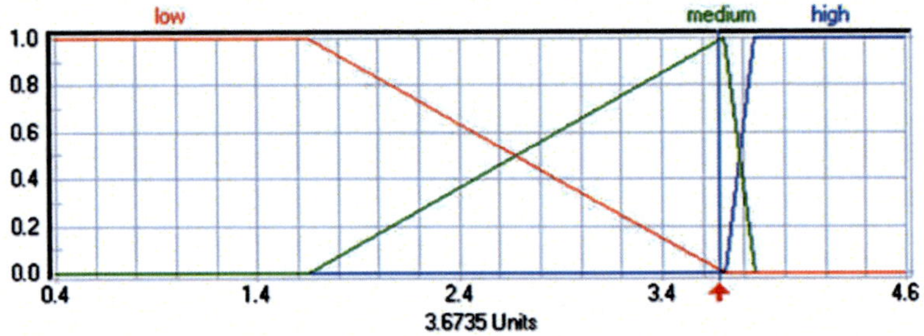

Fig. 42.1 Membership function of variable $d_{i,j}$ (From FuzzyTech 5,52)

(low, medium and high) with standard membership functions; the output variable is expressed by five terms to emphasize the sensitivity to the variations of inputs. Neuro-fuzzy tool, a hybrid that synergizes the human-like reasoning style of fuzzy systems with the learning and connectionist structure of neural networks, is used to improve the shape of membership function and the degree of support of the rules in the rule block, so the soft estimation of $GM_{i,j}$ is as much as possible close to the real value of the gross migration. We have seen that used membership functions become asymmetric, fuzzy variable time spending distance $d_{i,j}$ has the most relevant asymmetry. Figure 42.1 shows the shapes of the membership functions and that the most important definition point is the point 3.7 (distance of 40 h), where all three membership functions coincide by changing direction and the term "high" terminates its influence to the decision process. Therefore we divided the population of flows to those, having time spending distance shorter than 40 h and others

We see that important definition point is at d(t) = 40. The results of regression output for first part of population (d(t) < 40 h) at 825 observations gave adjusted R^2 equal to 0, 59 and the values of coefficients like presented in Table 42.3.

Here we can see that adjusted R^2 *is* nearly 60% (note that it can not be as high as the values at smooth regression analysis, where the total areas are supposed to

Table 42.3 The regression output for d(t) < 40 h

	Coefficients	P-value		Coefficients	P-value
α	0.000156	6E – 42	β_3	−0.5403	4E – 09
β_1	0.8914	9E – 83	γ_1	0.6714	9E – 19
β_2	0.8910	6E – 91	φ_1	1.1745	6E – 52

be more homogeneous). Using fuzzy approach R^2 was improved relatively for 7% and the absolute value of β_3 is 2.2 times higher and it has smaller standard error. While for the other 102 relations β_3 was −0.93, which is close to ESPON's value (−1), but P-value was too high (0.21), and also volatility of intercept and other factors have been very high. From here we can calculate the accessibility of regions and expected out-migration and in-migration as the sum of those from less remote regions (less than 40 h travel by car) and small participation of very volatile second part—the flows from the remote countries. The total out-flows from region i and the total in-flows to the region i are expected to be:

$$\begin{aligned} GM_{i,out} &= \sum_j GM_{i,j} = 1.56 \cdot 10^{-4} P_i^{0,89} K_{BDPp,i}^{0,67} \sum_j d_{i,j}^{-0,54}(t<40) P_j^{0,89} K_{BDPp,j}^{1,17} \\ &\quad + 3.54 \cdot P_i^{0,45} K_{BDPp,i}^{0,39} \sum_j d_{i,j}^{-0,93}(t \geq 40) P_j^{0,38} K_{BDPp,j}^{0,65} \\ GM_{in,i} &= \sum_j GM_{j,i} = 1.56 \cdot 10^{-4} P_i^{0,89} K_{BDPp,i}^{1,17} \sum_j d_{i,j}^{-0,54}(t<40) P_j^{0,89} K_{BDPp,j}^{0,67} \\ &\quad + 3.54 \cdot P_i^{0,38} K_{BDPp,i}^{0,65} \sum_j d_{i,j}^{-0,93}(t \geq 40) P_j^{0,45} K_{BDPp,j}^{0,39} \end{aligned} \tag{42.4}$$

The expected net migration NM_i in the region i is in this case:$NM_i = GM_{in,i} - GM_{i,out}$. Available human resources in this flow can be calculated from ESPON database, taking in account the structure of cohorts in the flow. If λ_i is the ratio of working cohort in the flows between the regions, than the expected availability of the new human resources would be: $NM_i(\textit{human resources}) = \lambda_i(GM_{in,i} - GM_{i,out})$ The human resources in the structure of net migration in Spain (λ_i **= 0, 6**) is there 120,3000 and for Romania ($\lambda_i = 0{,}87$) is −59,021. The other values are possible to calculate. Using this approach we can calculate what will be the change of human resources flows if the value of K_{BDPp} at the certain region i is changed.

42.5 Conclusion

Environmental capital, economic capital, anthropic capital, social & cultural capital, human capital and institutional capital which could be evaluated through different indicators make regions attractive or sticky. To be able to evaluate the number of available human resources in the region for the next years we have to evaluate these indicators in region and in all other regions around, from where and to where these flows could come or go. But very important is also the distance between the region of origin and the region of destination which very much influence the flow of human resources. But the migrant does not "fill" the time spending distance between regions 100 km apart and the time spending on the distance between 3010th and 3020th km between origin and destination the same, having the same accessibility value in ESPON databases. Therefore in the gravity models we have to be aware that the exponent β_3 is changing by distance. Using fuzzy approach the function $\beta_3(t)$ could be approximated with the stepwise function, where discontinuities—steps are best to be put at the definition points of membership functions. For the further improvement of the Lowry-like model which will be able to forecast the future spatial growth and availability of human resources in EU regions we need to elaborate the indicators and their dynamics which often depends on industrial, other economic and political decisions in the regions (competitiveness of the regions), but also more attention has to be put to investments in Trans-European transportation networks which are able to increase accessibility of European human resources which "fill" that the new destination is not so fare anymore. Their impact on the migration flows could be study by fuzzy approach to Lowry- like models. For detailed analysis of flows also the unit of origin or destination should be the region on NUTS3–NUTS5 level.

Acknowledgments The chapter and its dissemination was partly financed by ATTREG—ESPON applied research 2013/1/7 and ARRS:J5, 4279–2734.

References

Anjomani A (2002) *Regional Growth and Interstate Migration*–Social-Economic Planning Science 36:239–265

Bogataj L, Bogataj M, Drobne S, Vodopivec R (2004) "The influence of investments in roads and border crossings capacities on regional development after accession". Modern Traffic, Mostar (24):103–111

Bogataj M, Drobne S, Bogataj L (1995) "The influence of investment and fiscal policy on growth of spatial structure". Suvremeni promet, Zagreb 15(5):239–245

Chun J (1996) Interregional migration and development. Ashgate Pub. Co, Vermont

Drobne S, Bogataj M, Bogataj L (2008) "Spatial interactions influenced by European corridors and the shift of the Schengen border regime", KOI 2006: proceedings, The 11th international conference on operational research. Croatian operational research society, Zagreb, pp 185–198

ESPON (2009), "Territorial dynamics in Europe, trends in accessibility", Territorial observation No. 2, Luxembourg

Janssen R (2008)–*Flexi-security and the ETUC*–Etuc Review, Oviedo;(available on: http://www.eu-employment-observatory.net/resources/meetings/ETUC-Janssen.pdf

Lisec A, Drobne S, Bogataj M (2008), "The influence of the national development axes on the transaction value of rural land in Slovenia = Vpliv nacionalnih razvijnih osi na transakcijsko vrednost kmetijskih in gozdnih zemljišč v Sloveniji. *Geod. vestn.*, Vol 52, No. 1, pp 54–68. http://www.geodetski-vestnik.com/52/1/gv52-1_054-068.pdf Accessed 15 Aug 2010

Lowry IS (1966) Migration and metropolitan growth: two analytical models. Chandler Publishing Company, San Francisco

Russo AP, Bogataj M, Drobne S, Lisec A, Vodopivec R, Zavodnik Lamovšek A, Trobec B. (2010) *Attractiveness of European region and cities for residents and visitors—ATTREG:* Activity report PP8: Report 1. Ljubljana

Chapter 43
Parallel CUDA Architecture for Solving de VRP with ACO

Francisco Javier Diego, Eva María Gómez, Miguel Ortega-Mier and Álvaro García-Sánchez

Abstract Ant Colony Optimization (ACO) is an effective meta-heuristic for the solution of a wide variety of problems. Its computation is intrinsically massively parallel, and it is therefore theoretically well-suited for implementation on Graphics Processing Units (GPUs). In this paper, we propose a parallelization strategy to solve the VRP with ACO on the GPU.

43.1 Introduction

The Vehicle Routing Problem (VRP) deals with the distribution of goods from depots to customers. The VRP is a combinatorial optimization problem, highly important in different logistics environments; it is about designing the optimal set of routes for a given fleet of vehicles in order to serve a given group of customers. The interest in the VRP is motivated by its practical relevance, as well as by its considerable difficulty. To improve the resolution of high parallel implementation problems such as the VRP, NVIDIA® proposes a new technology to reduce the computing time, the GPU. We have implemented a new algorithm to take advantage of the new device properties.

F. J. Diego (✉) · E. M. Gómez · M. Ortega-Mier · Á. García-Sánchez
Escuela Técnica Superior de Ingenieros Industriales, Universidad Politécnica de Madrid, Madrid, Spain
e-mail: javier.diego@upm.es

E. M. Gómez
e-mail: emgomezjulio@gmail.com

M. Ortega-Mier
e-mail: miguel.ortega.mier@upm.es

Á. García-Sánchez
e-mail: alvaro.garcia@upm.es

S. P. Sethi et al. (eds.), *Industrial Engineering: Innovative Networks*,
DOI: 10.1007/978-1-4471-2321-7_43,

43.1.1 Related Work

The most commonly used techniques for solving Vehicle Routing Problems are exact approaches, heuristic methods and meta-heuristic methods such as ACO. The Capacitated Vehicle Routing Problem (CVRP) is the most basic form of a VRP, and it was the first VRP to be tackled by an ACO approach. Bullnheimer et al. (1999) presented an adaptation of Ant System Rank which was later improved by Reimann et al. (2004). Another VRP extension was solved with ACO: the Vehicle Routing Problem with Time Windows (VRPTW), where a time window is associated with each customer, defining an interval wherein the customer has to be supplied. ACO algorithms have also been applied to a number of other VRP variants, including a pickup and delivery problem under time window constraints in a hub network (Dawid et al. 2002), or the VRP with backhauls and time window constraints whose differences are the way the heuristic information and the pheromone trails are computed, or the details of the solution construction procedure. In recent times, Reimann et al. (2004) proposed an algorithm called Unified Ant System that applied to four VRP variants too. Dorigo and Sützle (2004) describe the simplest case of parallelization for the ACO algorithm, where independent instances of the algorithm are executed in different processors with no communication between them. The solution is taken as the best of all the independent executions. Recently, Chen et al. (2008) divided the population in sub-colonies executed in different processors, with each sub-colony searching for an optimal local solution, and information being exchanged between processors periodically. Lin et al. (2007) propose dividing the problem into sub-components, with each sub-graph assigned to a different processing unit. To explore a graph and find a complete solution, an ant moves from one processing unit to another, and messages are sent to update pheromone levels. The authors demonstrate that this approach reduces local complexity and memory requirements, thus improving overall efficiency.

GPUs are now emerging as powerful parallel processing devices replacing multi-core CPUs. They are well-suited to problems that could be expressed as data-parallel computations with high arithmetic intensity. Every thread executes the same instructions for each data element and the CPU is in charge of the flow control. Problem implementations which need large data sets can use a data-parallel programming model to reduce the computation time. In 3D rendering, large sets of pixels are mapped to parallel threads. This type of data processing could be used with no graphic-associated algorithms, thus making the GPU a General Purpose computing tool (GPGPU). Although GPUs can offer unprecedented performance gain, implementation of an algorithm over a GPU to take full advantage of this new technology involves a significant complexity of parallelizing across the multiple cores. Memory management over a GPU makes things even more challenging.

Compute Unified Device Architecture (CUDA) is a parallel computing architecture developed by NVIDIA. CUDA is the compute engine in NVIDIA's CUDA compatible GPUs, and it is accessible to software developers through industry standard programming languages such as C. CUDA is widely used for programming NVIDIA GPUs for general-purpose processing. You (2009) implemented an ACO for the Travelling Salesman Problem (TSP) using CUDA kernels in the construction and the pheromone update stage. Recently, Cecilia et al. (2011) proposed a new method where the pheromone is updated in different ways, but also to solve the TSP. Although these proposals offer a useful starting point to solve the TSP, there are no insights as to how to implement the VRP in any of its variants; in the next sections we propose a parallel implementation to resolve the CVRP with ACO and the NVIDIA GPU.

43.2 CVRP

The Vehicle Routing Problem has many variants depending on the constraints considered. In our research, we have studied the CVRP variant. In the CVRP, demands are deterministic and known, all vehicles are identical, only capacity constraints are considered, and there is a single central depot. The objective is to minimize the total cost to serve the customers. A solution is feasible if the total quantity assigned to each route does not exceed the capacity of the vehicle which services the route, and all customers are served.

43.2.1 CVRP Notation

For the linear programming formulation of the VRP whose ground set is the edges of a graph $G(V, E)$, where V is a vertex set of n cities, v_0 is the depot and A is the set of arcs, the VRP uses two indices in the formulation, where the variable x_{ij} is 1 if arc (i, j) belongs to the optimal solution, and 0 otherwise, and C is a matrix of non-negative costs or distances c_{ij} between customer v_i and customer v_j. The matrix C could be symmetric or not depending on whether we consider the Euclidean distance between two points or not. In the symmetric case the resulting problem is called Euclidean SCVRP. A general formulation of the linear programming can be given as follows:

$$\min \sum_{i \in V} \sum_{j \in V} c_{ij} x_{ij} \tag{43.1}$$

$$\sum_{i \in V} x_{ij} = 1 \quad \nabla j \in V \backslash \{0\} \tag{43.2}$$

$$\sum_{j \in V} x_{ij} = 1 \quad \nabla i \in V \backslash \{0\} \tag{43.3}$$

$$\sum_{i \in V} x_{i0} = K \tag{43.4}$$

$$\sum_{j \in V} x_{oj} = K \tag{43.5}$$

$$\sum_{i\notin S}\sum_{j\in S} x_{ij} \geq r(S) \quad \nabla S \subseteq V\{0\}, S \neq ø \tag{43.6}$$

$$x_{ij} \in \{0,1\} \quad \nabla i,j \in V \tag{43.7}$$

Where Eqs. 43.2 and 43.3 mean that the number of vehicles that enter and exit from a vertex equals 1, so there is only one vehicle which visits a customer, and Eqs. 43.4 and 43.5, where *K* is the number of vehicles, impose that the vehicles must depart from and arrive at the depot for their routes. Equation 43.6 is the Capacity-Cut Constraints (CCCs) which impose the connectivity of the solution and the capacity requirements of the vehicle, where *r(S)* is the minimum number of vehicles to serve a subset *S*. An alternative formulation could be given by Eq. 43.6 and the following ones (43.8):

$$\min \sum_{i\in S}\sum_{j\in S} x_{ij} \leq |S| - r(S) \quad \nabla S \subseteq V\{0\}, S \neq ø \tag{43.8}$$

Equations 43.6 and 43.8 grow exponentially with the number of cities. This means that it is practically impossible to solve directly the linear programming relaxation of the problem. A possible way to partially overcome this drawback is to consider only a limited subset of these constraints. Fisher (1994) and Miller propose the branch-and-cut method. Alternatively, a family of constraints equivalent to (43.6) and (43.8) can be considered as the equation presented for the TSP by Miller, Tucker and Zemlin and extended to the Asymmetric Capacity Vehicle Routing Problem (ACVRP) by Christofides et al. (1981):

$$u_i - u_j + Cx_{ij} \leq C - d_j \quad \nabla j \in V\{0\}, i \neq j, \backslash d_i + d_j \leq C \tag{43.9}$$

$$d_i \leq u_i \leq C \quad \nabla i \in V\{0\} \tag{43.10}$$

Where C is the maximum capacity of the vehicle, u_i is a continuous variable that represents the load of the vehicle after visiting the city i, and d_i represents the demand of the city i. Equations 43.7, 43.9, and 43.10 impose the connection and the weight constraints for the CVRP variant. The VRP belongs to the NP-hard type of problems. To address the complexity of these problems, heuristics and meta-heuristics are used; in this paper we used the ant colony algorithm.

43.3 Ant Colony Parallelization for the CVRP

ACO algorithms are stochastic search iterative processes, where the construction of solutions to the problem is carried out according to the probabilistic pheromone trails, which are numerical values that represent the experience of the algorithm search. These algorithms are composed of two basic processes: construction of solutions and pheromone deposit. The deposit made in each iteration pheromone guides the process of building better solutions in subsequent iterations. We implemented one of the simplest methods; the pheromone is initialized to 0.5, based on the recommendation of Dorigo and Sutzle (2004) and is evaporated in all the arcs. The best solution of the algorithm deposited pheromone in each iteration.

43.3.1 Graph Construction

ACO algorithms are based on a tree graph construction, to obtain the solution from the path followed by the ants. The artificial ants begin their journey at an initial node and continue their way by jumping to other nodes until they reach the final one. In VRP the nodes are the customers to be serviced (deliveries), and the tree consists of all the delivery nodes, the depot and the arcs which link the different nodes.

Algorithm 1. ACO algorithm for VRP
Require: A = (C, ξ, Π), n_a, C_m, ρ, T^{bs}: = NULL,
T^{ib}: = NULL

for all $e \in \Pi$ **do**

$\tau_e := 0.5$

end for
while Stop conditions don't satisfy **do**

for $k = 1$ to n_a **do**

$T^k = \{0\}$
for all $c \in C/c \neq a$ **do**

$a_c^k :=$ FALSE

end for
T^k: = Construct(A,a,C_m,T^k)

end for
$T^{ib} :=$ argmin(f(T^1),…,f(T^n))
if T^{bs} = NULL or (f(T^{ib}) < f(T^{bs})) **then**

$T^{bs} := T^{ib}$
PheromoneActualization(F,T^{ib},ρ)

end if

end while
return T^{bs}

$C = \{a, E\}$ is the set of nodes, a is the depot node, E is the set of deliveries, and there are = $1 + |E|$ nodes in the graph. So we define the graph as $A = (C, \Pi)$, where Π represent the tree arcs. In the VRP, an arc exists between the two deliveries, and also between the deliveries and the depot. In ACO, the arcs contain the so-called total information, which is composed by the heuristic information that indicates in a numeric way how desirable the arc is; the heuristic information is calculated as the inverse of the cost; and the pheromone information T_e that informs about the solution explored by the ants, after initializing the variables T^{bs}

(the best of the algorithm), T^{ib} (the best of this iteration) and c_f, is fixed to 0.5. In each iteration, n_a artificial ants initialize their variables and construct the solutions through the tree graph A, and then the best solution deposits pheromone. Figure 43.1 explains the algorithm. A solution consists of supplies that are vehicles routes. Each route begins and ends at the storage node a. The steps to be taken to construct a solution are explained in Fig. 43.2.

In every step, a new node o_i is added to the solution, $T^j_{p+1} = T^j_p\ U\{o_i\}$, and the o_i load must be lower than the actual capacity of the vehicle (C_m is the maximum capacity). If there are no candidates, the vehicle will have to return to the depot. To pick a node, it follows stochastic rule 43.11.

$$p_{ij} = \frac{[\tau_{ij}]^{\alpha}[\eta_{ij}]^{\beta}}{\sum_{l \in N_i^k}[\tau_{il}]^{\alpha}[\eta_{il}]^{\beta}} \tag{43.11}$$

Algorithm 2. Building solutions
Require: tree A, depot a, Max Load Cm, empty solution T^j_0

Cost : = 0, Load : = 0, Assigned : = 0, p : = 1
$T^j_{p+1} \leftarrow T^j + a$
Repeat

  C^j_p : = EMPTY
  for all $o_i \in N(o_i)$ **do**

    if a^j_i = FALSE and p_i + Load $\leq$ Cm **then**

      $C^j_p \leftarrow C^j_p\ U\{o_i\}$

    end if
  end for
  if $|Cj| = 0$ **then**
    $o_i = a$
    Load: = 0
  else
    Selection $o_i \in C^j_p$
  end if
  $T^j_p \leftarrow T^j_p$ U$\{o_i\}$
  if $o_i \neq a$ **then**
    a^j_i: = TRUE
    Assigned ← Assigned +1
    Load ← Load + p_i
    Cost ← Load + dis(T^{j-1}_p, o_i)

  end if
  $p \leftarrow p + 1$

until Assigned |E|
return T^j

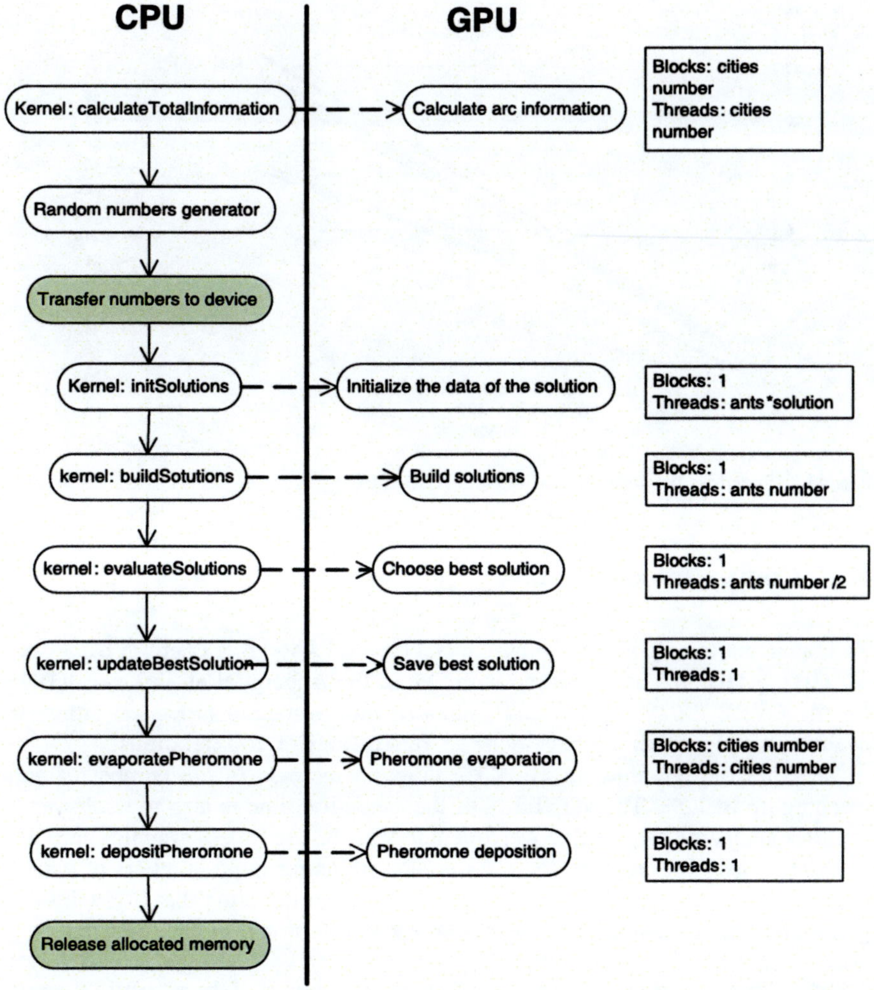

Fig. 43.1 Algorithm kernels implemented in the GPU

Before the best algorithm solution deposits pheromone, the pheromone depot is evaporated by the equation:

$$\tau_{ij} \leftarrow (1-\rho)\tau_{ij} \quad Vi, j \,/\, e_{ij} \in \tag{43.12}$$

Where $\rho \in (0,1]$ is a constant value. To solve the parallel algorithm with the NVIDIA device, several kernels have been implemented as indicated in Fig. 43.1. We have tried to implement as many steps as possible to resolve the exchange information time.

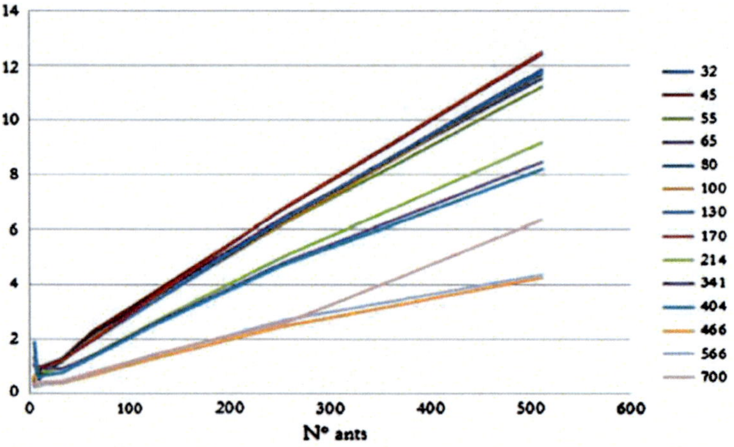

Fig. 43.2 Speed-up between ACO–GPU and ACO–CPU

43.4 Computational results

To implement the algorithm we used a GeForce GTX460 in a Pentium Dual Core, 2.7 GHz, 2 GB. We have worked with five of the Augerat et al. instances (1995) for the CVRP (36–80 cities) and generated new instances (100–700 cities); to evaluate the algorithm we have done 25 replications of 100 iterations.

The speed-up graphics in Fig. 43.2 show an increase in the number of ants, reaching up to 12%. The variability of the execution time is also much lower.

After analyzing the results, we conclude that the use of the graphics card as a calculation tool not only results in faster algorithms for large numbers of parallel computing, but also less run time variability. It is worth noting that if low level of parallelization is required, it is better, in terms of runtime, to use the CPU.

Acknowledgments This work partly stems from the participation of the authors in a research project funded by the Spanish Ministerio de Industria, Comercio y Turismo, Proyecto Avanza reference TSI-020100-2010-962, titled 'Arquitectura de servicios de supercomputación en la nube' (AMBU).

References

Augerat P, Belenguer J, Benavent E, Corbern A, Naddef D, Rinaldi G (1995) Computational results with a branch and cut code for the capacitated vehicle routing problem. Research report 949-m. Universite Joseph Fourier, Grenoble

Bullnheimer B, Hartl RF, Strauss C (1999) Applying the ant system to the vehicle routing problem. Ann Oper Res 89:319–328